Basic Confocal Microscopy

W. Gray (Jay) Jerome • Robert L. Price
Editors

Basic Confocal Microscopy

Second Edition

Springer

Editors
W. Gray (Jay) Jerome
Department of Pathology, Microbiology and Immunology
Vanderbilt University School of Medicine
Nashville, TN, USA

Robert L. Price
Department of Cell Biology and Anatomy
University of South Carolina School of Medicine
Columbia, SC, USA

ISBN 978-3-030-07358-9 ISBN 978-3-319-97454-5 (eBook)
https://doi.org/10.1007/978-3-319-97454-5

Softcover re-print of the Hardcover 2nd edition 2018

This Springer imprint is published by the registered company Springer Nature Switzerland AG
The registered company address is: Gewerbestrasse 11, 6330 Cham, Switzerland

Preface

Biological confocal microscopy is still a relatively young and rapidly advancing field. Since the first edition of this book was published in 2011, many significant advances have been made in confocal technology, including the introduction of highly sensitive detectors, improved software applications, several super-resolution instruments based on confocal technology, and more. In preparation of this second edition, it amazed us as to how far the field has advanced in a very short period of time. In this second edition, we introduce many of these advances while attempting to maintain the basic instructional nature of the text. Most chapters have been significantly updated with new information, and an entirely new chapter on analysis of fluorescence co-localization by Dr. Teng-Leong Chew has been added.

Most researchers in the field would date the modern era of biological confocal microscopy from the 1985 description of a particularly useful confocal design published by White and Amos in the *Journal of Cell Biology*. Since that time, the use of confocal microscopes by biologists has increased phenomenally, with new converts joining the ranks daily, many with little or no previous microscopy training. For that reason, in 2001, when we were asked to organize a 1-day session on basic confocal microscopy for attendees at the Southeastern Microscopy Society annual meeting in Clemson, SC, we decided not only to focus on the confocal microscope itself, but also on ancillary subjects that are critical for getting the most from confocal microscopy.

Our initial effort seemed to meet a growing need to train new students, technologists, and faculty wishing to use confocal microscopy in their research. Evidence for this need is that, each year since 2001, we have been invited by several meeting organizers and microscopy core facility directors to present our take on what is important to successfully using confocal microscopy for biological exploration. In 2005, we also began teaching a 5-day intensive, hands-on workshop at the University of South Carolina each year. As that course evolved, we invited various colleagues to help with the course. This book is a direct outgrowth of that course and follows the general structure of the didactic portion of the course. In line with the course philosophy, we have not attempted to cover each topic in depth. However, we have maintained a focus on basic information, and we have endeavored to completely cover information that is important for designing, carrying out, and interpreting the results of basic confocal microscopy-based biological experiments. We were very fortunate that two of the other course instructors, Drs. Ralph Albrecht and Tom Trusk, have provided chapters for this volume and have embraced the overall philosophy of presenting a basic knowledge base in a complete but concise manner.

Although the forums have been different and the course lengths have varied anywhere from 1 to 5 days, we have always based the workshops on the original concept that there is a group of core issues that must be understood before one can efficiently get the best results from the use of a confocal microscope. The early chapters in this book address these core issues, and it is not by accident that, after an initial introductory chapter on confocal microscopy, the chapters describing components of the confocal microscope and how to correctly set the various operating parameters are located toward the end of the book. Without a well-designed research plan and properly prepared specimen, the data collected by the microscope will not be optimum. Thus, we have devoted Chaps. 2 and 3 to fluorescence and understanding the use of fluorescence microscopy and Chaps. 4 and 5 to specimen preparation and labeling strategies. These chapters are essential since, regardless of the quality of the confocal microscope, if the sample is not prepared properly, the data collected will not be optimal.

Confocal microscope images are digital. Thus, many of the basic operating parameters for confocal microscopy involve setting up the analog to digital conversion of specimen information. It is essential that a confocal microscope operator has a thorough understanding of how digital images for scientific purposes should be collected and analyzed. For this reason, following the chapters on specimen preparation, Chaps. 6 and 7 discuss digital microscopy with respect to confocal imaging.

Although it might seem odd that a book on confocal microscopy contains only two chapters directly devoted to the actual operation of the confocal microscope, these chapters are packed with practical information and, taking advantage of the preliminary information presented in preceding chapters, they provide all that is necessary to begin doing confocal microscopy and optimizing the information obtained. After Chaps. 8 and 9, which discuss the types of confocal instruments and setting up proper operating parameters, the final set of chapters provides information on the three-dimensional analysis and reconstruction of data sets, analysis of co-localization, some ethical considerations in confocal imaging, and some resources we have found useful in our own use of confocal microscopes. After mastering the basic information presented in this book, these resources are great guides for continuing your education into more advanced forms of confocal microscopy.

This book has benefited from our association with numerous colleagues who have challenged and informed us. In particular, numerous debates with one of the course instructors, Dr. John MacKenzie, Jr., have helped hone the information on digital image processing to the most important concepts. We are also grateful to Drs. K. Sam Wells, David Piston, and John Fuseler for stimulating and challenging conversations that have made us better microscopists. We also owe a huge debt to the many students over the years whose enthusiasm and questions have guided our decisions regarding what to include and exclude from the workshops and chapters in this book and to the many readers of the first edition. We are thankful for the many positive comments we have received about the book and the encouragement colleagues have given us to provide a second edition with updated information. We are also thankful to the many companies that have provided resources and applications experts, which have significantly enhanced our hands-on workshops at the University of South Carolina.

Finally, we must thank our lab members and families for not only putting up with our obsession for microscopy, but also encouraging us in our pursuits.

Columbia, SC, USA Robert L. Price
Nashville, TN, USA W. Gray (Jay) Jerome

Contents

Contributors

Jesse Aaron Advanced Imaging Center, Howard Hughes Medical Institute Janelia Research Campus, Ashburn, VA, USA

R. M. Albrecht Department of Animal Sciences, Pediatrics, and Pharmaceutical Sciences, University of Wisconsin – Madison, Madison, WI, USA

Teng-Leong Chew Advanced Imaging Center, Howard Hughes Medical Institute Janelia Research Campus, Ashburn, VA, USA

John Fuseler Department of Pathology, Microbiology and Immunology, University of South Carolina School of Medicine, Columbia, SC, USA

Anna M. Harper Department of Cell Biology and Anatomy, School of Medicine, University of South Carolina, Columbia, SC, USA

W. Gray (Jay) Jerome Department of Pathology, Microbiology and Immunology, Vanderbilt University School of Medicine, Nashville, TN, USA

J. A. Oliver Department of Biological Sciences, University of Wisconsin – Milwaukee, Milwaukee, WI, USA

Caleb A. Padgett Department of Cell Biology and Anatomy, University of South Carolina School of Medicine, Columbia, SC, USA

Robert L. Price Department of Cell Biology and Anatomy, University of South Carolina School of Medicine, Columbia, SC, USA

Amy E. Rowley Department of Cell Biology and Anatomy, School of Medicine, University of South Carolina, Columbia, SC, USA

Thomas C. Trusk Department of Regenerative Medicine and Cell Biology, Medical University of South Carolina, Charleston, SC, USA

Chapter 1
Introduction and Historical Perspective

Robert L. Price and W. Gray (Jay) Jerome

1.1 Introduction to the Second Edition of *Basic Confocal Microscopy*

Since publication of the first edition of *Basic Confocal Microscopy* in 2011, a number of advances have occurred influencing several aspects of confocal microscopy technology, including a number of super- and enhanced resolution techniques, specimen preparation methods, lasers, detectors, and operating and image analysis software. For the purpose of discussions throughout this text, we will define enhanced resolution techniques as those that improve resolution from the historical Abbe defined limit of 200 nm down to approximately 140 nm and super-resolution techniques to those that provide resolution well below 100 nm. While techniques and technology have improved, the basics that must be understood to generate high-quality confocal images remain constant. In this second edition, we will address how these recent changes have improved the performance and expanded the research applications of confocal imaging but will also maintain the introductory concept to confocal imaging that made the first edition a success.

Because of advances that make possible imaging deep into thick specimens, we have added information on tissue clearing techniques that complements the deep imaging capability of some modern confocal systems. Tissue clearing, including techniques such as X-CLARITY®, benzyl alcohol/benzyl benzoate (BABB), 3Disco and iDisco (dichloromethane/dibenzylether), and several others, are mechanisms to

R. L. Price (✉)
Department of Cell Biology and Anatomy, University of South Carolina School of Medicine, Columbia, SC, USA
e-mail: Bob.Price@uscmed.sc.edu

W. G. Jerome
Department of Pathology, Microbiology and Immunology, Vanderbilt University School of Medicine, Nashville, TN, USA
e-mail: Jay.Jerome@Vanderbilt.edu

W. G. Jerome, R. L. Price (eds.), *Basic Confocal Microscopy*,
https://doi.org/10.1007/978-3-319-97454-5_1

increase the depth from which useful image information can be extracted. This provides exciting possibilities for improving our understanding of three-dimensional relationships between structures in large regions of tissue. The protocols and advantages and disadvantages of several tissue clearing techniques will be discussed. Also in the area of sample preparation, we will provide information on antigen retrieval protocols and extend the discussion of available fluorescent probes available for confocal microscopy.

There have been rapid and significant advances in the hardware configurations of confocal microscopes in recent years, and now, unlike 7–8 years ago, systems are sold with a complete configuration of diode lasers rather than gas lasers such as argon or helium neon. This range of new diode lasers, along with high-sensitivity detectors such as the gallium arsenide phosphide (GaAsP) detectors, has made it possible to detect very low signals in point scanning instruments. Likewise, significant advances in cooled charged coupled devices (cCCD) and scientific complementary metal oxide semiconductors (sCMOS) now allow faster and more sensitive image capture in microscopes where a full field is captured in a single process. These detector advances have greatly expanded our capability to examine samples with low levels of fluorescence and to attenuate probe intensity to minimize specimen damage in live cell imaging or samples that rapidly photobleach. High-sensitivity detectors have also contributed to the development of new technologies such as the enhanced resolution Zeiss Airyscan and Leica HyVolution instruments that can exceed the 200 nm resolution level obtainable with standard confocal configurations and can approach 140 nm resolution. While not the sub-100 nm resolution level of super-resolution techniques such as STORM (stochastic optical reconstruction microscopy), PALM (photoactivated localization microscopy), STED (stimulated emission depletion microscopy), and SIM (structured illumination microscopy), these enhanced resolution instruments provide a cost-effective mechanism to beat the diffraction limitations of light present in standard confocal configurations.

As often discussed, confocal microscopes enhance many research projects. However, it is not always the best instrument for some imaging situations. For example, imaging a thin monolayer of cells may be better served with an epifluorescence wide-field microscope than optical sectioning with a confocal microscope. Likewise, while super-resolution instruments have greatly enhanced our understanding of some biological principles, they typically serve a specific research purpose and may have limited depth of imaging capabilities (Wentao et al. 2016), require special fluorochromes (Dempsey et al. 2011), or have other difficult-to-meet requirements for optimal image quality (Ashdown et al. 2014). As a consequence, super-resolution instruments are often great at addressing specific research questions, but they typically are not suitable for other research applications better suited for standard confocal imaging. A discussion of these techniques, their applications, and comparisons to more traditional methods has been added to Chap. 8 where different types of confocal systems are presented.

A number of advances have also taken place in operating software and analysis of two-dimensional and three-dimensional images. Chapter 9 in the first edition described in detail the Zeiss AIM software used to operate the LSM 510 line of instruments. About the time the first edition of *Basic Confocal Microscopy* was published, Zeiss introduced a new operating system (ZEN) that had many additions to improve the user interface. Operating software for other confocal systems have also advanced and now include similar functions such as online libraries for protocol development. Chapter 9 from the first edition has been updated to include information from these new operating systems (the Leica LAS software will be described in Chap. 9 of this edition) while still covering the basic setup of the microscope for optimal image quality. For analysis of co-localization and correlation of two molecules in two-dimensional (2-D) space, we have also added a new chapter by Dr. Teng-Leong Chew that includes the implementation and interpretation of various co-localization and correlation coefficients.

Even though confocal technology has advanced, the basics of optical imaging and the principles for collection and analysis of publication quality images and data remain essentially the same. It is our hope that by presenting information on these updates in confocal imaging that we will be able to maintain our goal of providing the requisite basic information for confocal imaging in a well-organized manner that will assist novice users in understanding the basics of confocal imaging.

1.2 Why an Introductory Text on Confocal Microscopy?

The premise of the first edition was that during our combined 35 plus years of operating confocal microscopes and managing core microscopy facilities, and through teaching our Basic Confocal Microscopy Workshop at several venues, we found that students and technicians that are novice users of confocal microscopes are often instructed by their mentors to go to the confocal system and collect some images. Often the implied message is that it should be easy and quick since it is only a microscope. Unfortunately, all too often the advisor of the student or supervisor of the technician does not have a full understanding of the complexity of a confocal microscope. Unless these novice users are in a situation where others have the time and knowledge to properly train them, their initial efforts often amount to an exercise in futility because key parameters are not properly considered. This leads to specimens that are not prepared properly and a lack of understanding of how to operate the confocal microscope in a way that maintains the fidelity of the specimen information. In too many instances, this lack of user training is exacerbated further because there is little or no daily oversight of the setup and maintenance of the microscope. In this combined scenario, neither the experimental preparation nor microscopes are capable of producing the highest-quality information. Now with well over 45 years of combined experience in managing core microscopy facilities, we have unfortunately found that the premise of the first edition of *Basic Confocal Microscopy* (that many faculty, technicians, and students operating confocal microscopes do not adequately understand the technology) is still valid.

Good confocal microscopy is obviously dependent upon proper specimen preparation and the correct setup of various microscope parameters. However, even if an excellent confocal image is collected, there is often a poor understanding of how to properly display the full richness of the information contained in the image and how best to analyze two-dimensional (2-D) and 3-D confocal images. There is an abundance of good image processing and analysis software available to the user. However, these robust programs also provide the capability of inappropriately manipulating the data or inadvertently degrading the image information. A lack of understanding of basic digital imaging and image processing theory frequently results in improper image processing in 2-D programs such as Image J and FIJI (NIH freeware), Photoshop (Adobe Systems, Inc., San Jose, CA), MetaMorph (Molecular Devices, Sunnyvale, CA), or others and in more advanced 3-D volumetric programs such as AMIRA (Thermo Fisher Scientific, Hillsboro, OR), IMARIS (Bitplane, Concord, MA), or VoxBlast (VayTek, Inc., Fairfield, IA).

The goal of this book is to provide beginning and intermediate users of confocal microscopes a resource that can be used to address many of the frequently asked questions concerning confocal imaging and to provide a strong foundation for maximizing the data obtained from experiments involving confocal microscopy. While most of the information is directly relevant to single-photon scanning laser systems, much of the information also applies to spinning disk, multiphoton, and enhanced and super-resolution confocal systems. In several chapters specific comparisons of the technology that differentiates these systems will be made and advantages and disadvantages of each presented. The information presented will also provide the background information necessary when moving forward to complex imaging protocols such as Forster (or fluorescence) resonance energy transfer (FRET), fluorescence recovery after photobleaching (FRAP), fluorescence lifetime imaging (FLIM), and other advanced techniques.

1.3 Historical Perspective

It has long been recognized by microscopists that as the thickness of the specimen increases, light emerging from scattering objects above and below the focal plane of the microscope degrade the quality of the image. This occurs primarily because of reduced image contrast. The loss of contrast is caused by impinging light produced from the out-of-focus planes. Like turning on the lights in a movie theater, this stray light reduces the signal-to-noise ratio (SNR) and obscures important image details. The various factors affecting the axial (Z) resolution (ability to distinguish two small objects as separate and distinct along the axial axis) were explored by Berek in 1927 (Berek 1927). In Berek's analysis, the three key elements affecting image quality were (1) spreading of the light beam emerging from objects in the specimen, (2) the magnification of the image, and (3) the sensitivity of the detection system. For Berek, the detection system was the observer's eye. However, in the modern age of microscopy, the eye has been replaced with more sensitive detectors. With regard to Berek's item 2, microscopists have always worked with the highest magnification required for maintaining image data fidelity. This leaves the spread of out-of-focus light into the image plane as the last of Berek's parameters that needs to be minimized to obtain good axial resolution. Obviously, if one could limit the projection of out-of-focus light onto the image, then a significant gain in resolution should be achieved. The removal of the obscuring out-of-focus light is precisely what the confocal microscope is designed to do, and the subsequent gain in axial resolution remains the biggest advantage of confocal microscopy. However, as will be described in subsequent chapters, several other advantages accrue from the confocal design, including increases in lateral resolution.

The first confocal microscope is generally credited to Marvin Minsky (Minsky 1988). In his 1957 patent application, Minsky described a microscope in which the typical wide-field illumination arrangement was replaced with one in which a point source is focused to a small spot within the specimen. Light arising from the illuminated spot is focused by the objective lens to a small spot at the image plane. Thus, a point source of light is in conjugate focus (confocal) at the specimen and at the image plane (Fig. 1.1a). Placing a small pinhole aperture made of an opaque material at the image plane permits only the light coming from the focal point of the specimen to pass to the detector. In contrast, light coming from above and below the plane of focus will not be in focus at the image plane and will be rejected by the opaque material surrounding the pinhole. This confocal setup can also be achieved in an epi-illumination setup (Fig. 1.1b). The confocal arrangement dramatically improves contrast by removing the out-of-focus light originating above and below the focal plane. The arrangements diagramed in Fig. 1.1 are not the only possible designs. Since its inception, various other designs have been introduced for creating the required confocality of focus at the specimen and image planes.

Of course, a single point within a specimen does not provide much information about the specimen. In order to acquire full details across the lateral (X-Y) focal plane of the specimen, the spot must be scanned across the image and the image information collected sequentially. In Minsky's original design, the scanning was produced by translating the specimen laterally. This method was slow and prone to vibration, both of which presented problems for biological work. A notable advance for the use of point scanning instruments in biology was made in the 1980s with the development of the ability to raster the illumination across the specimen rather than translating the stage. This allowed for faster scan rates without the introduction of vibration. The publication of images of biological samples using the beam-scanning instrument (White et al. 1987) spurred an extreme interest in confocal microscopy for biological research.

Arguably, the development of beam scanning along with concurrent advancements in laser technology, fluorescent labels, lens design, and computer processing really set the stage for the rapid deployment of laser scanning confocal microscopy as a key tool for cell biological research. However, laser scanning instruments are not the only mechanism for implementing confocal microscopy. A parallel development occurred based on Paul Nipkow's invention of a method for converting an optical image into an electrical signal that could be transmitted over a cable (Nipkow 1884). Nipkow's technique converted the 2-D image information into a 1-D serial

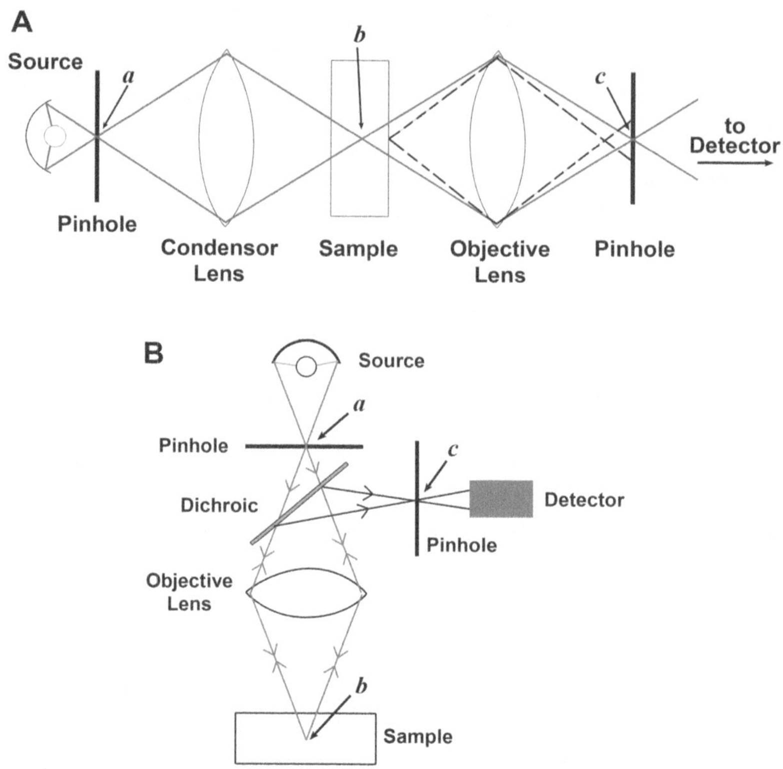

Fig. 1.1 Optical train for confocal microscope in conventional (**a**) and epi-illumination setups (**b**). The light path of the confocal beam is represented by the gray lines. In the conventional arrangement, light from the photon source is focused onto the entrance pinhole (a). This pinhole provides a bright focused point source. Light from this point source is collected by the condenser lens and focused to a spot (b) within the sample. The light emerging from the focused spot within the specimen is collected by the objective lens and focused at a second (exit) pinhole (c). Points a, b, and c are in conjugate focus (confocal). The path of light emerging outside of the focal point *B* is represented by the dotted black lines and arrives at the exit pinhole out of focus. Thus, most of this light is rejected and not transmitted to the detector

In an epi-illumination setup (**b**), the objective lens acts as both the condenser and objective lens. Light returning from the specimen is diverted by the dichroic (dichromatic beam splitter), and this diverted light (dark gray lines) is focused on the exit pinhole (dark gray lines). As with the conventional arrangement, light from above or below the focal point in the specimen arrives at the pinhole out of focus (not depicted) and so is rejected. Conventional wide-field fluorescence systems lack the pinhole so all out-of-focus light becomes a component of the final image as shown in Fig. 1.3

signal by scanning the image using a spinning wheel with precisely placed rectangular holes. The holes were arranged in a spiral pattern around the wheel such that when the wheel was spun, the small areas being sampled changed. The moving holes filled in the gaps between the initially sampled regions. In 1967 Eggar and Petráň (Eggar and Petráň 1967; Petráň et al., 1968) modified the design of the

Nipkow disk by including multiple spirals in a single wheel. They then used the spinning disk to provide both illuminating and imaging pinholes for a confocal microscope.

As with point scanning microscopes, over the years several different arrangements have been designed for spinning disk confocal microscopes. Figure 1.2 illustrates one such arrangement for an epi-illumination system. In this design, light is passed through the pinholes, directed onto the specimen, and the image light passes back through conjugate pinholes in the disk as it spins. By including sufficient numbers of pinholes and spinning the disk at a suitable speed, a real-time confocal image of the specimen can be obtained that can be viewed by eye or collected directly by a detector. One of the key benefits of this type of confocal microscope compared to laser scanning instruments is that spinning disks allow much faster image acquisition times. Further information on the design and use of spinning disk confocal systems is given in Chap. 8.

The Minsky and Petráň microscopes define the two principal implementations of confocal microscopy: the sequential scan (point scan) and spinning disk (multipoint scan, area scan) microscopes, respectively. As one might imagine, however, variations on these two schemes have been designed to overcome specific limitations of each for specific applications. A nice review of some of these implementations is provided by Shinya Inuoué (Inuoué 2006). Of course, the full power of imaging a

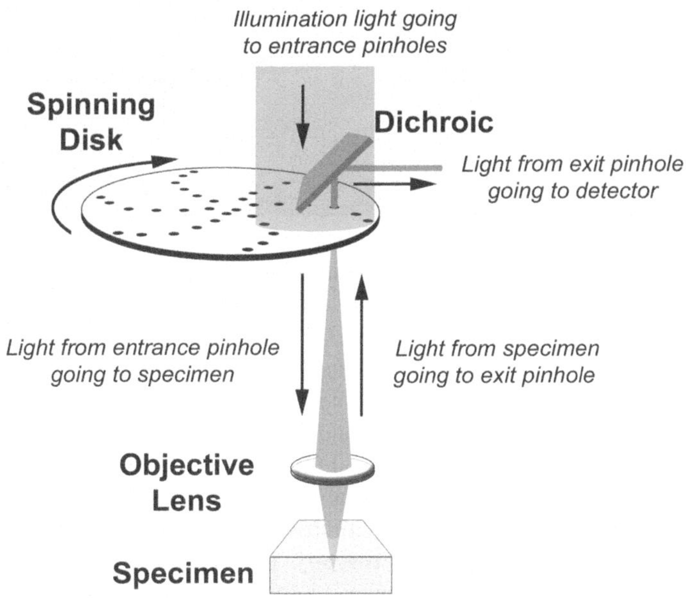

Fig. 1.2 Design of an epi-illumination spinning disk confocal microscope. Although multiple areas of the specimen will be illuminated at once, to simplify the diagram only light from one pinhole is depicted. As in Fig. 1.1 only focused light reaches the detector. Since light emitted from all pinholes reaches the detector simultaneously, image collection is rapid, but resolution and often overall signal is compromised in spinning disk systems as discussed in Chap. 8

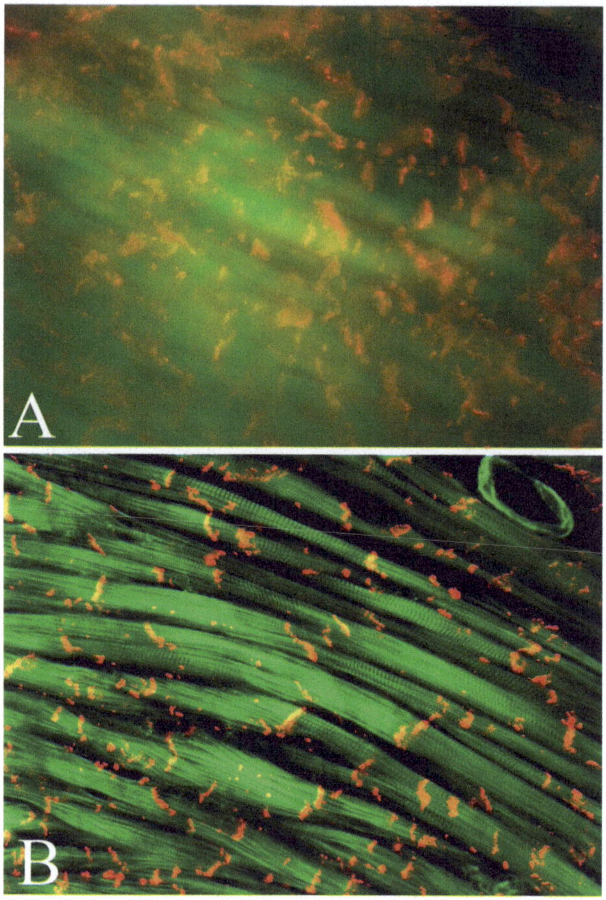

Fig. 1.3 Wide-field fluorescent (top) and single-photon confocal scanning laser microscope (CSLM) (bottom) images taken from a 100 m thick vibratome section of mouse heart stained for f-actin (green) and connexin 43 (red). In the wide-field image, out-of-focus light that contributes to the formation of the image significantly decreases the resolution and contrast of the image. Use of the pinhole in the confocal image to remove the out-of-focus light results in an image of much higher contrast and resolution as shown by the striated pattern of the myocyte sarcomeres and distinct cell: cell junctions labeled by the connexin 43 antibody

thin plane within a specimen is best exploited by scanning multiple thin planes in succession and reconstructing a high-resolution 3-D map of the specimen by stacking the 2-D images. As described in Chaps. 6, 7, 8, 9 and 10, key advances in digital imaging, detectors, and improved computer power over the last two decades now provide a convenient method of capturing, storing, and displaying sequentially acquired image information in both 2-D and 3-D formats.

While the above approaches to confocal imaging are still prominent, robust, and very important in today's research environment, more recently the development of enhanced and super-resolution confocal microscopes has significantly expanded the field of confocal and fluorescence microscopy. The importance of these techniques in furthering our understanding of many biological principles was evidenced by awarding of the 2014 Nobel Prize in Chemistry to Drs. Eric Betzig, Stefan Hell, and William Moerner. While the enhanced and super-resolution techniques may use different approaches for improving resolution, most use deconvolution, or mathematically reassigning signal generated by photons back to its point of origin, to improve resolution. A detailed discussion of deconvolution will be presented in Chap. 8, and examples of how some of these techniques are used to beat the 200 nm lateral resolution limits due to diffraction of light will be discussed in several chapters.

1.4 Is the Confocal Hype Legitimate?

Why has confocal microscopy revolutionized the way many laboratories image their samples? The simple answer is that the use of specific wavelengths of light, typically emitted from a laser, and the use of pinholes or some other mechanism to eliminate or reassign out-of-focus light as briefly mentioned above (and described in more detail in Chap. 9), has significantly increased our ability to resolve and co-localize small structures and molecules in high contrast images. An example of this is shown in Fig. 1.3. Wide-field images (Fig. 1.3a) contain large amounts of out-of-focus light that significantly deteriorates image resolution and contrast making it difficult to observe specific structures and detail. A confocal image (Fig. 1.3b) from the same region of the same sample clearly shows increased resolution and contrast making it much easier to discern the structures present in the section of heart muscle shown.

The improvement in image quality in Fig. 1.3 is obvious, but the confocal image in Fig. 1.3b remains limited by the diffraction of light and instrument configuration. As an example of further improvement in image quality available with enhanced resolution techniques, Fig. 1.4 shows a comparison of images from the same optical field of the hippocampus in a brain slice collected in confocal and enhanced (Airyscan) resolution mode with a Zeiss LSM 800 Airyscan confocal microscope. The improved resolution, as shown by the punctate staining in the Airyscan mode, is apparent, while the loss of resolution in the normal confocal mode is evidenced by the diffuse distribution of the green fluorescence in the tissue.

With the development of sensitive detectors, fast computing capabilities, and high-density media for storage, confocal imaging technology has grown rapidly. These advancements have made it possible to collect a large number of optical sections through a sample and to rapidly reconstruct them into a high-resolution high contrast projection of the sample where all detail is in focus (Fig. 1.5). Further advances in imaging software have made the use of 3-D data sets an important element in studying most biological systems. Many of these advances will be discussed in subsequent chapters of this book. However, both confocal imaging hardware and digital imaging software technologies are advancing at a very rapid pace making it essential that researchers stay vigilant in determining how confocal imaging may benefit their individual research programs.

The answer to the above question about confocal hype is obviously a resounding yes. Even though commercially available systems have only been available for about 30 years, and well-equipped confocal systems often cost $500 K or more and can be expensive to maintain, the thousands of publications that utilize confocal imaging and the large range of applications from biological to material samples

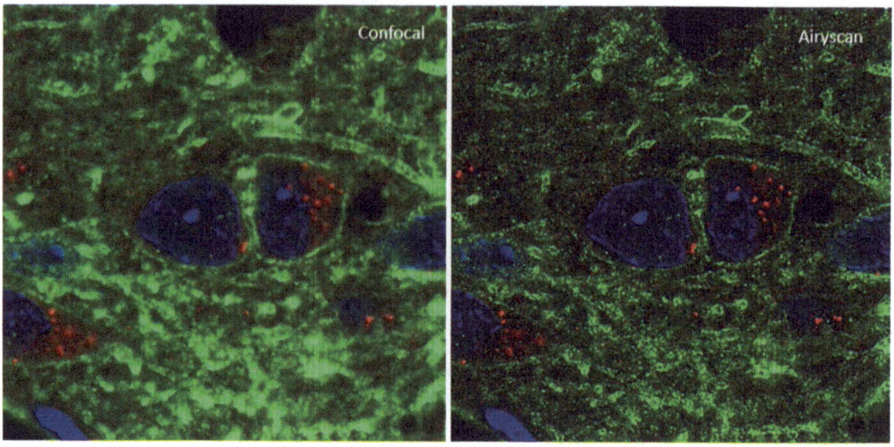

Fig. 1.4 Comparison of images collected on the Zeiss LSM 800 in normal confocal and Airyscan modes from the hippocampal region of a brain slice. Resolution of the Airyscan images is significantly improved over that seen in the image collected in confocal mode showing the punctate presynaptic terminals labeled with M2 mAChRs (red) and the GluN1 subunit of the NMDA receptor important for synaptic plasticity and learning

imaged clearly indicate that confocal microscopy has revolutionized the way many laboratories perform their research. Recent advances including spectral imaging, new fluorochromes and lasers, and increased imaging speed and resolution all indicate that confocal imaging will continue to be an important component of the imaging sciences in many fields of investigation.

1.5 The Ten Commandments of Confocal Imaging

As part of our Basic Confocal Microscopy Workshop, we often have students create a list of Confocal Commandments, which are comprised of statements we make that might be considered unequivocal in nature. The following is a list of some of these commandments that we have collected over the years that need to be considered by all undertaking the task of learning and using confocal microscopy as a research tool. These commandments establish some general guidelines to consider when using a confocal microscope, preparing a specimen, and handling digital images, which are all integral and equal parts of operating a confocal microscope. In fact, how we process and present the images we collect is every bit as important as how we do the initial data collection. The various chapters in this book will expand on the basic principles that lead to these commandments.

Our Ten Commandments of confocal imaging are as follows.

1.5.1 The Perfect Microscope and the Perfect Microscopist Do Not Exist

As we will discuss in great detail, physical factors inherent when using photons to produce microscopic images and the characteristics intrinsic to the design of many microscopes result in limitations in the amount of light that can be collected and restricts the obtainable resolution. While super-resolution systems have circumvented some of these defects, the confocal systems commonly available in core facilities and laboratories still must contend with these limitations. This makes it even more critical that the operator understands and adheres to proper preparation of specimens and knows how to appropriately set up the microscope before capturing images. Although some limitations can be minimized by selection of optimal microscope components, they cannot be totally eliminated. Even with the best microscope optics available, the physical nature of light and refractive index mismatch as the light passes through the several interfaces in the optical path of the microscope and specimen will result in image defects. These defects result in the loss of signal and resolution.

Moreover, even with optimal image quality, the human element of understanding image collection and data interpretation is often a limiting factor in getting the most out of a microscope. North (2006), in a feature article for the *Journal of Cell Biology*, noted that all data are subject to interpretation and that in microscopy a great number of errors are introduced in complete innocence. A common example is the frequent interpretation that the appearance of the color yellow in a sample stained with green and red emitting fluorophores indicates co-localization. However, many factors may affect this interpretation. Without a thorough understanding of sample preparation, optics, imaging parameters, and data analysis, an incorrect conclusion of co-localization may be reached in complete innocence. Several reasons why yellow in an image generated from a sample stained with green and red fluorophores may not represent true co-localization will be discussed in subsequent chapters.

1.5.2 Confocal Microscopy Is More Than a Confocal Microscope

To effectively use a confocal microscope, investigators must have an understanding of specimen fixation and processing, antigen-antibody interactions, fluorescence theory, microscope optics and hardware components, and the handling of digital images for both image enhancement and analysis protocols. Each of these topics will be addressed in subsequent commandments and discussed in detail throughout the text.

The fact that performing confocal microscopy is much more than operating a microscope is illustrated by the sequence of the following chapters. It is essential that information on specimen preparation, fluorescence theory, and the basics of

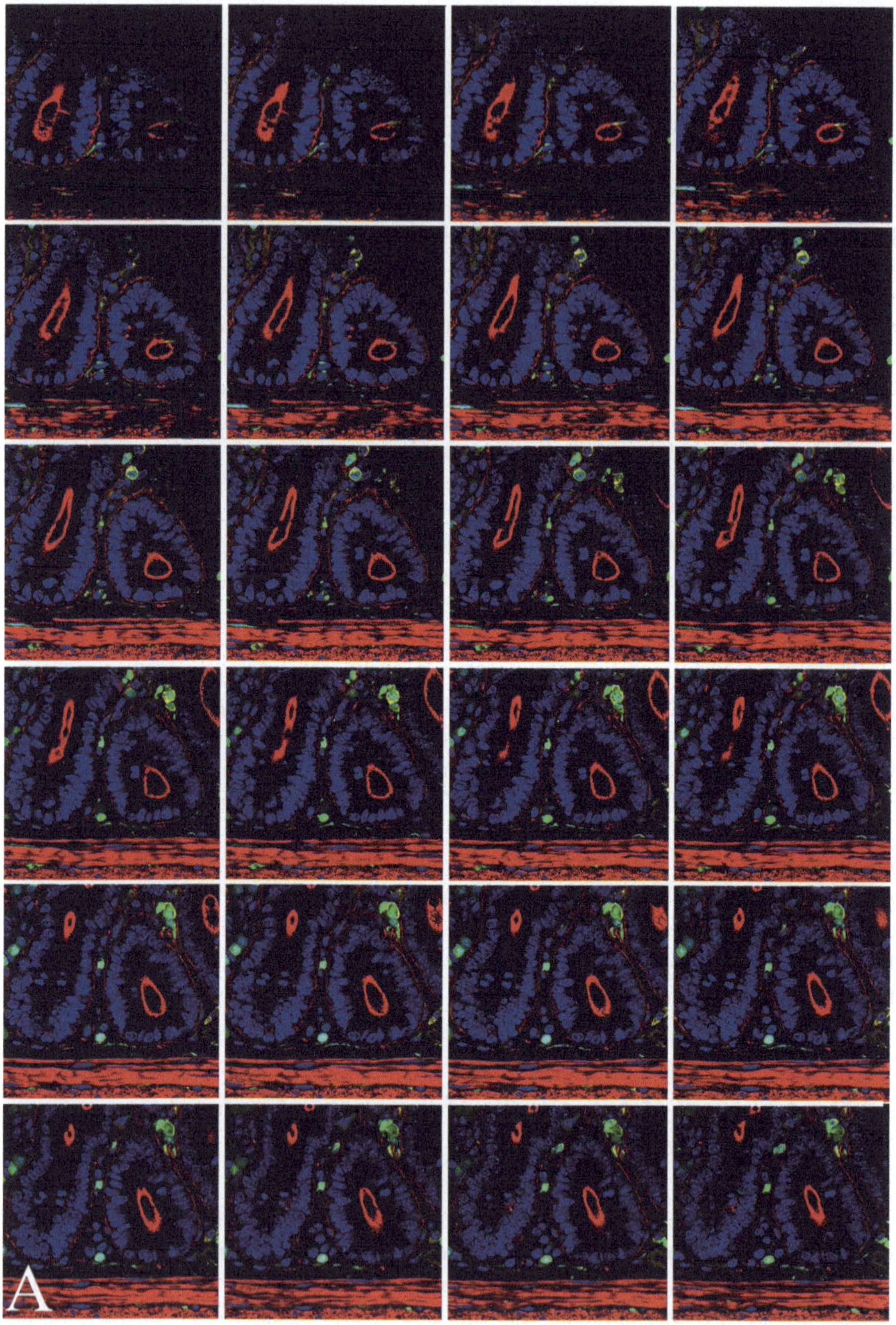

Fig. 1.5 Confocal optical sections (Z-series) through a section of intestine stained with multiple fluorescent dyes. Images were collected at 1 m intervals through a 50 m thick section of tissue, and every other section (2 m intervals) is shown in (**a**). All sections were then projected into a single composite image as shown in (**b**). The procedures for collection and projection of data sets are discussed in later chapters. Blue, DAPI stain for nuclei; red, f-actin stain; green, green fluorescent protein; yellow, mRNA-stabilizing protein

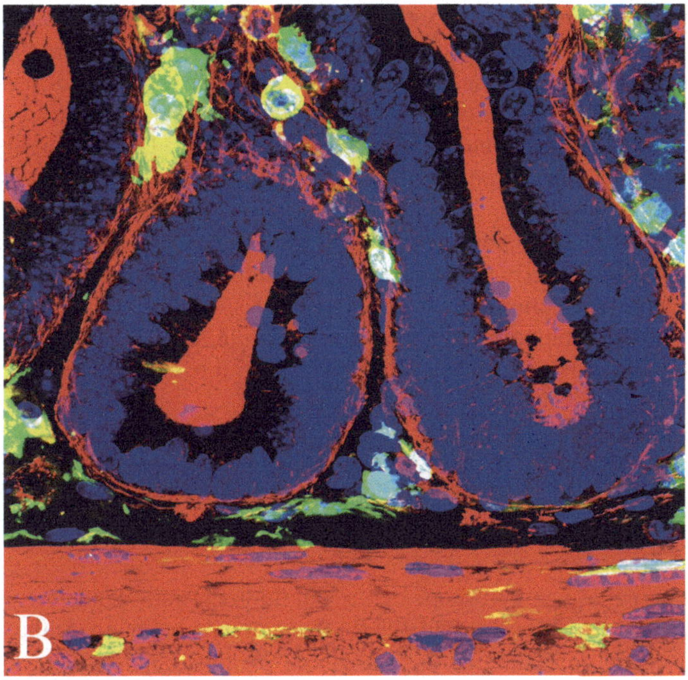

Fig. 1.5 (continued)

digital imaging be provided prior to material on confocal instrumentation if users are to understand the operation of a confocal microscope and be able to get the optimum amount of information from their samples.

1.5.3 During Specimen Processing the Integrity of the Specimen Must Be Maintained as Much as Possible

The integrity of the specimen includes the 3-D architecture. A major advantage of confocal imaging when compared to wide-field epifluorescence imaging is the acquisition of high-resolution, high contrast images which can be obtained through the Z-axis of a sample and the capability of software programs to reconstruct the 3-D nature of cells and tissues (Fig. 1.5).

Biological confocal microscopy often involves antigen staining to localize specific molecules and structures. It is essential that specimen fixation and subsequent processing maintain, as much as possible, the antigenicity of a specimen and the in vivo localization of cell and tissue antigens, analytes, structural components, etc. This may require extensive adjustment of protocols involving time, temperature, pH, and concentrations of fixatives and primary and secondary antibody solutions. Many of these issues, such as antigenicity and antibody penetration, have become more relevant as tissue clearing techniques have increased in popularity. Chapter 4 will address advantages and disadvantages of tissue clearing protocols, and Chap. 5 will address antigen-antibody interactions, labeling strategies, and potential problems that may arise during staining of samples with various fluorochromes.

Once successful processing protocols are developed, it is also essential that specimens be mounted properly to maintain the 3-D architecture of the sample. Chapter 4 also presents information on various aspects of specimen preparation including the use of various fixatives, buffers, mounting media, and strategies for mounting specimens to insure maintenance of the 3-D architecture of the specimen.

1.5.4 Photons Are Your Friends and Signal-to-Noise Ratio (SNR) Is King

Many factors including microscope optics and fluorochrome characteristics tend to reduce the number of photons available for formation of an image. At the same time that we are trying to maximize the number of photons (signal) collected, microscope hardware such as detectors and electronics introduce electronic noise that may result in a poor SNR. As a result, the operator must always be aware of the SNR in an image in an effort to establish operating parameters that maximize image quality while minimizing specimen damage. The SNR problem is being addressed by the development of new, high-sensitivity detectors such as the GaAsP and hybrid detectors discussed in Chap. 7. However, although these new detectors provide a better set of tools, their sensitivity is not infinite, and so sufficient signal-to-noise ratio remains a problem in many imaging protocols. In particular, a high SNR is critical for some forms of enhanced resolution microscopy. Thus, several chapters in this book discuss various aspects of fluorochrome and system properties that affect the SNR and provide suggestions on how to maximize the signal for optimal image quality.

1.5.5 Quantification of Fluorescence in a Confocal Micrograph Is a Challenge and at Best Is Only Semiquantitative

This is perhaps one of the most important commandments when dealing with today's competitive research environment and the need for quantitative data that is essential for funding opportunities and high-impact publications. Even though a large percentage of researchers using confocal microscopes report quantitative

Table 1.1 List of some factors (adapted from Pawley 2000) that may affect the quality and quantification of confocal images. The relevance of these and other factors will be discussed in subsequent chapters with the goal of improving the confocal imaging experience of students, technologists, and principal investigators

Microscope, specimen, or image component	Consideration that may affect quantitation
Laser unit	Alignment
	Instability with age
	Efficiency of optical coupling
Scanning system	Zoom magnification/Nyquist considerations
	Raster (pixel) size
	Distortions
	Environment (stray fields, vibrations)
Microscope objective characteristics	Numerical aperture
	Magnification
	Dipping/immersion lens
	Spherical/chromatic aberrations
	Cleanliness
Other optical components	Mirrors
	Excitation and emission filters
	Coverslips
	Immersion oil
Fluorochromes	Concentration
	Quantum efficiency
	Saturation state
	Loading
	Quenching
	Reaction rates
	Dye/dye interactions – FRET
Pinhole	Alignment
	Diameter
Detectors	Sensitivity
	Inherent noise
Digitization	Linearity – statistical noise

results from their studies, one must use caution when inferring numerical data from images collected with a confocal microscope. Pawley (2000) posed the question "does a fluorescent micrograph reveal the actual location and number of labeled molecules in a cell or tissue" to members of his well-known *3D Microscopy of Living Cells* course. Based on responses collected in the course, he published "The 39 Steps: A Cautionary Tale of Quantitative 3-D Fluorescence Microscopy" in *BioTechniques*. Table 1.1 is an abbreviated list of some of the factors that microscopists using confocal systems must be aware of during every imaging session. The conclusion of Pawley's paper is that "all you can really be sure of measuring with most laser-scanning confocal microscopes in the fluorescence mode is some feature of the number of photons collected at a particular time." Throughout the following chapters, we will discuss many of the issues that limit the effectiveness of confocal microscopes as a quantitative research tool and provide tips and suggestions for specimen preparation, imaging parameters, and handling digital images so that as much data as possible can be collected from each image data set.

1.5.6 Scientific Digital Imaging and Normal Digital Imaging (Family Photography) Are Not the Same

The greatest power of digital imaging is that exact copies of data can easily be made. This is excellent when archiving data and reverting to the original files when image processing does not result in the desired effect. However, while it may seem obvious that much of the processing we do on images collected with over-the-shelf digital cameras should not be done with scientific images, the innocence of the investigator again may be a problem. For example, when adjusting the contrast and brightness of a confocal image in programs such as Photoshop, the gamma function should always be used rather than the contrast and brightness functions. Gamma corrections should also be performed only after the histogram stretch functions are completed. While rules such as this are not important in family photography applications, not applying them correctly to digital images collected for scientific applications has the potential to alter the appearance of the data.

As discussed extensively in Chaps. 6 and 12, it is essential that an original, unaltered file of the data is archived for reference. All changes in the image should be made only on a copy of the original file. There are specific guidelines that have been published by several groups, including the Microscopy Society of America (http://www.microscopy.org), that specifically state how scientific digital images should be handled. More information concerning these guidelines and the ethics of handling digital images generated for scientific studies will be provided in Chaps. 6, 9, 10, 11 and 12 on processing of confocal images and the ethics associated with the presentation of the images.

Most hardware used for the collection and display of digital images utilizes software that includes some form of image processing prior to rendering the image. Frequently, manufacturers do not make this information available resulting in images that are collected without a full understanding of how they have been processed by the hardware used in image capture. While this is typically not a problem in recreational photography, processing of scientific data by collection devices prior to saving the information should always be a concern. Whenever possible, when working with images collected as scientific data, a thorough understanding of how the images are collected and processed by the system hardware is desirable. Unfortunately, this information is sometimes difficult to obtain from the manufacturer of the equipment or even worse, considered proprietary and so never revealed. We strongly feel that equipment and software manufacturers owe it to the scientific community to make critical information that can affect image fidelity readily available.

1.5.7 Your Image Is Your Data: Garbage in Will Result in Garbage Out

One should always be detail oriented in sample preparation, image collection, and handling digital images. The factors listed in Table 1.1 and by Pawley (2000) that affect quantitative confocal microscopy imaging are equally important in the acquisition of images for qualitative studies in which "pretty" pictures to demonstrate a scientific point are required. Without heeding each of the factors, it is unlikely that publication quality confocal images will be generated or that data collection from images will be maximized.

1.5.8 The Resolution and Bit Depth Present in a Digital Image Are a One-Way Street

After image capture the resolution of an image is set, and image processing protocols will not improve or increase the resolution of an image. While it may be possible through gamma, contrast, and brightness functions, and other types of algorithms such as sharpening filters to improve the aesthetic appearance of an image, as will be seen in Chaps. 6, 7, 10, and 11, once an image is collected with hardware and software available on a system, any structures that can be resolved in the image will be present. Using software to increase the number of pixels in a digital image will not improve the resolution, but only result in the creation of pixels by interpolation. These pixels are created by an algorithm such as averaging neighboring pixel values and appear as the computer "believes" they should look.

One may argue that image processing through deconvolution improves the resolution of a data set, but the limits of resolution have already been determined by the hardware present on the microscope and the physical properties of the light used to collect it. Deconvolution uses the point spread function (PSF) to mathematically reassign photons from the blur to their point of origin to reduce blur and improve resolution of an image. However, the ultimate resolution was set during collection of the image by factors such as the wavelength of light used and numerical aperture of the objective as discussed in detail in Chap. 7. Deconvolution may enable one to better define the data present, but the limits of resolution were set during collection of the image.

1.5.9 The JPEG (Joint Photographic Experts Group) Image File Format Is EVIL but Useful

This statement applies to any file format that compresses the data and does not allow full recovery of all of the information present in the original file. The JPEG format is the one encountered most often in imaging and so the one we chose to single out.

As noted above, resolution is a one-way street, and the original data should be stored as collected. Chapter 6 will show that saving files in the JPEG format results in significant loss of information, and especially damaging to scientific images, this loss is greatest along the edges. All original data should be stored in a lossless format such as a TIF (Tagged Image File) or a proprietary version of a TIF format as provided by the instrument manufacturer. JPEG and other compression formats may be used in situations where images either need to be shared electronically or inserted into formats for lectures, seminars, and posters. In these situations resolution may be sacrificed in favor of smaller file sizes to make handling of images more reasonable. However, these compressed images should never be used as the primary source of data. File format options will be discussed in detail in subsequent chapters.

1.5.10 Storage Media Is Essentially Free and Infinite

The message of this commandment from the first edition of *Basic Confocal Microscopy* is that it is essential that the original data sets be archived appropriately and that any image manipulation is performed on copies of the data that must also be archived. When the first edition was published, this essentially required several CDs or DVDs to accumulate and store data sets that were typically in the range of a few gigabytes at most. While essentially still true, this commandment from the first edition does need to be qualified to some degree. When compared to data sets from a few years ago, today's data sets have grown exponentially due to the speed of image collection, the types of imaging present such as light sheet microscopy, and resolution available. This has resulted in a single data set that may approach or even exceed several terabytes. It is also possible to collect several data sets in a fairly short period of time, and cumulative data sets are now approaching a petabyte in size. Thus the mechanism and potential cost of storage has changed significantly in the few short years since publication of the first edition and the term Big Data, and the cost and processing of Big Data, is becoming a concern for many laboratories.

Even though it may now be necessary to have servers with several nodes to store data, and these may be expensive, as noted in the first edition, the cost of data storage is still minimal compared to generating new experiments if the data is questioned and the original files are no longer available. In addition, archiving of data is now a requirement of many funding agencies such as NIH and NSF that have specific policies on data storage and accessibility for the scientific community (https://grants.nih.gov/grants/policy/data_sharing/ and https://www.nsf.gov/sbe/ses/common/archive.jsp), as do many other foundations and funding agencies. For example, NSF guidelines indicate that "for appropriate data sets, researchers should be prepared to place their data in fully cleaned and documented form in a data archive or library within one year after the expiration of an award. Before an award is made, investigators will be asked to specify in writing where they plan to deposit their data set(s)." Additional archiving guidelines will be discussed further in Chap. 12 on Ethics and Resources.

Thus it is no longer acceptable to simply archive data on personal storage devices such as external hard drives, thumb drives, etc., and there may be significant cost involved with storage on Cloud devices or servers supported by universities.

1.6 Summary

These Ten Commandments for confocal imaging provide a set of principles to guide users in a confocal microscopy laboratory. Other commandments have occasionally been added to the list during our workshops, but if close attention is paid to each of the above, and a detailed understanding of the importance of each is developed, users will have a strong understanding of confocal technology for use in their research.

In Chaps. 2, 3, 4, 5, 6, and 7, we present information on the topics of fluorescence, specimen preparation, and digital imaging which are essential for understanding confocal imaging. In subsequent chapters we present information on various types of confocal instruments, the proper setup of operating parameters for confocal imaging, and appropriate techniques for enhancing and analyzing confocal images. Topics pertinent to the various commandments as well as some frequently asked questions such as:

1. Are these fluorescent markers co-localized?
2. Can I quantify the amount of labeled material present based on the fluorescence intensity which is present?
3. Can I measure the size or area of these structures based on a confocal data set?
4. How deep can I image into my sample?

are addressed. Hopefully by learning the basic principles of confocal imaging, the quality of the confocal imaging experience of many beginning and intermediate users of the technology will be improved.

Literature Cited

Ashdown GW, Cope A, Wiseman PW, Owen DM (2014) Molecular Flow Quantified beyond the Diffraction Limit by Spatiotemporal Image Correlation of Structured Illumination Microscopy Data. Biophys J 107:L21–L23

Berek M (1927) Grundlagen der tiefenwahrnehmung im mikroskop. Marburg Sitzungs Ber 62:189–223

Dempsey GT, Vaughan JC, Chen KH, Bates M, Zhuang X (2011) Evaluation of fluorophores for optimal performance in localization-based super-resolution imaging. Nat Methods 8:1027–1036. https://doi.org/10.1038/nmeth.1768

Eggar MD, Petráň M (1967) New reflected light microscope for viewing unstained brain and ganglion cells. Science 157(786):305–307

Inoue S (2006) Foundations of Confocal Scanned Imaging in Light Microscopy. In: Pawley JP (ed) Handbook of Biological Confocal Microscopy, 3rd edn. Springer, New York 985 pp

Minsky, M. (1957) U.S. Patent #3013467. Microscopy Apparatus

Minsky M (1988) Memoir on inventing the confocal scanning microscope. Scanning 10:128–138
Nipkow, P. (1884) German Patent no. 30, 105. Germany
North AJ (2006) Seeing is believing? A beginners' guide to practical pitfalls in image acquisition. J Cell Biol 172:9–18
Pawley J (2000) The 39 steps: A cautionary tale of quantitative 3-D fluorescence microscopy. BioTechniques 28:884–888
Petráň M, Hadravsky M, Egger MD, Galambos R (1968) Tandem scanning reflectgd light microscope. J Opt Soc Am 58:661–664
Wentao Y, Ziheng J, Dashan D, Xusan Y, Yunfeng X, Qihuang G, Peng X, Kebin S (2016) Super-resolution deep imaging with hollow Bessel beam STED microscopy. Laser Photonics Rev 10(1):147–152. https://doi.org/10.1002/lpor.201500151
White JG, Amos WB, Fordham M (1987) An evaluation of confocal versus conventional imaging of biological structures by fluorescence light microscopy. J Cell Biol 105:41–48

Chapter 2
The Theory of Fluorescence

W. Gray (Jay) Jerome

2.1 Introduction

Confocal microscopy uses either transmission or reflection mode to look at nonfluorescent material. However, for most biological work, the confocal microscope is used in transmission mode with fluorescent samples. Fluorescence imaging provides a specific, high-contrast signal that maximally exploits the ability of the confocal microscope to remove out-of-focus light. For this reason, it is imperative for confocal microscopists to have a basic knowledge of fluorescence theory and imaging. In this chapter we review the fundamentals of fluorescence as applied to confocal imaging. In most cases, these basic principles are also applicable to wide-field fluorescence microscopy.

2.2 General Principals

Fluorescence microscopy usually involves observing light within the visible range, although detection systems such as charge-coupled device (CCD) cameras are available that will detect fluorescence in the ultraviolet (UV) and infrared (IR) range. Although UV and IR detectors have their uses, the current discussion will be limited mostly to detecting visible light, since this is the range of wavelengths most often used in biological confocal microscopy.

Visible light is that portion of the electromagnetic spectrum that can be detected with the human eye. The color that the brain perceives is a function of the specific wavelength of the light; this encompasses electromagnetic wavelengths in the range

W. G. Jerome (✉)
Department of Pathology, Microbiology and Immunology, Vanderbilt University School of Medicine, Nashville, TN, USA
e-mail: Jay.Jerome@vanderbilt.edu

W. G. Jerome, R. L. Price (eds.), *Basic Confocal Microscopy*,
https://doi.org/10.1007/978-3-319-97454-5_2

Fig. 2.1 The visible electromagnetic spectrum

from about 380 nm to 750 nm. Although the visible light spectrum is continuous, it can be somewhat arbitrarily divided up into discrete colors (Fig. 2.1) according to wavelength. These divisions are violet (380–425 nm), indigo (425–450 nm), blue (450–495 nm), green (495–570 nm), yellow (570–590 nm), orange (590–620 nm), and red (620–750 nm). Indigo is not well differentiated from violet by human eyes, so wavelengths in the indigo range are often included in the violet category. Near-UV light (300–380 nm) is useful in fluorescence microscopy primarily as a source for excitation photons, and the near-infrared wavelengths (750–1,400 nm) are useful for multiphoton excitation.

A fluorescent molecule (fluorochrome) is one that absorbs a photon of light of a particular wavelength and, after a brief interval, emits some of that energy in the form of a photon of a different wavelength. The delay in emission is called the fluorescence lifetime. Some of the absorbed energy is lost to non-radiation processes, and so the emitted photon has less energy than the absorbed photon. Planck's law indicates that the radiation energy of a photon is inversely proportional to its wavelength. Since the emitted photon will have less energy, it will have a longer wavelength and thus a different color than the absorbed photon. The difference in wavelength between the absorbed and emitted light is called the Stokes shift. Although the shift will always be to a longer wavelength, the degree of shift is highly dependent upon the molecule being excited. Table 2.1 lists the Stokes shift and the excitation and emission maxima for some fluorochromes often used in confocal microscopy. It is important to know these values for each fluorochrome with which you work. These values are critical guides to setting up a microscope to insure that the detected fluorescence is in fact coming from a specific fluorochrome. Being aware of the excitation and emission curves and Stokes shift is particularly critical when designing experiments with multiple fluorochromes to prevent overlap in the excitation and emission spectra of fluorochromes in the sample and resultant misinterpretation of the image. This is discussed in more detail in Chaps. 3 and 9.

Energy absorption by a fluorochrome is the result of interactions of the oscillating electric field vector of the light wave with electrons in the fluorochrome. For a given electron, a certain amount of energy is required for an electronic transition. Only energies close to this transition energy will be absorbed; the fluorochrome is essentially transparent to photons of other energies. Since energy and wavelength are related, this means that for each fluorochrome there are only certain wavelengths

Table 2.1 Excitation and emission wavelengths for some common fluorochromes

Dye	Excitation max (nm)	Emission max (nm)	Difference (Stokes Shift)
Hoechst 33342	352	461	109
DAPI	358	461	103
Lucifer Yellow VS	430	535	105
FITC	490	523	33
Alexa Fluor ® 488 (0.92)	495	519	24
Cy 2 (0.12)	492	510	18
Alexa Fluor 546 (0.79)			
Propidium Iodide	536	617	81
Cy 3 (0.15)	548	562	14
DiI	549	565	16
TRITC	557	576	19
Lissamine Rhodamine B	575	595	20
Alexa Fluor 594 (0.66)	590	617	27
Texas Red	596	615	19
Cy 5 (0.28)	649	669	20
Alexa Fluor ® 647 (0.33)	650	668	18

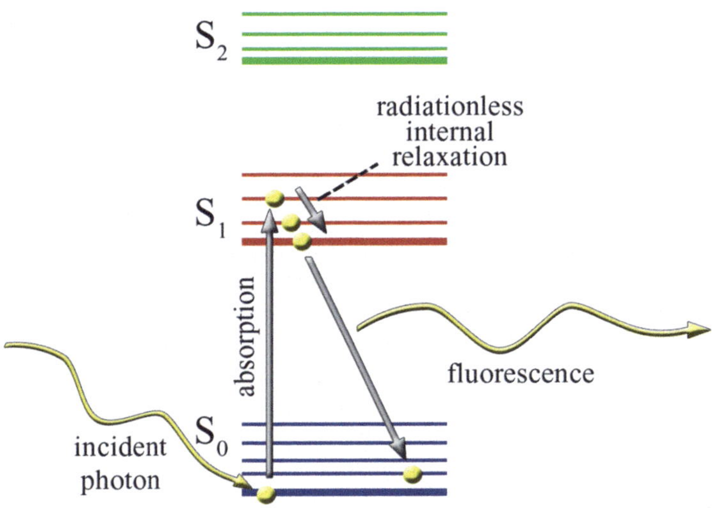

Fig. 2.2 Jablonski diagram of excitation and fluorescence

capable of a productive interaction that generates fluorescence. This provides specificity to the fluorescence process.

The absorption and emission of photons by a fluorochrome are best illustrated using a Jablonski energy diagram (Fig. 2.2), named for Aleksander Jablonski, the Polish physicist who introduced the diagram in 1953. In this diagram, increasing energy is depicted along the vertical axis from bottom to top. At room temperature, most electrons in the fluorochromes will only have enough energy to exist in the lowest energy (ground) state (S_0). However, the absorption of a photon transfers the photon's energy to a ground state electron and raises the electron's energy to a higher (excited) state, usually the S_1 state (first excited singlet state), although higher singlet states are possible (e.g., S_2, etc.).

Absorption occurs within femtoseconds. Within each state (S_0, S_1, S_n), several vibrational sublevels exist, and the transition can be to any of the possible sublevels, depending upon the energy absorbed. Soon after absorption (~ picoseconds), the electron's energy relaxes to the lowest vibrational level within the excited state. This process is termed internal conversion and does not result in the generation of a photon. Rather, the energy is released in non-photon-generating reactions. Most often this loss is in the form of heat absorbed by the surrounding material. Eventually, the electron reaches the lowest S_1 sublevel (lowest excited singlet state). From the S_1 state, the electron's energy can undergo a slower (~nanoseconds) decay back to the S_0 state. The energy loss during the decay is in the form of a photon that has lower energy (longer wavelength) than the excitation photon. This is the observed fluorescence. For good fluorochromes, most of the return to the S_0 state involves this fluorescence pathway. Importantly, only the decay from S_1 to the ground state is capable of producing fluorescence. Thus, for any given fluorochrome, the emission spectrum is predictable.

In addition to fluorescence, there are other competing methods by which the excited molecule can lose energy. The most common are heat generation and transfer of energy to a neighbor molecule. These alternative processes can reduce the fluorescence emitted from a population of fluorochromes.

Molecules can exist in more than one excited state. For organic fluorophores the singlet state is most common. In the singlet state, all electrons are spin paired. However, molecules can also exist in a triplet state, where one set of electron spins is unpaired. There is an increasing likelihood of a molecule undergoing an intersystem crossing from the singlet to triplet state when the energy absorbed excites the molecule to states above the S_1 level. From the triplet state, the energy can be lost as internal conversion to heat allowing decay back to the S_1 state generating a photon (albeit with delayed timing). This is called phosphorescence.

For an individual molecule, absorption is an all-or-none phenomenon; it only occurs if the incident light has sufficient energy for an electronic transition of the fluorochrome to a higher energy state. These states are determined by the atomic organization of the molecule. The limitation on allowable wavelengths that will excite a fluorochrome is due to the fact that the transition from ground state to excited state happens so quickly that the molecules cannot move. Thus, the only transitions allowed are those in which the electron positions in the ground and excited state overlap. This restricts the wavelengths (energies) which can produce a transition. This is a key property for the use of fluorochromes in microscopy because it allows the identification of specific fluorochromes in a sample based on their excitation and emission frequencies.

Because within each of a molecule's excitation states there are a number of closely spaced energy sublevels, there is not a single energy but a range of energies that can excite the fluorochrome. The emitted photons also have a distinct range of energies. For this reason, a particular fluorochrome will have a range of excitation and emission energies with maxima at the most likely energies. For instance, Texas Red absorbs wavelengths in a range from about 470 nm to 630 nm with a maximum at about 596 nm. In comparison, the absorption spectrum for Lissamine Rhodamine B ranges from about 470 nm to 610 nm and has a maximum at 575 nm (Table 2.1).

For most fluorochromes, the photon absorption and fluorochrome excitation spectra are matched; i.e., the molecule is maximally excited when photons are maximally absorbed. Since it is easier to measure the excitation of the molecule, this is the information that is usually provided. Figure 2.3 shows the excitation and emission spectra for tetramethylrhodamine isothiocyanate (TRITC) in methanol. Although TRITC is maximally excited at a wavelength of 557 nm, absorbance of wavelengths slightly less than or more than 557 nm can also result in substantial fluorescence. Obviously, since more than one wavelength can excite a specific fluorochrome and photons with a range of wavelengths are emitted, when designing fluorescence experiments, it is important to know the range of wavelengths applicable for a particular fluorochrome, not just the excitation and emission maxima.

Excitation and emission characteristics for fluorochromes are typically provided in the Material Safety Data Sheet (MSDS) supplied when a fluorochrome is purchased. If the spectral information for the fluorochrome is not readily available, it is usually possible to find it on websites such as the Invitrogen Spectra Viewer (http://www.invitrogen.com/site/us/en/home/support/Research-Tools/Fluorescence-SpectraViewer.html). Websites of this nature are also valuable when selecting

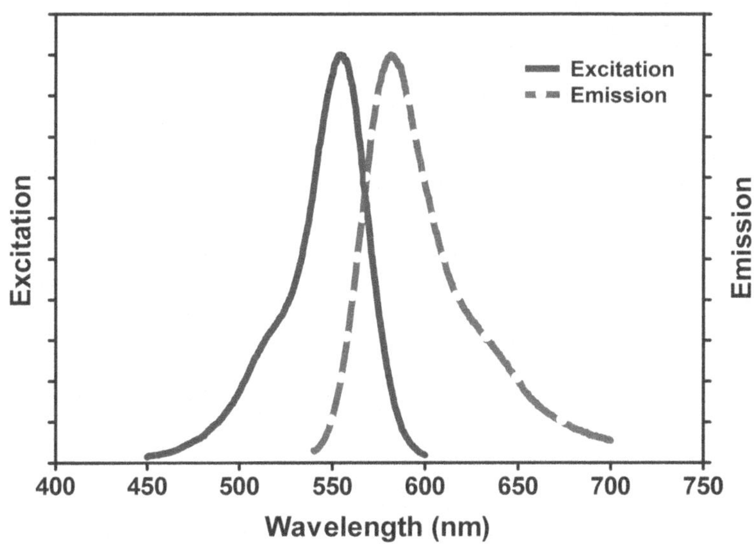

Fig. 2.3 Excitation and emission spectra for tetramethylrhodamine isothiocyanate (TRITC) in methanol

multiple fluorochromes to insure minimal overlap of excitation and emission as discussed in Chap. 3.

The fact that the emitted wavelength is longer than the absorbed wavelength is the fundamental property that permits fluorescence microscopy because it allows for the excitation light to be filtered out from the specific wavelengths emitted from the specimen. In this way, the final image contains only the information from the fluorescence emitted from the specimen. This provides a very high-contrast image where only the emitted light is observed. The bright, high-contrast image produced by fluorescence microscopy is ideally suited for confocal microscopy.

An additional important piece of information about a fluorochrome is its quantum yield. This is the ratio of the number of photons emitted relative to the number of photons absorbed (Eq. 2.1) and is, therefore, a measure of the efficiency of fluorescence:

$$Y(\Phi) = \frac{\text{photons emitted}}{\text{photons absorbed}} = \frac{k_f}{(k_f + k_{\text{nr}})} \tag{2.1}$$

where QY = quantum yield

k_f = rate constant for fluorescence decay

k_{n} = sum of rate constants for non-radiactive effects (e.g. internal conversion, quenching)

A perfect emitter would produce one photon of longer wavelength for every photon absorbed. However, in most situations non-radiative phenomena such as heating and photobleaching reduce the energy leading to quantum yields that are less than 1. Table 2.2 lists the quantum yields for several typical fluorochromes. Another important, and related, fluorescence parameter is the molar extinction coefficient. The extinction coefficient is the quantity of light absorbed as a function of the path

Table 2.2 Quantum yields

Fluorophore	Solution	QY
Bodipy ® FL	Water	0.90[a]
Cy 5 ™	PBS	0.27[b]
Fluorescein	Aqueous 0.1 N NaOH	0.94[c]
JOE ™	Water	0.60[a]
Oregon Green ® 488	Water	0.90[a]
Rhodamine B	Water	0.31[d]
Tetramethylrhodamine	Water	0.20[a]
Texas Red	Water	0.90[a]

[a]VL. Singer, ID. Johnson. http://www.promega.com/geneticidproc/ussymp8proc/21.html
[b]Mujumdar et al. (1993)
[c]Magde et al. (2002)
[d]Magde et al. (1999)

length of the excitation light and the concentration of the fluorochrome. It is usually measured at a path length of 1 cm and at a one molar concentration. As a direct measure of the ability of a molecule to absorb light, it is obvious that those molecules with high molar extinction coefficients are more likely to exhibit fluorescence.

2.3 Factors Affecting Fluorescence Emission

While it may seem intuitive that fluorochromes with the highest quantum yield, such as fluorescein isothiocyanate (FITC) at 0.94, should be used for confocal microscopy, the probability of fluorescence emission is greatly dependent upon the surroundings of the fluorochrome. Contact with any adjacent particle that has the ability to accept transferred energy from the fluorochrome can rob the fluorochrome of sufficient energy for a fluorescent event to occur. For instance, binding of a fluorochrome to another molecule, such as an antibody, usually reduces the probability of fluorescence. The loss of energy due to transfer to surrounding molecules is known as vibrational relaxation. It is very common in fluids or gasses because in these situations the solvent molecules are more likely to collide with the fluorochrome. The resulting vibrational relaxation produces a reduction in fluorescence intensity from a population of fluorochromes. It is important to remember that these interactions do not alter the shape of the emission spectrum; only the height of the curve (intensity) is reduced indicating decreased emission of photons.

Energy absorption, particularly of excess amounts, can also promote chemical modifications of the molecule that render it no longer capable of fluorescence. This is called photobleaching. Photobleaching is a critical concept to successful fluorescence microscopy. Although most fluorochromes can go through multiple excitation/relaxation cycles, chemical reactions will eventually alter the molecule such that it no longer fluoresces. It is estimated that a single FITC molecule can only undergo about 30,000 cycles before it can no longer absorb an incident photon. At this point, the FITC molecule is said to be “bleached.” This is not a long time given that the cycles are nanoseconds in length. Thus, for confocal microscopy the bleaching process for FITC occurs very rapidly, and even though fluorescein has a very high quantum yield, its bleaching characteristics when excited with a high-intensity laser generally make it an unsuitable fluorochrome for confocal microscopy. Luckily, as described in Sect. 2.6, the popularity of confocal microscopy has spurred the development of new fluorochromes with greater photostability.

We described earlier the intersystem crossing to the triplet state. Transition back to the singlet state is not chemically favored, and so when a molecule enters the triplet state, most of its energy is lost to nonfluorescent radiation such as photobleaching or phosphorescence. A major cause of photobleaching is the reaction of the excited molecule with oxygen. Thus, antioxidants are often included in the mounting medium or surrounding fluid to help retard photobleaching of fluorochromes during microscopy (Chap. 4).

Instead of generating a photon, under favorable conditions the excitation energy of an excited fluorochrome can instead be transferred to a nearby fluorochrome in the ground state. This is called Förster resonance energy transfer (FRET). The transfer of energy from the donor fluorochrome leads to the excitation of the acceptor fluorochrome, and this energy can then return back to the ground state producing a photon with a different wavelength (Fig. 2.4). Thus, a donor fluorochrome sensitive to blue light can transfer its energy (which is now reduced because of internal conversion and relaxation within the S1 state) to an acceptor fluorochrome sensitive to this lower energy level (i.e., that of a green photon). The green-sensitive fluorochrome becomes excited, and the decay of its energy back to the ground state will produce a photon with a wavelength in the red spectrum. Thus, a blue incident

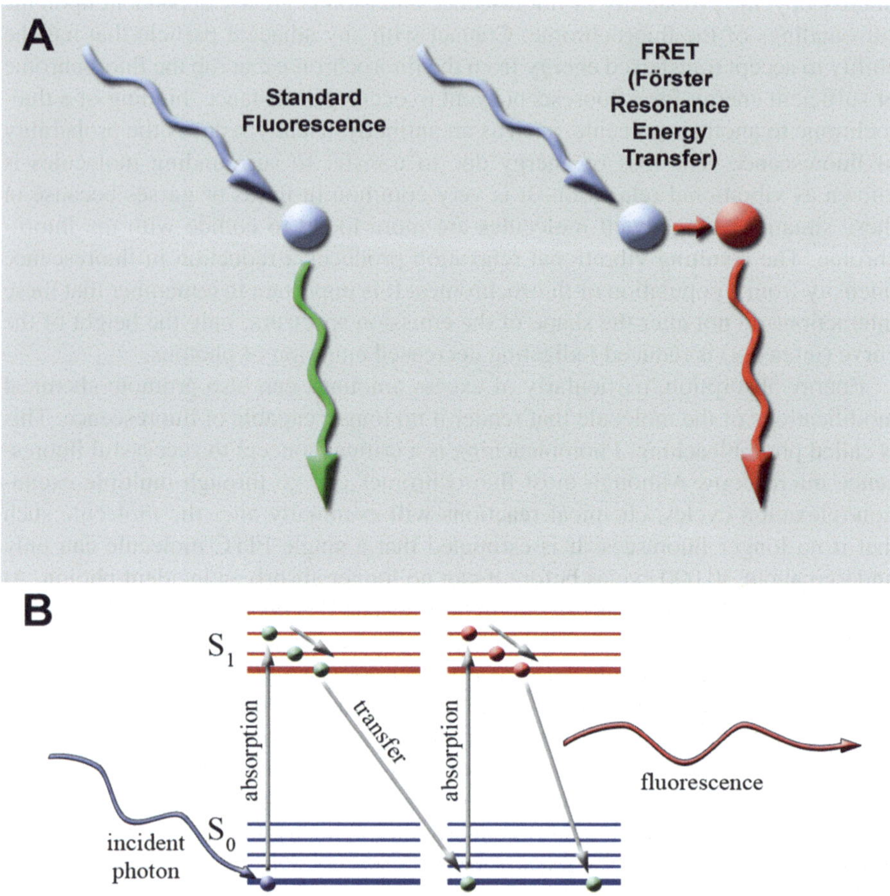

Fig. 2.4 Förster resonance energy transfer (FRET). (**a**) Schematic diagram comparing standard fluorescence emission to that occurring by Förster resonance energy transfer. (**b**) Jablonski diagram of energy during excitation, transfer, and emission occurring during Förster resonance energy transfer

photon yields red rather than green fluorescence. FRET is very sensitive to distance, and the efficiency of transfer is usually reduced by greater than 50% at distances larger than 5 nm and is nonexistent beyond about 10 nm separation. For this reason, FRET is a useful method for determining if two molecules are closely associated within a structure. Choosing the proper FRET pairs for labeling two molecules and determining if FRET has occurred requires care, since many factors besides distance can influence energy transfer. For those interested in further information, the review by Paul Selvin provides a nice introduction (Selvin 2000). FRET and other methods for microscopically co-localizing molecules are discussed further in Chap. 11.

2.4 Nonlinear Optical Methods

As described above, under most conditions a fluorescent event involves absorption of a single photon of relatively high energy resulting in the emission of a single photon of relatively lower energy. The single nature of the event is because absorption must occur very rapidly. However, if a large number of photons are packed into a very small space, the probability of two or more photons transferring their energy simultaneously, or nearly simultaneously, to a fluorochrome increases to the point that such an event can occur. This is known as multiphoton excitation. When the energy from multiple photons is absorbed, it is additive resulting in the transfer of enough energy to a fluorochrome molecule to raise it from the resting S_0 state to the excited S_1 state (Fig. 2.5). Thus, if a fluorophore is normally excited by photons with

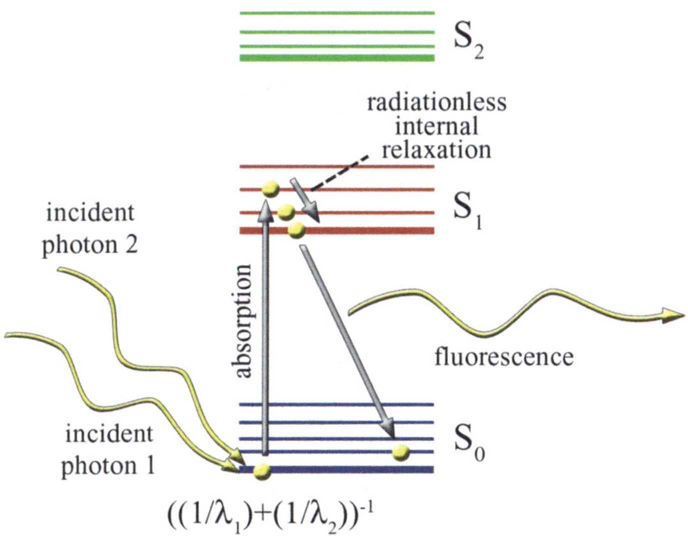

Fig. 2.5 Jablonski diagram for two photon excitation and fluorescence

a wavelength of 425 nm, it will also be excited if it absorbs two photons each of around 850 nm wavelength. Thus, in a multiphoton event, unlike single photon excitation, the emitted photon is of higher energy than the two or more low energy excitation photons.

Since the probability of two photons colliding with a fluorochrome at precisely the same time decreases rapidly away from the focal point of a light beam (the point where large numbers of photons are packed into a small space), in multiphoton microscopy the excited volume of a sample is confined to the area where the beam is focused. Fluorochrome molecules away from the focal point remain quiescent. The confinement of the excitation volume to the focal point, and the fact that longer excitation wavelengths are employed, provides advantages for fluorescence imaging of living material and increases the specimen depth from which useful information can be obtained. More details on the phenomenon and the use of multiphoton excitation in biological imaging are provided in Chaps. 8 and 9.

A separate but related nonlinear optical phenomenon useful for microscopy is second harmonic generation. To generate a second harmonic signal, two photons of the same energy interact with the specimen to generate photons with exactly twice the energy and therefore half the wavelength (Fig. 2.6). Unlike in two-photon microscopy, no energy is lost. The frequency doubling (production of a second harmonic wave) requires the illuminated sample to have very specific properties. Biological materials such as collagen and a number of cytoskeletal elements meet these criteria and so can be visualized by detecting the second harmonic signal coming from samples containing these materials (Fig. 2.7). Since it does not involve excitation of molecules, a second harmonic signal does not lead to photobleaching or phototoxicity. Moreover, since it exploits the optical properties of native materials, no exogenous label is required.

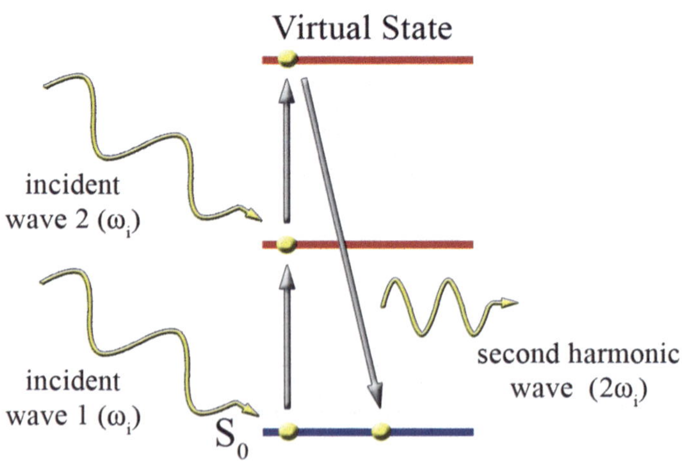

Fig. 2.6 Frequency doubling in second harmonic generation

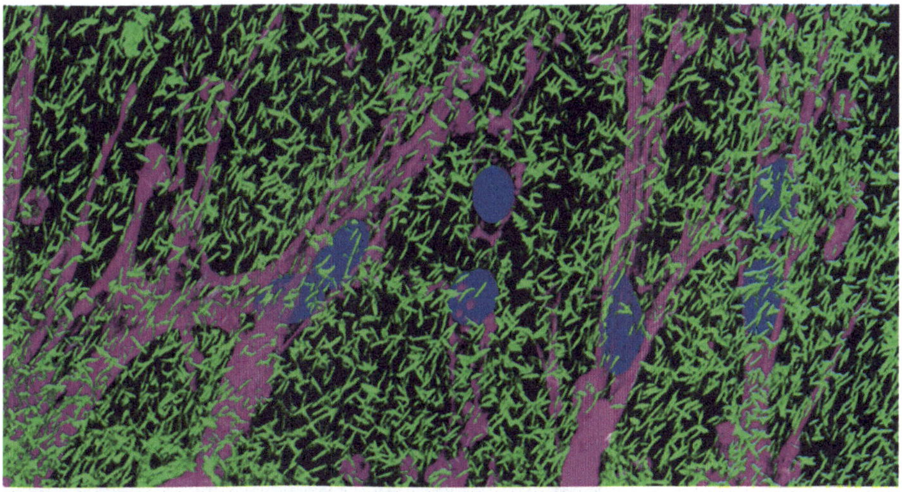

Fig. 2.7 Three-dimensional reconstruction of second harmonic generation images of collagen (green) from a 3D collagen gel culture of fibroblasts stained with DAPI (blue) and phalloidin (lavender)

2.5 Biological Specificity of Labeling

Fluorescence in subcellular imaging is most useful when there is specificity to what is fluorescently labeled. There are a number of mechanisms by which specificity can be assured. The most often employed method is to link the fluorochrome to an antibody molecule and use the specificity of the antibody binding to localize an antigen of interest. Details of this technique are provided in Chap. 5. However, this is not the only means of providing specificity. All types of ligand-receptor interactions can be exploited to bind a reporter molecule to a molecule of interest for microscopic localization. For instance, phalloidin has a high affinity for f-actin molecules. Fluorescent analogs of phalloidin can be used to label the actin cytoskeleton of cells (Fig. 2.8). In a similar manner, the binding of fluorescently labeled cDNA sequences to complementary sequences on DNA or RNA can be used to localize these sequences in cells or on isolated chromosomes. A fluorochrome linked to a biologically active substance is called a fluorophore.

Another method of identifying specific molecules, particularly large complex molecules, involves providing a small fluorescent precursor and allowing the cell to assemble the more complex molecule within the cell. Since the molecule contains the fluorochrome, it can be visualized microscopically. In a similar manner, the gene for a fluorescent protein, such as green fluorescent protein (GFP), can be linked to the gene for a known cellular molecule. Cells transfected with this construct will then synthesize a molecule of interest with its fluorescent reporter molecule already attached (Fig. 1.5) (for a review see Tsien 1998). The 2008 Nobel Prize in Chemistry was awarded to Osamu Shimomura, Martin Chalfie, and Roger Tsien for the isolation of GFP and the development and modifications of this protein into useful cytological tools.

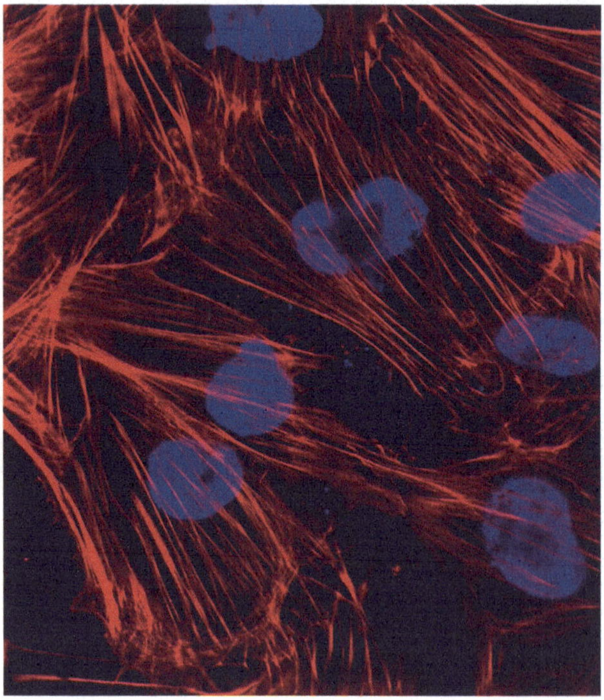

Fig. 2.8 Actin cytoskeleton stained with phalloidin (red) and nuclei stained with DAPI (blue)

Localized environmental conditions within a cell can also be probed using fluorophores that fluoresce only in a specific environment. These fluorochromes change their fluorescence properties in conjunction with changes in their environment. LysoSensor Yellow/Blue DND-160™ (Thermo Fisher Invitrogen) is one such molecule. This dye fluoresces blue at a neutral pH, but emission is shifted to longer wavelengths as the pH becomes more acidic. Thus, this fluorochrome can be used to follow the acidification of organelles (Diwu et al. 1999). Another such dye is HCS LipidTOX™ Red Neutral Lipid. This dye permeates throughout the cell but only fluoresces when surrounded by neutral lipids. It is useful for monitoring the accumulation of lipids within cells (Fig. 2.9). Similar dyes for imaging calcium fluxes (Fura) as well as other environmental physiological changes are also available for examining various cellular processes (O'Connor and Silver 1998).

Finally, cells contain naturally occurring fluorochromes that can be probed by fluorescence microscopy. The fluorescence properties of some of these are sensitive to their biochemical state. The reduced form of nicotinamide adenine dinucleotide (NADH) is fluorescent, while the oxidized form (NAD^+) is not. Because of this shift, one can use NAD's fluorescent properties to monitor the redox state of NAD (Williams et al. 1994; Piston and Knobel 1999).

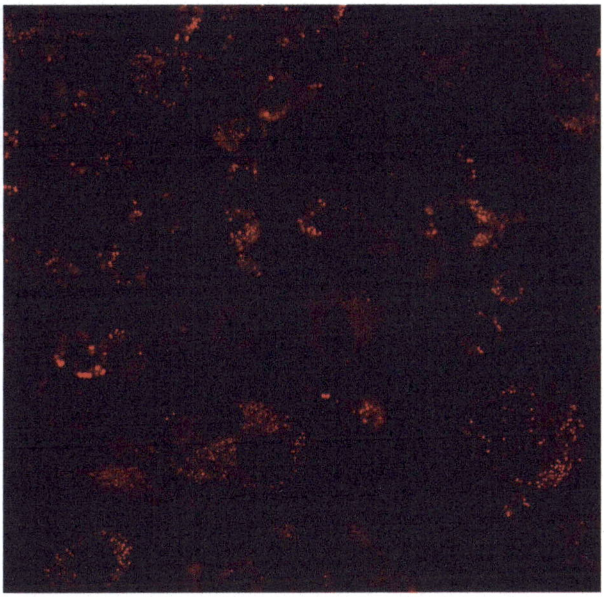

Fig. 2.9 LipidTOX red staining of neutral lipid droplets in cultured human macrophages

2.6 Recently Developed Fluorescent Probes

The use of fluorescence microscopy has greatly expanded in the past 15 years, fueled in part by an increase in the number of fluorochromes available. Many of the newest generation of fluorochromes, such as the cyanine and Alexa Fluor ® family of dyes, have been specifically engineered to have greater quantum yield and narrower excitation and emission spectra that match the demands of confocal imaging (Tables 2.1 and 2.2).

Most of the newer probes have relied on standard organic fluorochrome chemistry. However, some nontraditional approaches to fluorescence are now being developed. Among these are quantum dots. These are fluorescent semiconductor nanocrystals. Like organic fluorochromes, they emit light at a longer wavelength than that of the absorbed light. Several properties of quantum dots make them attractive for biological imaging; primary among these properties are their high quantum yields and extreme photostability. Quantum dots are very bright and exhibit very little photobleaching. Moreover, for a given composition of quantum dot, different sizes of the same crystal will emit at different wavelengths (Fig. 2.10). This means that a single type of material, for instance, CdSe, can produce an array of colors with similar brightness and similar chemical properties, such as binding to ligands. In addition, quantum dots absorb a broad spectrum of wavelengths below their emission wavelengths; thus a single wavelength can be used to excite multiple fluorophores. This is ideal for experiments in which two or more structures are analyzed using different color fluorochromes. Multicolor experiments are also simplified with quantum dots because the crystals emit within a very narrow spectrum

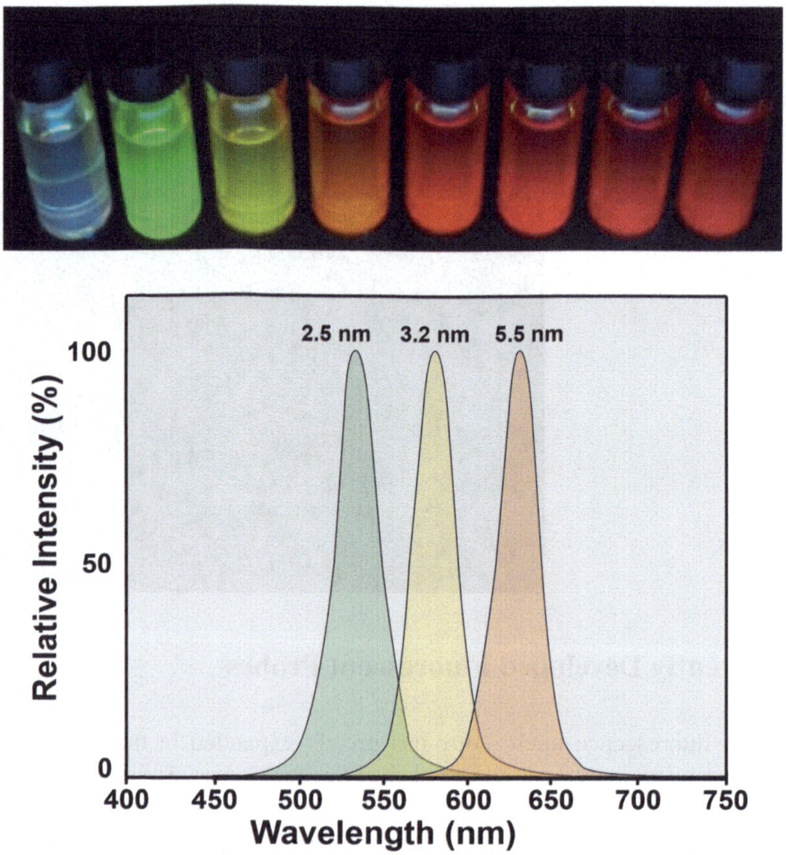

Fig. 2.10 Quantum dot fluorescence. The top image shows fluorescence obtained from solutions of CdSe nanocrystals (quantum dots) of diameters 2.0, 2.5, 3.0, 3.2, 3.9, 4.4, 5.5, and 7.3 nanometers, respectively. All vials are excited at 375 nm. The graphs below show the emission spectra for CdSe nanocrystals excited at 375 nm for sizes 3.2 nm, 3.9 nm, and 5.5 nm

(Fig. 2.10). This allows a large number of different color reporters to be separated visually.

Two key problems currently limit the usefulness of quantum dots for biological assays. These are the large size of quantum dots and their toxicity. Quantum dots are not water soluble so, for biological use, they must have some type of coating to make them water soluble. This additional layer overlays the inorganic shell (such as ZnS) that is applied to the nanocrystal to reduce non-radiative transitions and boost the quantum yield. The need for multiple coatings produces complexes that are much larger (usually >10 nm) than reporters labeled with organic fluorochromes. Figure 2.11 compares the size of quantum dots to some other useful biological labels. In addition to size limitations, if quantum dots are to be used for imaging living material, care must be taken that the surface coat resists leakage. During

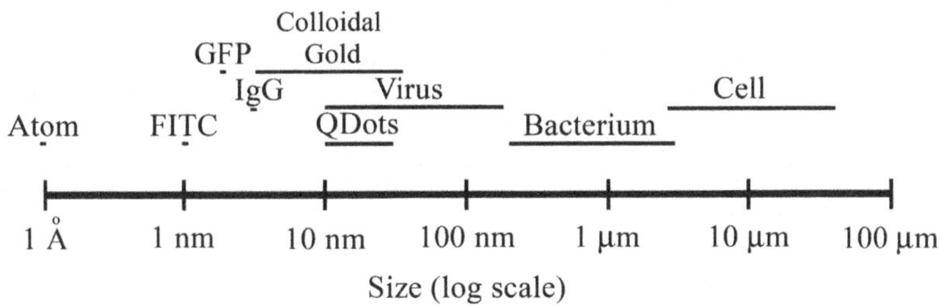

Fig. 2.11 Diameters of various labeling molecules

irradiation, the excitation photons can dissolve the core particle releasing toxic materials such as Cd. Despite these drawbacks, the exceptional fluorescent properties of quantum dots have spurred research in overcoming these limitations, and progress has been made in making quantum dots more biologically friendly. In addition to quantum dots, other inorganic and organic-inorganic composites are being explored for fluorescence properties. For instance, other types of manufactured nanoparticles with fluorescent properties are under development (Braeken et al. 2017; De-La-Cuesta et al. 2017). All of these developments will expand the choices of molecules available for biological fluorescent labeling studies.

A key problem for all current fluorescent molecules used for subcellular imaging is the need to get the reporter molecule (and any attached binding molecules) to the site to be labeled. Even a relatively small complex such as an IgG Fab fragment attached to Alexa 488® (Invitrogen) is about 5 nm in diameter (Fig. 2.11). The size and hydrophilic nature of most reporter-ligand complexes make these compounds difficult to get across lipid bilayers into the cell, and once inside the cell, their size can restrict their movement and limit their accessibility to the site to be labeled. This is not a critical issue for staining of thin histological sections for confocal microscopy. However, where the interest is generally in whole cells or thick tissue samples, the penetration of probes becomes an important concern. A lot of effort is being put into making probes smaller and in providing coatings that will facilitate entry into the cell and subsequent trafficking. None of these have yet been totally successful. Suffice it to say, however, that the diffusion property of a reporter molecule needs to be considered in any confocal experiment. It should not be assumed that the reporter molecule will have access to a ligand or that two different fluorophores will have equal access to their respective ligands. Lack of access may result in false-negative labeling experiments, and unequal access to ligands may result in incorrect interpretation of co-localization studies.

As described earlier, an alternate approach to the problem of label penetration into a sample has been provided by the elucidation of the gene for GFP. This breakthrough, when coupled with modern molecular biology techniques, has provided a mechanism to provide a fluorophore that is synthesized inside the cell instead of being provided exogenously. The coordinated cellular expression of the GFP protein covalently fused to another gene product of interest allows the synthesis of a fluorescently labeled molecule targeted to specific sites or with specific fluorescent properties. This can simplify the labeling of a molecule and circumvents the need for the fluorophore to cross the plasma membrane to get into the cell.

The introduction of GFP technology (and the development of fluorescent proteins of differing colors such as mCherry and red fluorescent protein (RFP)), the development of fluorochromes with improved fluorescent properties, and introduction of engineered fluorochromes with unique chemical properties expands the range of problems that can be attacked by confocal microscopy. Despite this, there are several important constraints that must be considered when using the growing number of fluorescent probes. For fluorescent proteins, for instance, one must control for ectopic expression of the gene and the possibility that fusion of the two proteins may interfere with function. For all fluorochromes, one must consider the relative size of the probe and its ability to disperse uniformly within the cell. Nonetheless, new and creative variations on the use of fluorescent probes for biological labeling are being published at an accelerated rate indicating that there will be continued improvements in this exciting technology.

Literature Cited

Braeken Y, Cheruku S, Ethirajan A, Maes W (2017) Conjugated polymer nanoparticles for bioimaging. Materials 2017(10):1420. https://doi.org/10.3390/ma10121420

De-La-Cuesta J, Gonzalez E, Pomposo JA (2017) Advances in fluorescent single-chain nanoparticles. Molecules 2017(22):1819

Diwu Z, Chen C, Zhang C, Klaubert D, Haugland R (1999) A novel acidotropic pH indicator and its potential application in labeling acidic organelles of live cells. Chem Biol 6:411–418

Magde DM, Rojas GE, Seybold P (1999) Solvent dependence of the fluorescence lifetimes of xanthene dyes. Photochem Photobiol 70:737–744

Magde D, Wong R, Seybold PG (2002) Fluorescence quantum yields and their relation to lifetime of rhodamine 6G and fluorescein in nine solvents: Improved absolute standards for quantum yields. Photochem Photobiol 75:327–436

Mujumdar RB, Ernst LA, Mujumdar SR, Lewis CJ, Waggoner AS (1993) Cyanine dye labeling reagents: sulfoindocyanine succinimidyl esters. Bioconj Chem 4:105–111

O'Connor N, Silver R (1998) Ratio imaging: practical consideration for measuring intracellular calcium and pH in living tissue. Methods Cell Biol 81:415–433

Piston D, Knobel S (1999) Real-time analysis of glucose metabolism by microscopy. Trends Endocrinol Metab 10:413–417

Selvin P (2000) The renaissance of fluorescence resonance energy transfer. Nat Struct Biol 7:730–734

Tsien R (1998) The green fluorescent protein. Annu Rev Biochem 67:509–544

Williams R, Piston D, Webb W (1994) Two-photon molecular excitation provides intrinsic 3-dimensional resolution for laser-based microscopy and microphotochemistry. FASEB J 8:804–813

Chapter 3
Fluorescence Microscopy

W. Gray (Jay) Jerome and Robert L. Price

3.1 Introduction

To harness the power of fluorescence for biological microscopic imaging, a number of additional components must be introduced to the standard light microscope. The most important modifications are (1) a strong illumination system that provides suitable wavelengths for exciting a specific fluorochrome, (2) some mechanism to limit the illumination beam to specific wavelengths so that only the fluorochrome of choice is excited, and (3) a means to image only the light emitted from the fluorochrome so that the excitation light (and other stray wavelengths) are not imaged. Inclusion of light from anything other than the fluorochrome would degrade the image fidelity and signal to noise ratio (SNR). In this chapter we will discuss a basic set up that meets the criteria for good fluorescent imaging. Subsequent chapters will expand on this theme and discuss additional modifications required specifically for laser scanning and spinning disk confocal fluorescence.

W. G. Jerome (✉)
Department of Pathology, Microbiology and Immunology, Vanderbilt University School of Medicine, Nashville, TN, USA
e-mail: Jay.Jerome@vanderbilt.edu

R. L. Price
Department of Cell Biology and Anatomy, University of South Carolina School of Medicine, Columbia, SC, USA
e-mail: Bob.Price@uscmed.sc.edu

W. G. Jerome, R. L. Price (eds.), *Basic Confocal Microscopy*,
https://doi.org/10.1007/978-3-319-97454-5_3

3.2 The Optical Path in a Fluorescence Microscope

In standard transmitted light microscopy, a beam of light is focused onto the specimen by a condenser located below the specimen (transillumination light path in Fig. 3.1). The light is altered as it traverses the specimen and this altered light is collected by the objective lens and transmitted ultimately to the imaging device (eye, camera, etc.). While this design can be used for fluorescence, there are significant problems involved in separating the excitation light from the light emitted by the fluorophore. Obviously, it is critical to image only the emitted light, since all other wavelengths are false signals that degrade image quality and might be incorrectly interpreted.

The separation of emitted light from other wavelengths is much easier when the emitted light is traveling in the opposite direction to the excitation light. For this

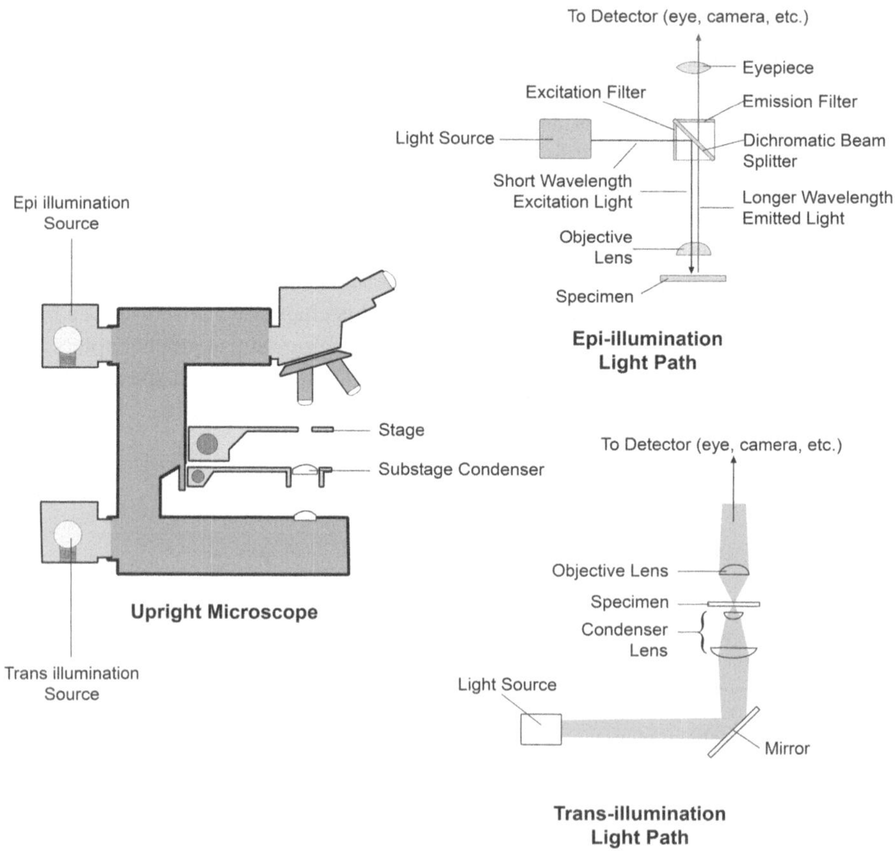

Fig. 3.1 Optical path in epi-illumination and transillumination microscope designs. Depicted here is the setup for an upright microscope. However, the trans- or epi-illumination design can be implemented just as easily with an inverted microscope

reason, most fluorescence is done using a type of illumination called epi-illumination (Fig. 3.1). In this design, the excitation illumination is focused onto the specimen by the objective lens rather than a condenser lens (i.e., the objective lens acts as a high-quality condenser lens). The objective lens is also used to collect and focus the light emitted by the fluorophore, just as it would be in a transilluminated light setup. By illuminating the specimen with incident light, the only excitation light in the imaging path is that which is reflected from the specimen or reflected from glass surfaces. When effectively minimized, the reflected light has a very minor intensity compared to the light emitted from the excited fluorochrome. Limiting reflection is a key component of successful epifluorescence microscopy. The incident light is further reduced by appropriate filters or other means that block its transmission to the imaging device (see Sect. 3.5). This epifluorescence strategy is useful for both wide-field and confocal fluorescence microscopy.

3.3 Light Sources

As indicated in Fig. 3.1, the light path for epi-illumination begins with the light source. A suitable light source must be able to provide a high photon flux at a wavelength that will excite the fluorochrome of interest. High levels of excitation are necessary because the fluorescence emission is usually low due to fluorochrome and specimen characteristics such as low quantum yield, limited labeling, and loss of light within the optical path of the microscope. These limitations are discussed later in this chapter and in the chapters on specimen preparation, immunolabeling, and system components. The bottom line is that one usually needs to maximally excite the fluorochrome in order to get a signal that is distinguishable from the general background noise level.

A second requirement of a fluorescence lamp is that the intensity of the illumination should be uniform at the back aperture of the objective lens. If it is not, uneven sample excitation and uneven brightness in the final image may result. This can also be a problem if the illuminating source is not properly aligned. Checking the source alignment on a regular basis should be a routine procedure.

Traditionally, two illumination sources that meet the above criteria of uniformity, intensity, and specificity have been employed: mercury arc lamps and xenon arc lamps. Both of these produce photons as the result of a voltaic arc running through an ionized gas. The mercury bulb produces photons with wavelengths that cover the full visible spectrum and the UV range, as well. For this reason, they are well suited for exciting fluorochromes with absorbance maxima below 380 nm as well as those with absorbance in the visible range. However, the intensity at different wavelengths is not uniform. Mercury arc lamps show peak intensities at 313, 334, 365, 406, 435, 546, and 578 nanometers (Fig. 3.2a). The intensities at other wavelengths are much lower. For this reason, when using a mercury lamp, it is best to carefully select fluorochromes that have good absorbance at, or close to, the peak wavelengths.

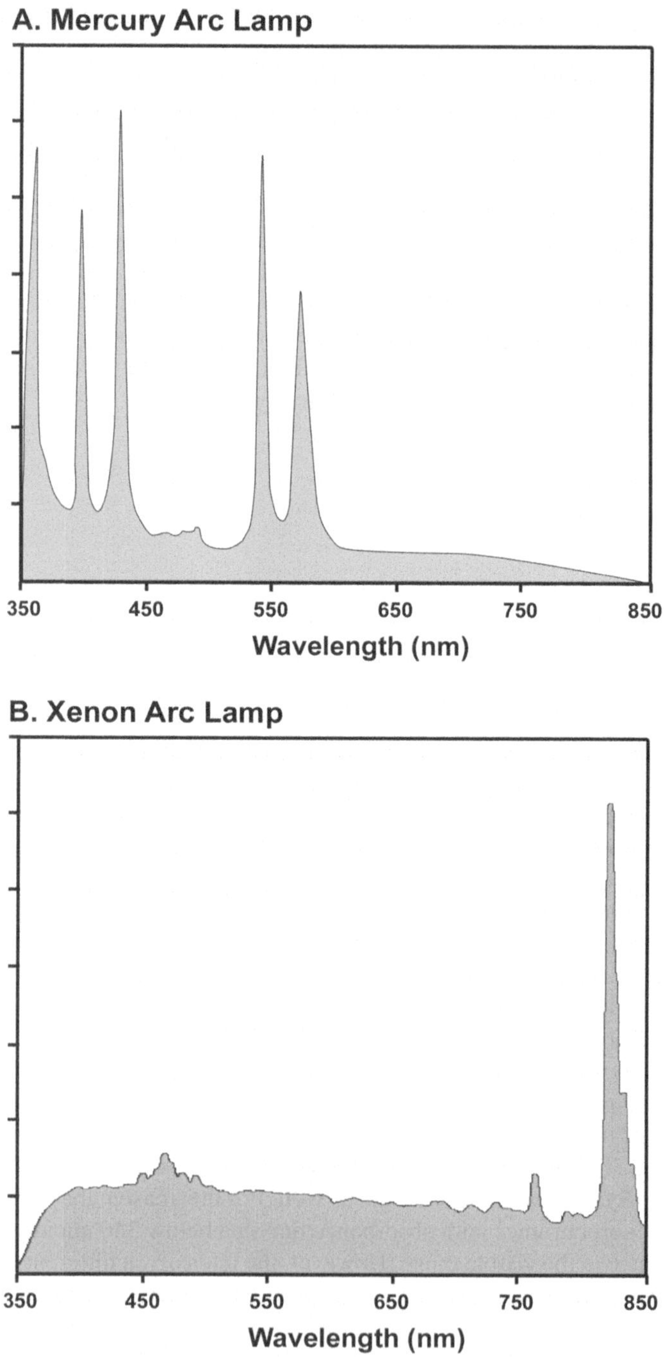

Fig. 3.2 Spectrum of light emitted from mercury arc (**a**) and xenon arc (**b**) lamps

Like mercury lamps, xenon lamps are arc lamps using an ionized gas to generate light. Xenon lamps have a more uniform distribution of intensities across the visible range, but their intensities fall off rapidly at wavelengths below 400 nm and so are less suitable for exciting fluorochromes with UV absorption (Fig. 3.2b). The intensities at peak wavelengths are also not as high as those from the most intense wavelengths of mercury lamps. Another drawback of xenon lamps is that they give off significant radiation in the infrared range which can produce excessive heat. This heat must be efficiently dissipated because excessive heat can damage specimens and can affect the characteristics of environmentally sensitive fluorochromes.

With both xenon and mercury lamps, one must keep track of the number of hours the bulbs have been used. Beyond the useful life of the bulb, the spectral emissions can change dramatically, altering the microscopy results. The increase in use of fluorescence microscopy, however, has spurred innovative approaches to improving arc lamp design. One such advancement has been the introduction of mercury arc lamps, such as the X-Cite ® series of lamps from EXFO (EXFO Photonic Solutions Inc. Ontario, Canada). These lamps include a cleaning cycle technology that removes burned material that tends to build up on the walls of arc lamps. This can greatly extend the useful life of the lamp by as much as tenfold.

Although arc lamps are very useful for fluorescence microscopy, they have their drawbacks. Chief among these are the relatively short life span, instability, and the need to filter out unwanted wavelengths. In addition, the heat from xenon lamps can be problematic for live cell imaging. In recent years, newer light sources have been introduced that overcome some of the issues with arc lamps. A particularly exciting change has been the development of high-performance light-emitting diodes (LEDs). The newest breeds of LEDs have several advantages over mercury or xenon arc lamps for fluorescence microscopy. Like arc lamps, LEDs have a high and stable output. However, they have very low energy consumption and generate less heat. They are also relatively inexpensive and have a very long useable life span. LEDs can produce light within narrow wavelengths, so filtering non-specific wavelengths is simplified. An array of LEDs can be employed to cover the spectrum from near UV through red; consequently the choice of fluorochromes is not limited. The drawback is that their light output is less intense than the output from arc lamps, but this parameter is improving rapidly, and newer LED systems now provide suitable photon flux for most fluorescent microscopy needs. A number of LED-based sources are now available for fluorescent confocal microscopy.

The use of arc burners and LED arrays in confocal microscopy are generally limited to spinning disk confocal microscopes. For single point scanning confocal microscopes, a laser is usually the light source of choice because of its high photon flux and physical properties. Laser is an acronym for *Light Amplification by Stimulated Emission of Radiation*. Chapter 2 discussed spontaneous emission of light. Laser light represents stimulated emission. This is a situation where an atom or molecule retains its excess energy until stimulated to emit the energy as light. Lasers both produce and amplify this stimulated form of light. Like spontaneous fluorescence, the emission is restricted to specific wavelengths that are dependent upon the material used for creating the light. Thus, most lasers produce an intense, monochromatic beam that is coherent and highly collimated. The coherence of the beam introduces problems when used in standard wide-field microscopy because scattering and interfering diffraction patterns are produced at all of the surfaces in the optical path, including any dirt in the system. This is much less of a problem, however, when a tightly focused beam is rastered point-by-point across the specimen and the image information collected sequentially from each illuminated point. This is exactly the procedure utilized in single point scanning confocal microscopy.

At its simplest, a laser consists of a gain medium contained in a reflective optical cavity with an energy supply (Fig. 3.3). Light is generated within the cavity, and, by placing mirrors within the cavity, the light is reflected back and forth within the gain medium. This leads to an amplification of the energy. The energy for amplification (pumping) is usually supplied as an electrical current or light of a different wavelength.

The gain medium is a material capable of stimulated emission. It can be a gas, liquid, plasma, or solid. The gain medium absorbs the pumped energy resulting in electrons in the medium becoming excited. In spontaneous emission this energy would be rapidly lost, and the electrons would return to the ground state. In stimulated emission, the electrons maintain their higher energy state. When the number of excited state electrons exceeds those in a lower state, this is called population inversion. It is a requisite for laser emission. In this state, when a photon is absorbed, the energy released as the atom returns to the ground state will exceed the energy of the photons absorbed. Thus, the light is amplified. Because the injected light passes back and forth through the medium multiple times, the emission becomes greatly amplified.

Although there are a variety of materials that can be used as the gain medium in lasers, for general biological laser scanning confocal microscopy, gas lasers have

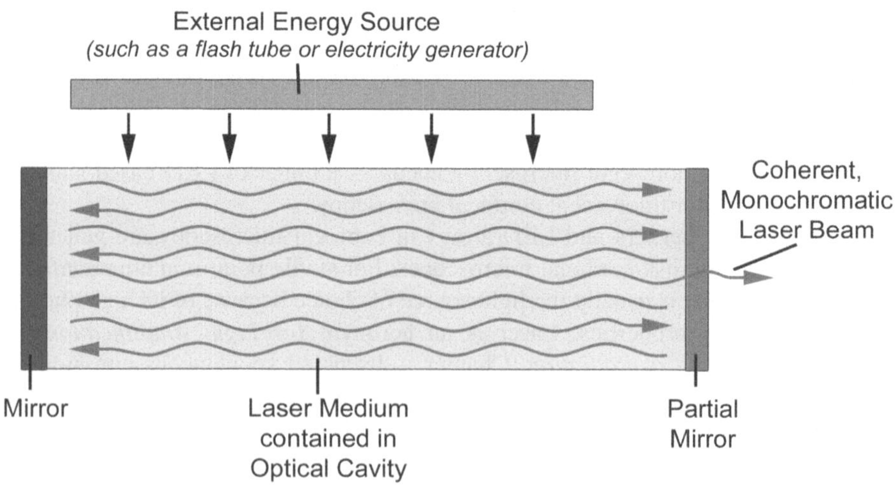

Fig. 3.3 Diagram of general laser principle

historically predominated. The most popular, because of available wavelengths, cost, and ease of operation, have been argon-ion and helium-neon lasers. Recently, however, the technology involved in other forms of lasers has advanced sufficiently to make them useful for laser scanning microscopy. In particular, titanium-doped sapphire (Ti:sapphire) solid state lasers have gained prominence for multiphoton excitation. For these lasers, sapphire is doped with titanium to provide the appropriate energy states for stimulated emission. Ti:sapphire lasers are tunable to a variety of wavelengths in the IR range and can be engineered to produce ultrashort pulses. These properties make them ideal for multiphoton excitation (see Chap. 2, Sect. 2.4).

Another form of laser is the diode laser. Diode lasers are semiconductor devices where a crystal substrate is doped with two very thin layers of material that will form positive (P) and negative (N) diode junctions (diodes). The difference in chemical potential between the n- and p-type semiconductors produces a depletion region at the interface. An electrical bias causes charge carriers (holes and electrons) to be injected into the depletion region from the p-doped (holes) and n-doped (electrons) sides. The recombination of electrons and holes in the depletion region produces photons which are stimulated by incident photons. The stimulation leads to amplification. The first diode lasers that were developed emitted mostly in the IR region. However, advances have increased the power output of diode lasers, and diodes that emit in the blue and red regions have been developed, making them useful for fluorescence microscopy. Diode lasers are now becoming common as the illumination source for newer confocal microscopes.

One of the strengths of lasers is the monochromatic nature of the emitted photons. For microscopy, the fact that the excitation light exists essentially as a single wavelength simplifies filtering of non-specific light. However, just like with LEDs, separate lasers are required for each excitation wavelength of interest. At best, a single laser may provide several suitable wavelengths (Fig. 3.4). Unlike LEDs, though, practical considerations, such as cost, usually limit the number of lasers provided with laser scanning confocal microscope systems. This means that a particular instrument usually is not equipped to cover the full visible spectrum and so

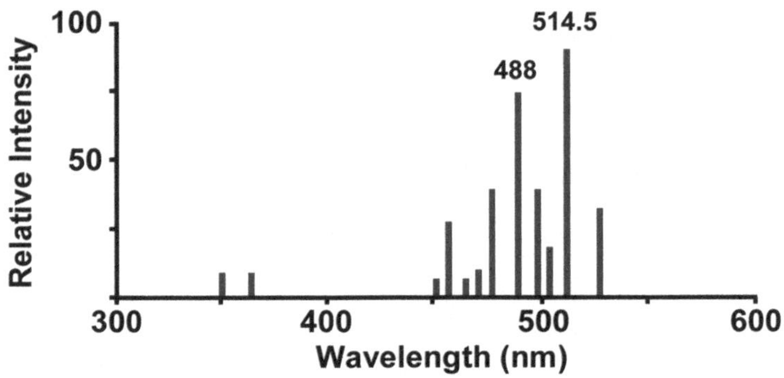

Fig. 3.4 Spectrum of light emitted from argon gas laser

one is limited in the choice of fluorochromes that can be employed. Luckily, fluorochrome manufacturers have realized an opportunity, and a great deal of work has been done in the last two decades to develop fluorochromes with properties that match the standard laser lines available with most laser scanning confocal microscopes. Thus the presence of limited wavelengths is no longer as restrictive to biological experiments as it was previously. Very recently, white light lasers have been introduced. These use a low energy-pulsed infrared fiber laser that is then amplified by a diode laser to produce wavelengths in the 450 to 670 range. An acoustic optical device is then employed to select specific narrow wavelengths for imaging. At present, these are very expensive but as the technology is developed and cost reduced their utility in confocal microscopy will increase.

3.4 Objective Lens Characteristics and Terminology

All components in the optical path are important in the collection of a high-quality confocal image. However, the objective lens may be the most critical element since the initial image is formed, the resolution determined, and most of the magnification of the final image is obtained via the objective. Even with an optimum specimen and the operating parameters of the confocal system set up properly, if the objective lens is of poor quality, or an improper lens used, the image will be inferior. While everyday users may not have control over the choice of available objectives on a core or established instrument, it is important to understand the optical characteristics of each available objective and how these characteristics may affect the collection and interpretation of the data. Only with this knowledge will you be able to select from your available choices a lens that is most suitable for your experimental needs. For example, if a limited number of objective lenses are available, it may be necessary to use a standard 40X oil immersion objective to collect data from both a cell culture sample and a 100 μm thick section of tissue labeled with multiple fluorochromes. However, this objective is a poor choice for imaging thick specimens if optional 40X water immersion optics are available.

So, how does one choose the optimal lens? In this section we give a number of examples demonstrating how the selection of an optimum objective lens can improve the acquisition, reconstruction, and interpretation of confocal data sets. However, before giving the examples, we will first provide several definitions and information about objectives that are essential in understanding why it is essential to have the correct optics on a confocal microscope for the acquisition of a given data set. Much of the discussion relates to a fundamental problem that we face when imaging; the optical components of our systems and the specimens we are trying to image have fundamentally different physical and chemical properties. This can result in refractive index mismatch as various wavelengths of light are affected differently as they pass through each component (Table 3.1).

With regard to confocal microscopy, refractive index refers to a measurement of how the wavelength of photons is affected as they pass through a substrate. This is

Table 3.1 Refractive index depends on concentration and molecular weight of proteins and nucleic acids (Voros 2004; Samoc et al. 2007). Mismatch of refractive indices of various components in the optical path results in significant loss of signal in confocal imaging

Substrate	Refractive index
Air	1.0
Water	1.32–1.34
Immersion oil	1.515
Glycerol (75%)	1.44
Polyvinyl alcohol	1.52–1.55
Protein[1]	1.36–1.5
Nucleic Acids[2]	1.5–1.6

Fig. 3.5 Effect of refractive index differences on the visualization of a straight pencil

important for two reasons. First, light rays change direction when they cross the interface from the objective to air or the immersion fluid used for immersion and dipping lenses. Second, light reflects partially from surfaces that have a refractive index different from that of their surroundings. Figure 3.5 shows a simple example of how refractive index mismatch can distort an image. Although the pencil is straight, its perceived image is dramatically distorted by the refractive index differences between air, glass, and water. Thus, it is critical to understand, minimize, and correct for refractive index differences.

The refractive index of each component in the optical path, including the optics of the objective lens, the filters used to separate the various wavelengths of photons, the immersion fluid or air the lens is designed to image through, the coverglass, the mounting medium of the specimen, and the specimen itself is specific and based on the molecular composition of the material. As photons, both excitatory and emitted, pass through each component, their path is affected. If the refractive indices of each component of the microscope and sample are not matched, refractive index mismatch occurs resulting in image aberrations and loss of specimen signal reaching the detector (Dunn and Wang 2000; Diaspro et al. 2002). Some common refractive indices for various materials are given in Table 3.1. Optimizing microscope performance includes accounting for and minimizing the effects of refractive index differences in your image.

3.4.1 Objective Lens Inscriptions

Inscribed on all objectives is information concerning the objective specifications and information about the type of imaging for which the objective is optimized. Figure 3.6 shows several Zeiss and Nikon objectives. Inscribed on each is information concerning the type of lens; the magnification; the focal length; the numerical aperture (NA); if it is an immersion lens (Fig. 3.6b and c); whether it is corrected for use with water, oil, or glycerol; and the coverglass thickness for which the objective is corrected. Definitions and considerations important in lens selection for confocal imaging are presented below. Several images are also shown in the following sections which illustrate the importance of matching the correct objective with the goals of the experiment.

Plan objectives are flat-field objectives corrected to eliminate effects caused by curvature of the lens. Without correction, light from peripheral regions of the image will not be focused to the same point as light from more central regions. This would result in loss of focus in peripheral regions. Since the purpose of the confocal pinhole is to remove out-of-focus light, this may also result in loss of signal on peripheral regions of the image. If Plan lenses are not available, which may be the case for low magnification objectives, the confocal software can be used to zoom the image and limit image acquisition to the central area of the objective.

Numerical aperture provides an indication of both the light gathering capacity and resolution of a lens. A brief introduction to resolution with respect to NA is presented at this point to introduce the topic and provide the necessary information to understand the inscriptions on the microscope lenses. A more thorough theoretical discussion of resolution is presented in Chap. 7 which deals with the process of image capture. As NA increases, the number of photons that can be collected by the lens increases, and the resolution the lens is capable of producing is also improved. Key components in determining the NA, as shown in the equation:

$$\mathrm{NA} = n \sin \alpha$$

are refractive index (n) of the medium (air, immersion oil, glycerol or water) between the lens and the coverglass, and the half angle (α) of the cone of light that enters the objective. The significance of the refractive index and examples of refractive index mismatch between the components of the optical path will be discussed in detail in Sects. 3.4.2 and 3.4.3. As discussed in several sections, one of the limiting factors in confocal imaging is the acquisition of an adequate number of photons.

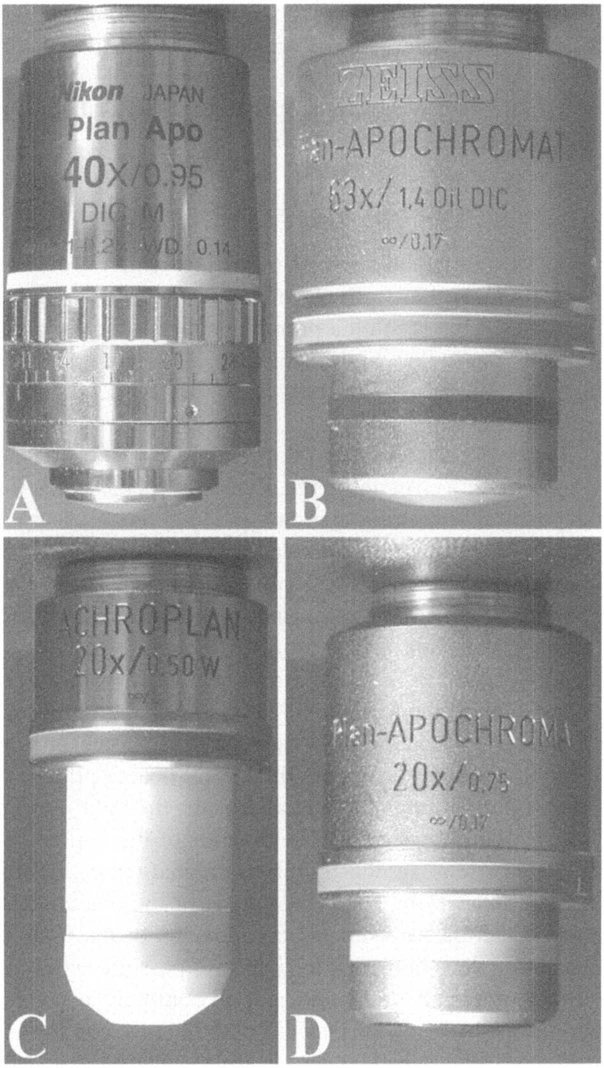

Fig. 3.6 Various types of objective lenses showing inscriptions on the barrel of the objective. (**a**) 40X Plan Apo with correction collar for adjustment to correct for coverglass thickness and refractive index mismatch; (**b**) 63X Plan Apo oil immersion; (**c**) 20X Plan Achromat water immersion; (**d**) 20X Plan Achromat

High NA lenses collect more photons. High NA lenses are also essential for high-resolution confocal imaging as shown in the equation for resolution:

$$d = 0.61\lambda \mathrm{NA}$$

where d is the resolution and λ is the wavelength of the photons. As numerical aperture increases, resolution also improves.

Depth of field (Z) refers to the distance along the vertical or z-axis of the optical path that the specimen, or an object within the specimen, remains in focus. As shown by the equation below, similar to resolution, depth of field is highly dependent on the NA, wavelength of light (λ) used as the imaging source, and the medium between the lens and object (*n*):

$$Z = n\lambda \mathrm{NA}^2$$

As the numerical aperture increases, which is desirable for high-resolution confocal images, the depth through which focused images can be obtained rapidly decreases. This is often a problem with confocal experiments where it is desirable to be able to collect high-resolution 3D z-stacks of images, but the required use of high NA oil immersion objectives limits the depth from which images can be collected.

Depth of focus refers to the thickness of the image plane, or depth of the area around an object in an image, that is in focus. Since this refers to an image plane, the position of the object in the specimen is not changed. If the confocal pinhole is set properly and image collected appropriately, the depth of focus in confocal imaging is not as critical as the depth of field.

Infinity-corrected optics are included on nearly all new optical microscopes (1980s and beyond). In infinity-corrected objectives, light leaving the back of the objective does so in parallel rays which do not come to a focal point to form an image until a tube lens (Telan lens) brings the light to a focal point. The advantage of infinity-corrected optics, especially in fluorescence microscopy, is that multiple components such as barrier filters and beam splitters can be placed in the optical path without affecting the parallel paths the light rays take out of the objective lens. After passing through the components of the optical path, the light is then focused by the tube lens. In non-infinity-corrected optics, light is focused to an image after passing through the back focal plane of the objective. This then requires considerable correction to insure all components of the optical path are matched so light diffraction does not occur as light rays pass through filter cubes, beam splitters, and other components of the optics. If all components are not optically matched, a number of image aberrations, as discussed below, can occur.

Immersion objectives are designed to minimize the refractive index mismatch between the components of the lens and other components within the optical path. For example, the optical components of oil immersion objectives are closely matched with that of the coverglass used to mount specimens. This minimizes the refractive index mismatch as excitation photons pass through the objective and coverglass prior to exciting fluorochromes in the specimen. This is also advantageous as emitted photons must also pass through the same glass components of the coverglass and objective lens. Typically oil immersion objectives have a very high numerical aperture (1.3–1.4) which also results in a high-resolution objective.

Other types of immersion objectives also exist. Water immersion objectives closely match the refractive index of the biological components (proteins, nucleic acids, water) by having a drop of water between the objective and the coverglass rather than using air or oil as the intermediate medium. While water immersion objectives are excellent at minimizing refractive index mismatch between water in the sample and the surrounding media and, thus, allow deeper imaging into tissues, the numerical aperture of these lenses is not as high as those available for oil immersion optics. Objectives are also available for immersion in glycerol to match the mounting medium used for many biological specimens. In addition, some objectives are classified as multi-immersion objectives, and these can be used for various mounting media or dipping lenses which can be placed directly into the buffer mounting medium of the specimen without the use of a coverglass. Immersion and dipping objectives are typically expensive, and cost increases as the numerical aperture and versatility of the lens increases.

The *working distance* of a lens is the distance between the objective and the coverglass (or specimen if using a dipping lens) when the specimen is in focus. Low magnification objective lenses and water immersion lenses typically have a long working distance, while oil immersion lenses have very short working distances. Care must be taken when focusing all objectives, but one must be diligent in making sure short working distance oil immersion objectives do not run into the coverglass of the specimen during focusing. This can damage the specimen but, more importantly, it can damage the objective. A potential problem that exists with automated microscope systems is the use of samples of varying thicknesses with objectives of different working distances. One should always focus a new sample at low magnification (i.e., 10X) prior to using a short working distance lens. An example of a potential problem with the fully automated Zeiss microscope in our core facility is that a number of investigators utilize the 63X NA 1.4 oil lens to image co-localization in cells cultured on coverglasses. Other investigators may use the same objective to image the surface of sectioned samples that are several microns thick. Since the automated system brings the focal point back to the one previously used, it is essential that those using thick specimens first focus their sample at low magnification. Otherwise, the automatic setting might crash the short working distance objective into the sample.

Wavelength-Corrected Objectives and Chromatic and Spherical Aberrations One of the most important characteristics of an objective used for acquisition of confocal images is the range of wavelengths of light for which it is optimized to collect. Since wavelengths of photons emitted from different fluorochromes vary, the focal point of each after it passes through the objective lens will also differ. It is essential to use objectives corrected for the various fluorochromes being used so that all are focused in the same optical plane. If the fluorochromes are focused in different optical planes, chromatic and spherical aberrations are introduced into the image which can affect data interpretation as discussed in Sects. 3.4.2 and 3.4.3.

On the barrel of the objective lens, inscriptions indicate the level of correction. Achromat lenses are designed to minimize chromatic aberrations in the blue and red wavelengths and spherical aberration in green wavelengths of light. Fluor objectives are also corrected for chromatic aberration in blue and red but are also corrected for spherical aberration in these ranges. Apochromat objectives are corrected for chromatic aberration for red, green, blue, and possibly UV light and spherical aberration for red and green light. In addition to these basic classifications, there are specialized objectives such as the C-Apochromats for up to six color correction in water immersion lenses which have been introduced for specific uses. These are critical for confocal work involving collection of multiple wavelengths, particularly within the red part of the spectrum. However, even with advanced correction, it is essential that care be taken when interpreting multicolor data as shown in the next several examples.

3.4.2 Effects of Chromatic Aberration

Use of lenses not specifically corrected for collection of multiple wavelengths of light may result in the introduction of chromatic aberration in the image. This can result in image artifacts and improper interpretation of data. Figure 3.7 diagrammatically illustrates the effects of chromatic aberration on the imaging of photons of different wavelengths. In objectives that have longitudinal (axial) chromatic aberration (Fig. 3.7a), photons of shorter emission wavelengths (blue) are focused closer to the objective than those of longer wavelengths (green or red). As shown in Fig. 3.7b, lateral chromatic aberration may result in a change in magnification with shorter wavelengths of light being magnified more than long wavelengths by 1–2%. The correction of chromatic aberration is typically accomplished in the design of the objective by inclusion of multiple lens elements with different optical characteristics. This results in different wavelengths of light being brought into focus in the same plane.

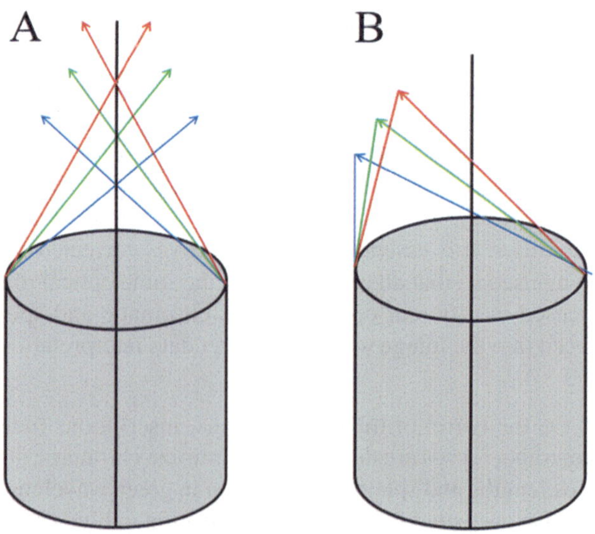

Fig. 3.7 Diagrams of the effects on focal plane and magnification of longitudinal (**a**) and lateral (**b**) chromatic aberrations

Chromatic aberrations may result in two types of defects in the image. First, since out-of-focus photons are removed from the image by a pinhole, the resulting signal will be significantly reduced in the wavelengths that are out of focus. The second artifact may result in a misinterpretation of co-localization data since photons of different wavelengths that are imaged from a single point in the specimen will be out of register in the X-, Y-, and Z-directions and observed as being in different points in the specimen. If working at low magnifications where Z-resolution is relatively poor, this may not be significant. This is why many low magnification objectives are not highly corrected for chromatic aberrations. However, in high-resolution confocal imaging, it is essential that highly corrected objectives be used for collection of images from samples labeled with multiple fluorochromes.

The effects of chromatic aberration on the interpretation of co-localization of structures in a sample have been illustrated by Dunn and Wang (2000) as shown in Fig. 3.8. In this set of experiments, both a mirror and fluorescent beads were imaged with 40X Fluor and 100X Plan Apo objectives at photon emission values of 520 nm, 600 nm, and 680 nm (Fig. 3.8a). With the Plan Fluor objective corrected for red and green emission, these two colors were co-localized, while the emission at 680 nm was displaced by 1.2 μm along the Z-axis. With the Plan Apo 100X objective, which was corrected for all three wavelengths of light, the signals coincided. To demonstrate this effect in a biological sample, the authors simultaneously labeled transferrin (Tf) molecules in cell endosomes with fluorescein, rhodamine, or Cy5. Since there is a very high expression of Tf sites available in these cells, a high probability for double label co-localization exists. In this case, true co-localization of the structures should be observed in the images if no longitudinal chromatic aberration is present. In Fig. 3.8b–d the Plan Apo 100X objective was used to image fluorescein (520 nm emission) and Cy5 (680 nm emission). With this objective, images correctly show a high coincidence in signal localization and intensity indicating that the fluorochromes are co-localized as indicated by the yellow color in Fig. 3.8b and the arrows in Fig. 3.8c and d.

The same sample was then imaged with the 40X Plan Fluor objective corrected for green (fluorescein) and red (rhodamine) emission. Fig. 3.8e again shows the co-localization of these two signals collected with an objective corrected for these fluorochromes. However, when the 40X Plan Fluor objective was used to collect fluorescein and Cy5 signals and the images overlaid, the Tf molecules no longer appeared co-localized. With this objective the image had distinct red and green colors indicating no co-localization (Fig. 3.8f). When a Z-series was collected (Fig. 3.8g) and projected, the signals again appeared co-localized indicating a shift in the Z-positioning of the emitted photons. When the individual fluorescein images collected in Fig. 3.8f were shifted 1.2 μm in the Z-direction, the fluorescein and Cy5 signals correctly appear co-localized (Fig. 3.8h). Fig. 3.8i and j shows that if the same focal plane is collected in the fluorescein and Cy5 channels and shown side-by-side, poor image registration of the Tf labeling results. However, if the fluorescein image is collected at a depth of 1.2 μm greater, there is perfect registration of the signal.

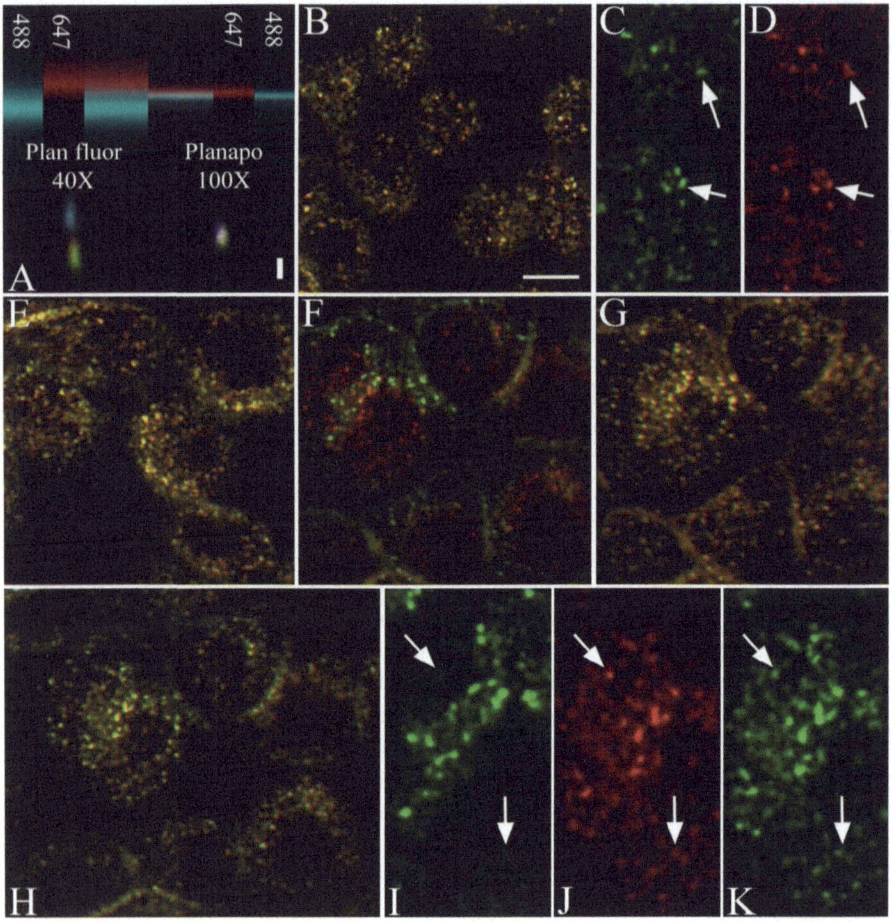

Fig. 3.8 Chromatic aberration in confocal microscopy. (**a**) XZ-sections of glass reflection (top) and microsphere fluorescent (bottom) images collected with a poorly color-corrected Plan Fluor 40x objective (left) and a well-corrected Plan Apochromat 100x objective (right). In the reflection images at the top of this panel, 647 nm light is shown in red, and 488 nm light is shown in blue. In the fluorescence images of fluorescent beads, 520 nm emissions are shown in green, 600 nm emissions are shown in red, and 680 nm emissions are shown in blue. For all images, the focal axis is oriented vertically, with the sample bar indicating a distance of 1 μm. (**b**) An image of a field of cells labeled with both fluorescein-Tf and Cy5-Tf collected with the 100X Plan Apochromat objective. Higher magnification images of the component green and far-red fluorescence are shown in Panels **c** and **d**, respectively. (**e**) An image of a field of cells labeled with fluorescein-Tf and rhodamine-Tf collected with a 40x Plan Fluor objective. (**f**) An image of a field of cells labeled with both fluorescein-Tf and Cy5-Tf collected with the Plan Fluor objective. (**g**) The projection of the vertical series of images from which Panel **f** was obtained. (**h**) The same field shown in **f** but combining green and far-red planes collected at 1.2 μm apart. Higher magnification images of the green and far-red fluorescence from the same focal plane are shown in Panels **i** and **j**, while Panel **k** shows the green fluorescence from a focal plane 1.2 μm lower. Scale bar in **b** represents a 10 μm length for Panels B–H, 5 μm in length for Panels **i–k.** (Used with permission of the authors and Biotechniques, 2008)

In the above set of experiments, a 100X Plan Apo objective corrected for common fluorochromes in the 500 nm through 680 nm emission range was used. With the development and addition of new lasers that have extended the range of fluorochromes that can be used in confocal microscopy, additional care must be taken when selecting fluorochromes for co-localization studies. In a set of experiments similar to those above, we have used a Zeiss Plan Apo 63X objective to image F-actin in cardiac myocytes labeled with 405 nm and 633 nm phalloidin. A similar assumption to the transferrin model above is made here in that both fluorochromes should be co-localized on the F-actin. As above, this Apochromat objective is corrected for fluorochromes emitting in the 500 nm through 700 nm range, it is not well corrected for fluorochromes such as the 405 nm fluorochrome excited with the 405 nm diode laser which emits in the 475 nm range. As a result, the two signals are not co-localized in the image even though they are labeling the same structures (Fig. 3.9). As near-UV lasers become more common on many confocal systems,

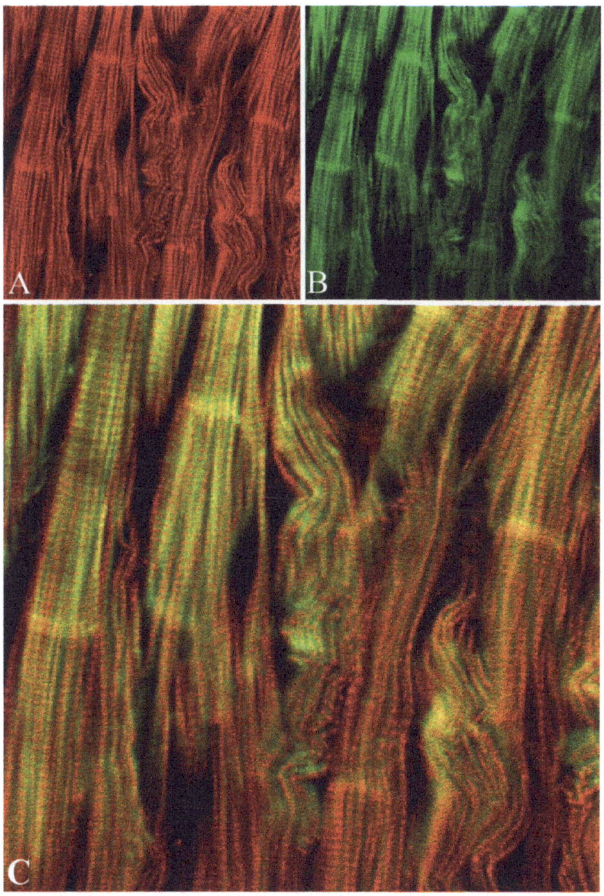

Fig. 3.9 A section of adult mouse heart labeled with 633 phalloidin (**a**) and 405 phalloidin (**b**). Both are labeling f-actin in the cardiac myocytes, but due to chromatic aberrations in the system, the red and green signals do not show 100% overlap

care must be taken in selection of fluorochromes used in co-localization studies to insure they match the lasers and optics of the available confocal microscope.

These experiments clearly show the importance of knowing both the objective lens and fluorochrome characteristics when designing an experiment. When dealing with co-localization data, it is important to be aware of the possibility of aberration and select fluorochromes for which objectives are available to adequately correct for focal shifts due to differing wavelengths.

3.4.3 Refractive Index Mismatch and Spherical Aberrations

As illustrated previously, refractive index mismatch is a primary cause of spherical aberration in microscope systems. This type of aberration results in peripheral rays of light that enter an objective being focused at a different point than the central rays of light that enter (Fig. 3.10). This may result in blurring of the image and significant loss of signal as the confocal pinhole effectively removes the out-of-focus light from the image. Much of the refractive index mismatch that occurs in a confocal system results from differences in the refractive indices of air or immersion oil and common specimen mounting media that are primarily water based, such as the 1:1 PBS:Glycerol mix used in many laboratories.

To address problems with refractive index mismatch, immersion lenses may be used. By utilizing immersion fluids rather than air as the interface between the lens and the specimen, a closer match in the refractive indices in the optical path occurs. This can significantly improve resolution as in the case of high NA oil immersion objectives or depth of imaging in the case of water immersion optics. Similar to

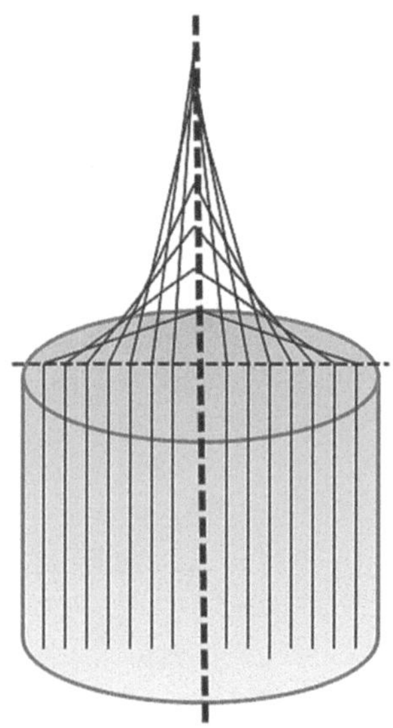

Fig. 3.10 Diagram showing the effects of spherical aberration resulting in peripheral rays of light being focused closer to the objective lens surface than rays of light coming through the central regions of the lens

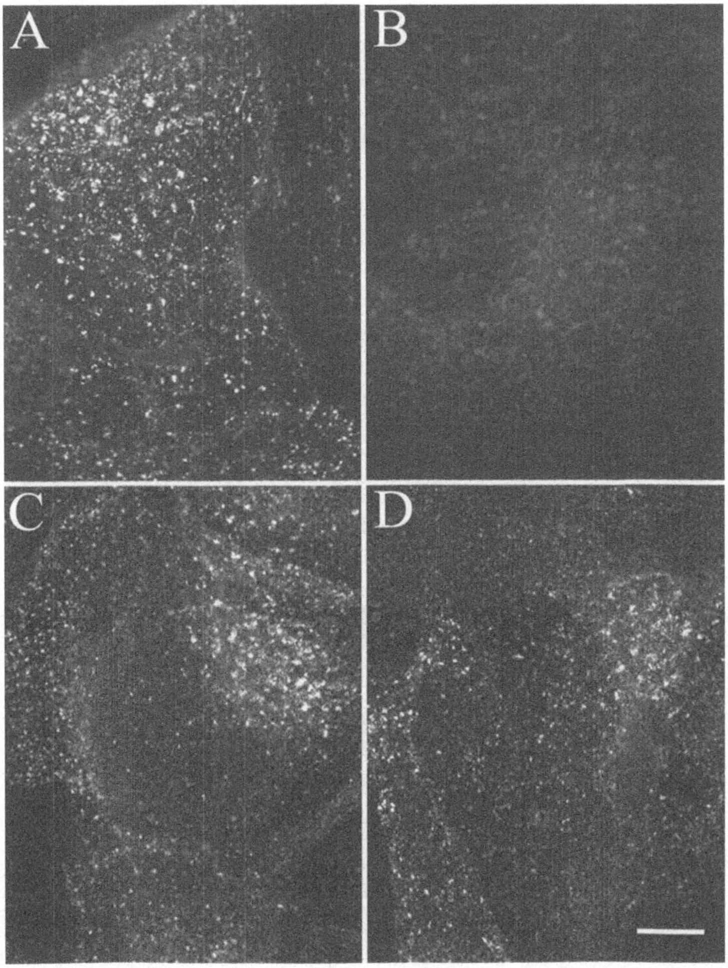

Fig. 3.11 Spherical aberration in confocal microscopy. Images of cells labeled with F-Tf and imaged with either the 100x Plan Apochromat oil immersion objective (**a** and **b**) or the 60X Plan Apochromat water immersion objective (**c** and **d**). Endosomes were imaged at the surface of the coverslip Panels **a** and **c** or at a depth of 35 μm (Panel **b**) or 66 μm (Panel **d**) into the aqueous sample medium. Scale bar is 10 μm in length. (Used with permission of the authors and Biotechniques, 2008)

their experiments demonstrating the effects of chromatic aberrations, Dunn and Wang (2000) have also used fluorescein-Tf-labeled endosomes to demonstrate the effects of refractive index mismatch by imaging the same sample with a 100X Plan Apochromat oil immersion objective and a 60X Plan Apochromat water immersion objective (Fig. 3.11). When imaging near the surface with the 100X Plan Apochromat oil immersion objective, a strong signal can be obtained. However, due to refractive index mismatch between the immersion oil and the water-based mounting medium of the sample, signal is rapidly lost due to spherical aberration as imaging depth is increased (Fig. 3.11 a–b). At a depth of only 35 μm, essentially no signal was available for imaging. When using the 60X Plan Apochromat water immersion lens, which minimizes refractive index mismatch between the immersion and mounting media, excellent images were obtained 66 μm deep in the specimen (Fig. 3.11 c–d).

In addition to spherical aberrations associated with refractive index mismatch between the immersion and mounting media, depth of imaging, even with water immersion optics, can be dramatically affected by the specimen itself. One of the common questions asked by investigators beginning confocal imaging is how deep can I image into my specimen? As seen above, knowing the specifications of the objective being used is important in answering this question. Equally important is a basic knowledge of the characteristics of the specimen being imaged. Figures 3.12 and 3.13 illustrate the effects of specimen (protein?) density on the depth of imaging. In Fig. 3.12 Cy3-phalloidin was used to label the somites of a chicken embryo, which was then imaged with 20X NA 0.75 air and 20X NA 0.5 water immersion objectives. With the 20x air objective, signal was significantly decreased at a depth of 100 μm in the specimen. However, with the water immersion objective, a strong

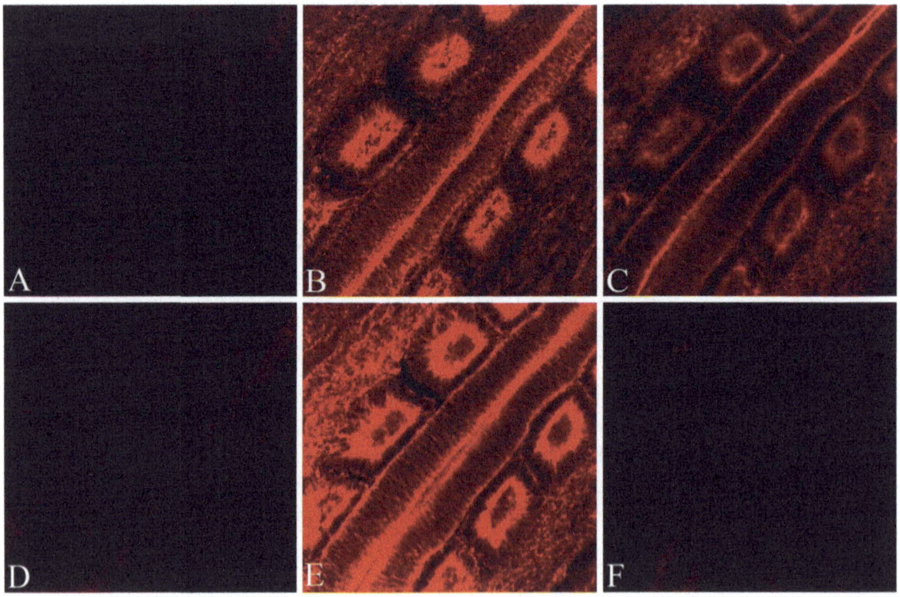

Fig. 3.12 Effects of refractive index mismatch on depth of imaging. A 200 mm thick section of mouse embryo was labeled with Alexa 543-phalloidin (red channel) and somites imaged with a 20x NA 0.5 water immersion lens (**a–c**) and a 20X NA 0.75 dry lens (**d–f**). A Z-series was collected with image acquisition optimized for collection at 50 μm deep in the sample. Images **a** and **d** were collected at the surface of the sample, images **b** and **d** 50 μm deep, and **c** and **f** 100 μm deep With the 20X dry objective, all signal was lost at 100 μm into the section, while a strong signal was still present at this depth with the 20X water immersion objective. Signal was not lost with this objective until a depth of 150 μm was imaged

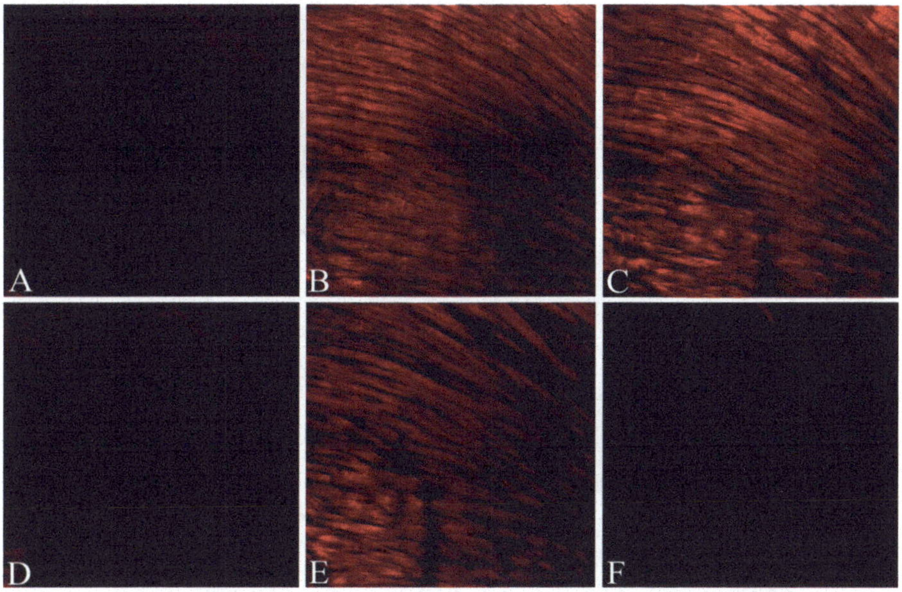

Fig. 3.13 Effects of refractive index mismatch and tissue density on depth of imaging. A 100 μm thick section of mouse heart was labeled with Alexa 543-phalloidin (red channel) and myocytes imaged with a 20x NA 0.5 water immersion lens (**a–c**) and a 20X NA 0.75 dry lens (**d–f**). A Z-series was collected with image acquisition optimized for collection at 15 μm deep in the sample. Images **a** and **d** were collected at the surface of the sample, images **b** and **d** 14 μm deep, and **c** and **f** 28 μm, deep. With the 20X dry objective, all signal was lost at 28 μm into the section, while a signal was still present at this depth with the 20X water immersion objective. Signal was not lost with this objective until a depth of 50 μm was imaged. Compare this to the results of imaging depth with embryonic tissue which could be imaged much deeper into the specimen

signal was still available at this depth, and signal did not decrease until a depth of 150 μm was tested.

Using the same 20X optics and Cy2-phalloidin label, we also imaged adult mouse heart, which is a much denser sample than the embryonic tissue imaged above. With this sample signal was lost at a depth of less than 25 μm with the 20X dry lens but was still strong with the water immersion lens at this depth (Fig. 3.13). Imaging depth of dense adult cardiac tissue was approximately 25% of that obtainable with embryonic tissue using the same preparation protocols and optics illustrating the importance of knowing specimen characteristics in addition to optical specifications when determining imaging parameters.

In response to the need to image as deep as possible into tissue, a number of techniques have recently been developed to clear tissues, and these will be discussed in detail in Chap. 4. To enhance one of these techniques, X-CLARITY, Leica has also developed an objective (Fig. 3.14) to match the refractive index of the mounting medium used in the process, which also closely matches the refractive index of biological tissues. An example of the increased imaging depth possible with this technique is shown in Fig. 3.15 where it was possible to image mCherry-

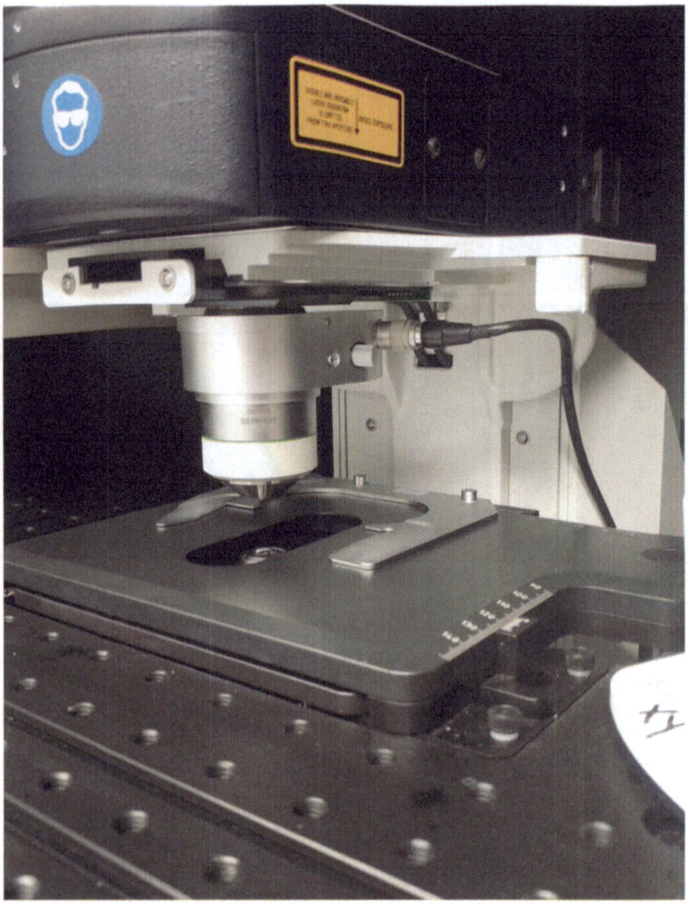

Fig. 3.14 Leica CLARITY objective mounted on an SP8 upright confocal system

labeled cells through 960 μm of tissue. Without clearing it is difficult to image through more than 50 μm of adult brain.

An additional factor that affects spherical aberration is the thickness of the coverglass since this affects the length of the optical path through the glass. Most modern microscope lenses are corrected for a number 1.5 coverglass, and this is the thickness of coverglass that should be employed unless a specific lens indicates otherwise. However, even within a batch of #1.5 coverglasses, there can be small but important differences in thickness. While it may not be practical to measure the actual thickness of every coverglass used, one must be aware that minimal variations in the thickness of a coverglass can affect image quality and interpretation. Manufacturers have addressed this problem, as well as the above factors affecting spherical aberrations, by providing objectives with correction collars. Correction collars can be adjusted to accommodate small variations in coverglass thickness and

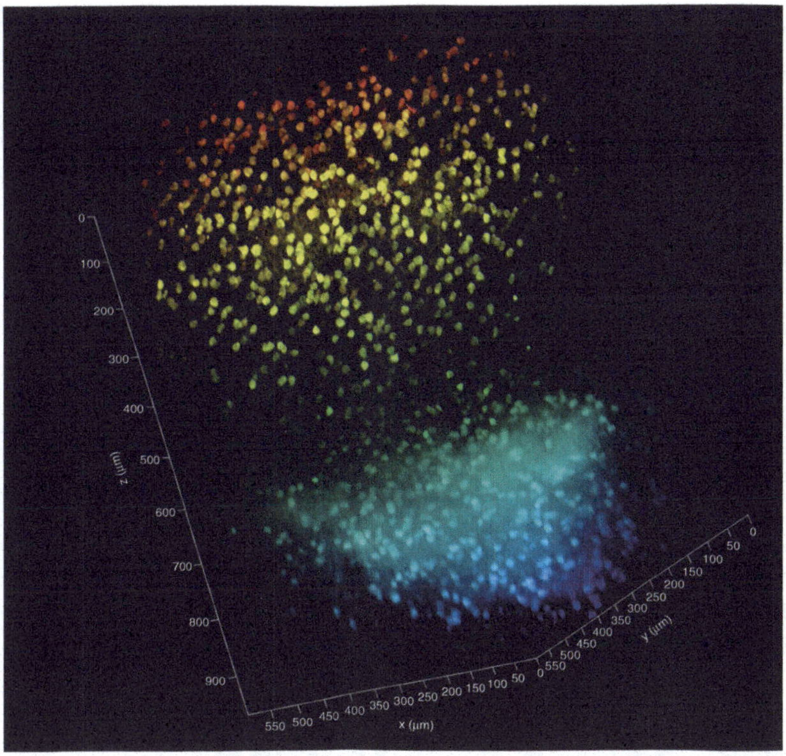

Fig. 3.15 Cleared slice of adult mouse brain stained with mCherry. It was possible to obtain signal 960 µm into the slice although spherical aberration started affecting resolution at approximately 500 µm as shown by the reduced sharpness (resolution) at that level. Signal also began to deteriorate as shown by the increased background level at approximately 600 µm deep

minimize image aberrations. While fairly expensive, these objectives are ideal for imaging deep into a specimen with minimal spherical and chromatic aberration as shown by another set of experiments performed by Dunn and Wang (2000). Using a 60X water immersion objective with a correction collar, they first collected a set of cross-sectional images from a glass surface at 488 nm and 647 nm at both the surface and at a depth of 63 µm in an aqueous medium (Fig. 3.16a). The signal from these two wavelengths differed by a depth of 0.6 µm in both cases indicating the shift in Z-direction was not due to imaging depth.

They then collected images from a triple labeled bead with emissions at 520 nm (green), 600 nm (red), and 680 nm (blue). With the correction collar adjusted for a measured coverglass thickness of 174 µm, the green and red signals coincided, while the blue signal was offset, as expected from the measurements made from glass. When the correction collar was purposefully misadjusted for a 150 µm coverglass to introduce spherical aberration, the red and green signals no longer coincided, and image intensity was significantly reduced. This type of error is often

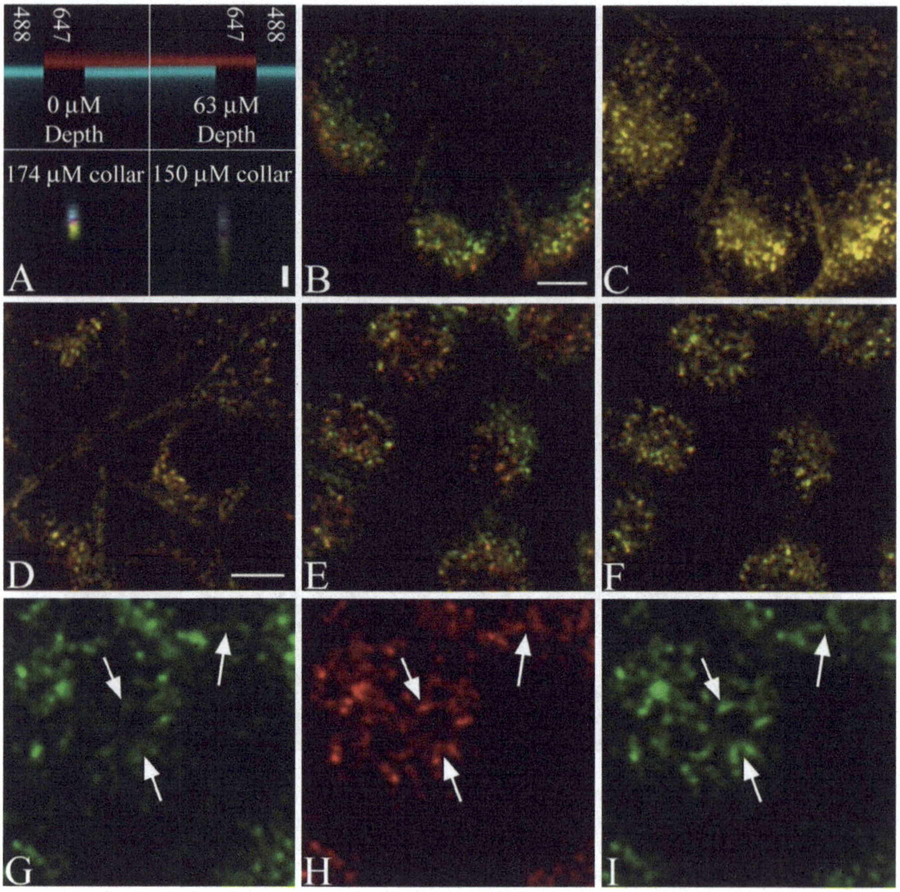

Fig. 3.16 Chromatic aberration in the 60X Plan Apochromat water immersion objective. (**a**) XZ-sections of glass reflection (top) and microsphere fluorescence (bottom) images. Reflection images are shown for depths of either 0 or 63 μm into an aqueous medium, with 657 nm light shown in red and 488 nm light shown in blue. Fluorescence images of fluorescent beads are shown for an objective with correct (left) and incorrect (right) coverslip-thickness collar setting. With 520 nm emissions shown in green, 600 nm emissions shown in red and 680 nm emissions shown in blue. For all images, the focal axis is oriented vertically, with the scale bar indicating a distance of 1 μm. (**b**) An image of a field of cells labeled with F-Tf and Cy5-Tf collected at a depth of 63 μm into an aqueous medium. (**c**) The projection of the vertical series of images from which Panel B was obtained. (**d**) An image of a field of cells labeled with fluorescein-TF and rhodamine-Tf collected at the coverslip surface. (**e**) An image of a field of cells labeled with both fluorescein-Tf and Cy5-Tf collected at the coverslip surface. (**f**) The same field shown in Panel E but combining green and far-red planes collected 0.6 μm apart. Higher magnification images of the green and far-red fluorescence from the same focal plane are shown in Panel **g** and Panel **h**, while Panel **i** shows the green fluorescence from a focal plane 0.6 μm lower. Scale bar in Panel **b** represents 10 μm length in Panels **b** and **c**. The scale bar in Panel **d** represents 10 μm length in Panels **d**–**f** and 5 μm in Panels **g**–**i** (Used with permission of the authors and Biotechniques, 2008)

encountered since the thickness of coverglasses from a single batch may vary by up to 40 μm around the thickness value at which they are sold.

To further demonstrate how these optical effects might lead to misinterpretation of co-localization data, Dunn and Wang (2000) again collected a series of images from Tf-labeled endosomes. In this set of experiments, they utilized the 60X water immersion lens and imaged through aqueous media at a depth of 63 μm. In Fig. 3.16b, a chromatic shift was present when imaging with fluorescein- and Cy5-labeled Tf. However, by projecting a Z-series of the images (Fig. 3.16c), the chromatic shift is no longer apparent, and the Tf molecules properly appear co-localized.

Consistent with data presented in Fig. 3.16a, when fluorescein- and rhodamine-labeled Tf molecules are imaged at the surface of the sample, they appear co-localized (Fig. 3.16d). However, even though imaging is at the surface of the sample, when fluorescein- and Cy5-labeled Tf molecules are imaged, there again is a chromatic shift indicating labeling of discrete populations of endosomes (Fig. 3.16e). If the fluorescein image is again collected 0.6 μm deeper, the co-localization is again apparent (Fig. 3.16f). Figures 3.16g–i show separate channels for fluorescein (Fig. 3.16g), Cy5 (Fig. 3.16h), and fluorescein collected 0.6 μm deeper (Fig. 3.16i) to further illustrate the mis-registration of images due to spherical aberration.

The examples above demonstrate the critical importance of carefully selecting and knowing the specifications of the objective lenses used in confocal and fluorescence imaging. Co-localization studies are common in confocal microscope experiments, and misinterpretation of co-localization data can easily occur if lenses are not corrected for the wavelengths of fluorochromes being used. Another common application of confocal systems is three-dimensional reconstruction of data sets. To optimize these experiments, minimizing spherical aberration in the optical path is essential in maximizing the usefulness of the confocal system.

3.5 Filters

In fluorescence microscopy filters are usually necessary for insuring that only light emitted by a fluorochrome in the sample is passed on to the detector and for blocking any stray light in the system. In addition, filters are needed to separate the various wavelengths of light emitted from different fluorochromes when using multiple fluorochromes emitting at different wavelengths. Finally, in situations where non-monochromatic light is used for excitation, filters are necessary for restricting the excitation wavelengths to only those necessary for excitation of the fluorochrome and restriction of the wavelengths that might be mistaken for emitted photons. Even when the emitted wavelengths have a narrow range, such as with lasers or LED arrays, it is usually desirable to include emission filters to restrict stray light entering the system. Also, since lasers generally have multiple narrow emitted wavelengths (Fig. 3.4), filtering helps insure that only the laser line of interest is part of the light path.

Basic filters can be classified roughly as short-pass filters (also known as low-pass filters), long-pass filters (also called high-pass filters), and band-pass filters. A long-pass filter (Fig. 3.17a) blocks all light below a certain wavelength and transmits wavelengths longer than that cutoff. Conversely, short-pass filters (Fig. 3.17b) transmit wavelengths shorter than the cutoff and block longer wavelengths. Band-pass filters (Fig. 3.17c) pass light within a certain range of wavelengths and block all wavelengths above and below that range.

For a long-pass filter, the filter is named by the wavelength at which 50% of its peak transmittance is passed. Figure 3.17a shows the transmission curve for a 459 nm long-pass filter. The filter transmits only 1% of 450 nm light that reaches the filter. However, 92% of 500 nm light passes through the filter. This is the maximum transmittance for this filter. The half maximum value is 46%, and this value is attained with light having a 459 nm wavelength. The filter would thus be designated a 459 nm long-pass filter. Similarly, short-pass filters are named for the wavelength at which 50% of their peak transmittance is blocked.

Band-pass filters transmit a range of wavelengths. The range can be very narrow, as is the case for the filter depicted in Fig. 3.17c, or the spectrum can be broad. Band-pass filters are designated by their central wavelength and the range of wavelengths they transmit. The lower and upper boundaries of the range are determined as the wavelength at which the half maximal values are achieved. This is often referred to as the full width half maximal (FWHM) value (double arrow in Fig. 3.17c). The central wavelength is the arithmetic mean of the upper and lower boundaries. For the filter depicted in Fig. 3.17c, the peak transmittance is 87%. Half of this value, 43.5% is attained at 528 nm and 554 nm. The arithmetic mean of 528 and 554 is 541 and the range from 528 nm to 554 nm is 26 nm. The filter would thus be designated as a 541/26 nm band-pass filter.

Fluorescence microscopy also requires that the excitation beam and emission beam are ultimately separated from each other. The most common method is to employ a specialized type of filter called a dichromatic beam splitter (also known as a dichroic mirror). Most filters are designed to have a zero angle of incidence. This is the angle between the optical axis of the incident light and the angle normal to the filter surface. The dichromatic beam splitter is designed for high incidence angles, usually 45 degrees. The beam splitter reflects light of short wavelengths and passes light of longer wavelengths, thus separating higher and lower wavelengths.

Figure 3.1 shows the filter arrangement in a standard epifluorescence microscope using an arc lamp source. An excitation filter (usually a band-pass or, sometimes, a short-pass filter) limits the wavelengths of light that illuminates the specimen. This helps make sure that only the desired fluorophore is excited. The dichromatic beam splitter reflects this light to the back focal plane of the objective lens which focuses the excitation beam onto the specimen. In a well-designed microscope, very little of this excitation light is reflected back. In this way, only the light emitted from excited fluorophores is collected and focused by the objective lens. The emitted photons travel back to the dichromatic beam splitter. Since these have a longer wavelength than the excitation light, they are transmitted through the dichromatic beam splitter, while any stray excitation light, because of its shorter wavelength, is reflected by the

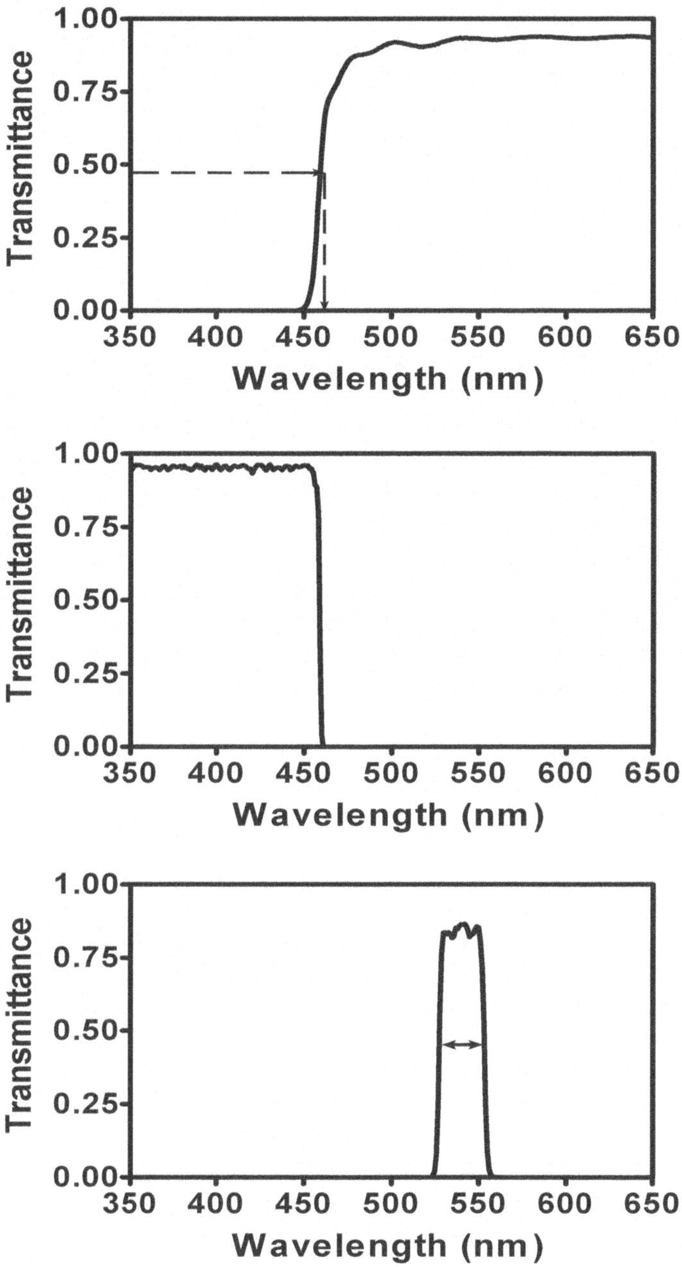

Fig. 3.17 Spectra for typical long-pass (**a**), short-pass (**b**), and band-pass (**c**) filters

dichroic back toward the light source. A filter on the other side of the dichromatic beam splitter helps to further limit the wavelengths that reach the detector to only those emitted by the fluorochrome. This emission filter (also called a barrier filter) is usually a long-pass or band-pass filter.

3.6 Types of Filters

3.6.1 Glass Filters

A variety of devices are available that discriminate and separate different wavelengths. The most common are colored glass filters. These are inexpensive, have long useful lives, and they are relatively insensitive to incidence angle. However, there are limitations to glass filters. Chief among them are low transmittance and high autofluorescence at longer wavelengths. Thin film interference coatings are an alternate method of making filters. These have high transmittance and can be designed to provide a wide variety of filtering parameters. The drawbacks of thin film coatings are that (1) their blocking performance holds for only specific wavelengths so other blockers must often be added, (2) they are very sensitive to angle of incidence, and (3) coatings that work well for visible light are usually not ideal for UV and so UV performance is compromised.

Filters that attenuate the intensity of the signal (called neutral density filters) are also useful for controlling the intensity of light on the specimen. This is particularly important for lasers, since they are high-intensity sources. While the high intensity of lasers can be an advantage in exciting fluorochromes and generating high signal levels, in many cases the intensity is so high that it is desirable to attenuate or decrease the intensity of light entering the optical path. Control of the intensity of light on the excitation side of the optical path is necessary to minimize photobleaching of the sample and to prevent cell death when imaging living samples. In early generation instruments, this was accomplished by placing a neutral density filter in front of the laser to decrease the amount of light entering the scan head. For instance, in the BioRad MRC 1024 line of instruments, neutral density filters that allowed 0.1%, 0.3%, 3%, 10%, 30%, or 100% (no filter) of the laser light through were options when imaging a sample. When samples were either very bright or susceptible to photobleaching or death, filters that allowed small percentages of laser light through to the sample could be selected. When samples generated a relatively poor signal, neutral density filters that allowed more excitation photons into the optical path could be used to improve the signal generated from the sample. However, this method of controlling the amount of excitation light contacting the specimen was limited in the selection of intensity level and also resulted in excitation of the entire sample area; no imaging could be done on smaller regions of interest (ROI).

3.6.2 Acousto-optical Tunable Filters (AOTF)

Most newer laser confocal systems utilize acousto-optical tunable filters (AOTF) to provide precise control of the laser intensity and the region of interest (ROI) in the sample that is excited. AOTF assemblies consist of a quartz filter subjected to various levels of ultrasound. By adjusting the intensity of the ultrasonic wave, the intensity of the laser light passing through the quartz can be adjusted between 0% and 100% in 0.1% increments. This provides very precise control over the intensity of light entering the optical path. In addition, selection of single or multiple excitation wavelengths of light for excitation is much easier with an AOTF than with a filter wheel containing multiple excitation filters. This is an important advantage when imaging multiple fluorochromes either simultaneously or sequentially.

An additional advantage of AOTF systems is that specific ROIs can be selected for excitation while the remaining areas of the sample are not exposed to excitation photons (Fig. 3.18). This is often important in samples that are sensitive to intense beams, such as living cells, since regions of the specimen outside the area of

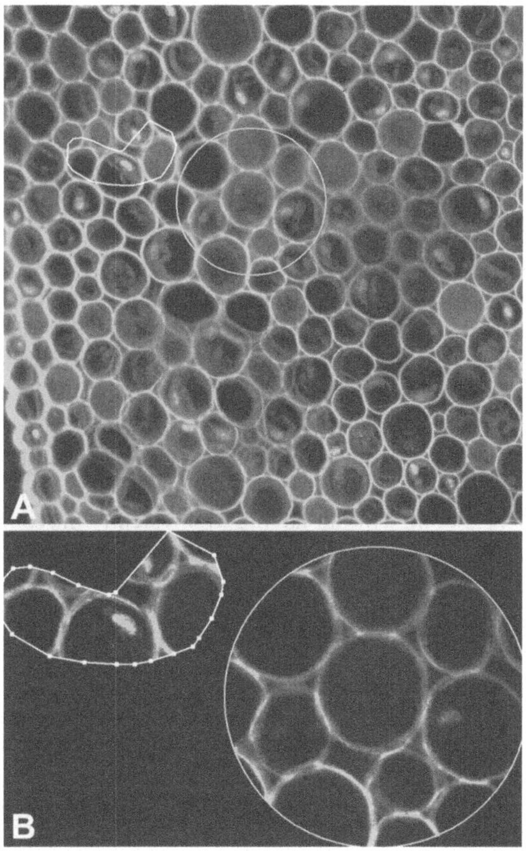

Fig. 3.18 Illustration of selection of ROI areas for scanning to increase the speed of scanning in select regions and to preserve the specimen in areas outside of the ROI. In **a** the entire specimen (Convallaria root tip) has been scanned and then a standard circle ROI and a hand-drawn ROI selected and scanned as shown in **b**

excitation are not illuminated and thus escape any damage. Region of interest imaging is also very useful for applications that involve rapid processes or restricted illumination of only small areas of the sample such as Förster resonance energy transfer (FRET) or fluorescence recovery after photobleaching (FRAP).

3.6.3 Acousto-optical Beam Splitters (AOBS)

There are several potential problems with the dichroic filters that are commonly used to sort the excitation and emission photons in wide-field fluorescence and confocal microscopes. In many confocal studies, multiple fluorochromes are used, and this requires the use of a series of filters in order to adequately separate the excitation and emission wavelengths of the multiple fluorochromes. Typically, the number of filters available in a system is limited, and the ones available may not provide the optimal transmission of photons for the range of fluorochrome wavelengths being generated. Even if multiple dichroic filters are available on a system, it is necessary to mechanically shift between these beam splitters as different fluorochrome excitation and emission wavelengths are generated. Although the filter changes are software driven and relatively fast, they do slow acquisition times when imaging multiple channels. Moreover, the mechanical shifting of filters can introduce vibrations. Finally, glass filters are not absolutely efficient, and so not all incident photons are passed through the filter. This decreases the signal and may adversely affect the signal to noise ratio in the final image.

In 2002, Leica introduced the AOBS as an alternative to dichroic mirrors in their confocal systems (Borlinghaus et al. 2006). Similar to dichroics, the AOBS transmits short excitation wavelengths of light to the specimen and also directs the transmission of longer wavelength photons emitted from the specimen to the detector. However, the AOBS utilizes an electronically tunable acoustic crystal in place of the dichroic mirror and is a fixed device. The ability to electronically tune the AOBS eliminates the need for dichroic mirrors to be mechanically inserted or removed from the optical path. This allows for faster switching between laser excitation wavelengths (lines), the use of up to eight illumination lines simultaneously, and the selection of narrow bands of light (0.6–2.0 nm) for excitation of fluorochromes. All this is done without having to physically move filters.

3.7 Determining an Optimum Filter Combination

Choosing the correct filter set for a given fluorophore is simply a matter of matching the characteristics of the fluorochrome with those of the filters. In Chap. 2, Fig. 2.3 depicts the excitation and emission curves for TRITC. This fluorochrome has a peak excitation at 555 nm and peak emission at 580 nm. The range of excitation wavelengths at FWHM is 533 nm through 571 nm. To optimize fluorescence the microscopist would like an excitation beam that is within this range and as close to the excitation maximum as possible. However, overlap with the range of emission wavelengths should be avoided. For TRITC, the emission range at FWHM is 563–610 nm. Thus, there is an overlap of the excitation and emission FWHM values between 563 and 571 nm.

A 540/25 nm band-pass filter would limit the excitation light to below the TRITC emission range. The 543 wavelength of a helium-neon laser could also be used for excitation of TRITC. This wavelength is well out of the emission range of TRITC but would promote sufficient absorption. A 565 nm dichromatic beam splitter would reflect the excitation light and pass most of the light emitted from TRITC. Finally, a 575 nm long-pass filter would insure that most of the signal hitting the detector was specifically from the excited TRITC. This of course is not the only filter combination that could be used with TRITC. For instance, a 610/75 nm band-pass filter would also work as a TRITC emission filter as would an optimally tuned AOBS.

For detecting fluorophores of a single wavelength (single color detection), the choice of the filters to use is relatively simple since you do not need to worry about specific excitation of or emission from competing fluorophores. This is not the case when you want to detect two or more fluorophores in a single sample. The presence of multiple fluorophores increases the probability of spectral overlaps. This is a problem because in fluorescence microscopy the detectors most often used do not discriminate between wavelengths (i.e., they do not recognize color); they only record the presence or absence of photons (the standard detectors used in confocal microscopy are described further in Chap. 7). The color specificity assigned to photons is arbitrary but generally based on assumptions about the wavelengths that are transmitted to the detector by the filters that are placed in the light path. If the microscopist has selected these filters incorrectly, spurious interpretations can occur.

Two principal types of artifacts can occur with imperfect filtering when two or more colors of fluorescence are analyzed. The first artifact results when the emission from the first fluorochrome has the proper wavelength to excite the second fluorochrome. This situation is depicted in Fig. 3.19. The emission peak of fluorochrome I falls within the excitation wavelengths of fluorochrome II. Without attention to proper filtration, this scenario could result in the fluorescence coming from both fluorochromes I and II being incorrectly interpreted as coming only from fluorochrome I based on the excitation wavelength employed. However, with a suitable emission band-pass filter, the emission from fluorochrome I can be discriminated from that arising from fluorochrome II as long as there is no substantial overlap of the two emission spectra. Figure 3.20 depicts two hypothetical emission peaks and indicates a band-pass filter which would allow detection of emission from only fluorochrome I. An alternative would be to choose different fluorophores where the emission spectrum of one does not overlap the excitation spectrum of the other. This is the safer alternative.

The second problem in assuring specificity of detection arises when two emission spectra have substantial overlap as shown in Fig. 3.21. If both signals are

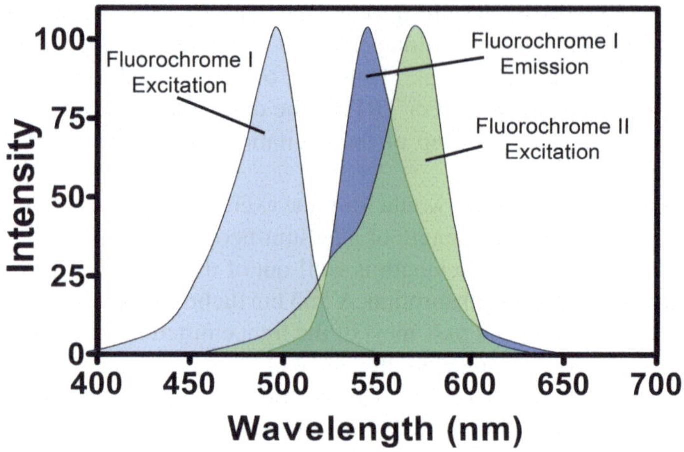

Fig. 3.19 Spectral overlap of emission from one fluorophore with excitation of a second fluorophore

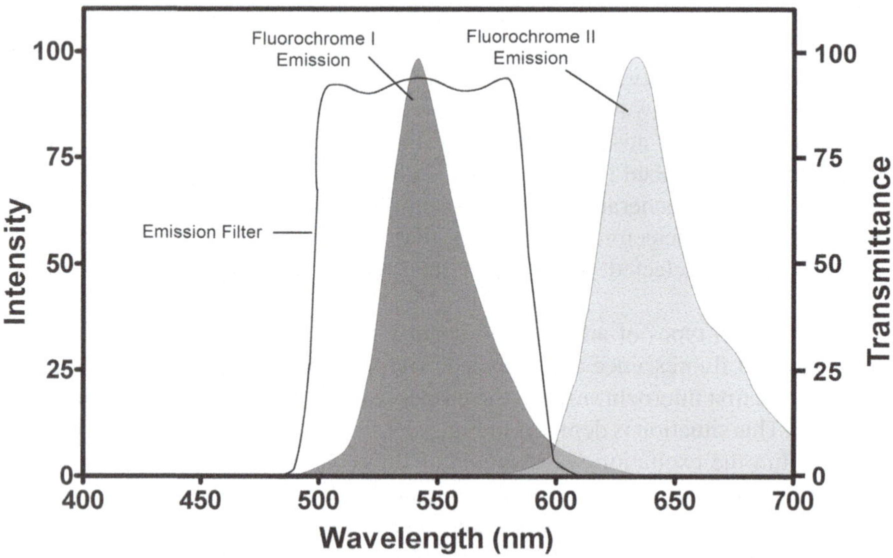

Fig. 3.20 Use of band-pass filter to separate emission of one fluorophore from emission of a second fluorophore when the two fluorochromes have similar excitation spectra

strong, as depicted in Fig. 3.21a, it would not appear that the fluorochrome 1 emission (stars) makes up a substantial component of the emission assigned to fluorochrome II (circles) using a band-pass filter designed for fluorochrome II (in this case a 600/50 band-pass filter). However, if the signal from fluorochrome II is low compared to fluorochrome I, the emission assigned to fluorochrome II could be substantially overestimated (Fig. 3.21b). Moreover, even if fluorochrome II was not present,

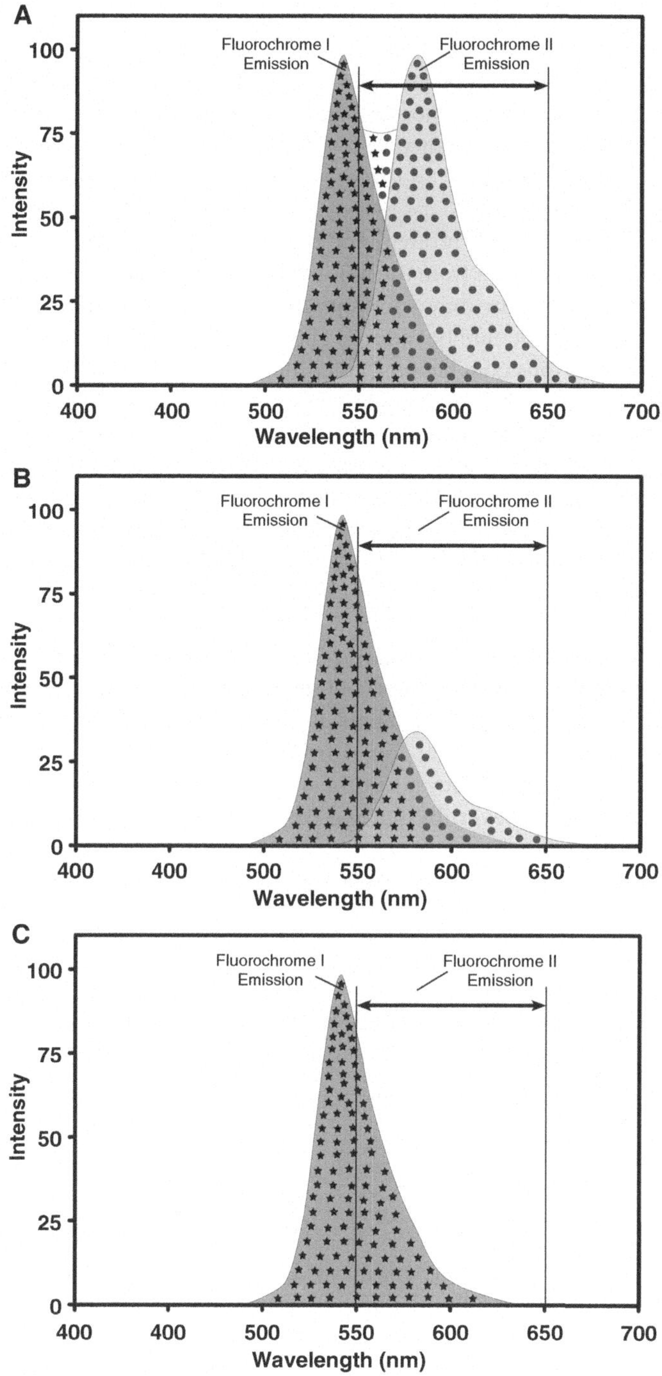

Fig. 3.21 Simultaneous two-color fluorochrome detection and the effect of bleed through of signal assigned to the higher wavelength fluorochrome. (**a**) When the signal from fluorochrome II (circles) is strong, the bleed through of fluorochrome I (stars) has a limited but still significant effect on the signal detected using a 600/50 band-pass filter. (**b**) When the signal strength of fluorochrome II is weak, the bleed through of fluorochrome I becomes the dominant signal. (**c**) When fluorochrome II is not present, bleed through produces a signal that could be spuriously interpreted as the presence of fluorochrome II

a signal would be detected and spuriously assigned to fluorochrome II (Fig. 3.21c). Thus this overlap of emissions (termed bleed through) must be carefully guarded against and proper controls initiated to test for possible overlap. Eliminating bleed through is always important but, as discussed in Chap. 11, it is critical when undertaking co-localization studies.

Of course, to avoid bleed through, one could use narrower band-pass emission filters to collect only photons from the non-overlapping regions, but, as in the example shown in Fig. 3.22, this may substantially reduce the number of photons collected. In fact, the signal could be reduced enough that it is not distinguishable from noise leading to spurious conclusions about the presence or amount of fluorophore present. A better approach is to sequentially excite one fluorochrome and then the other instead of attempting to collect the two signals simultaneously. Sequential collection reduces the necessity to discriminate between two simultaneously excited fluorochromes because it can optimize collection conditions for each fluorochrome separately. Advancements in automation of functions like filter switching and the AOBS make sequential imaging of two or more fluorochromes much faster and easier. Most modern confocal microscopes are ideal instruments for such a sequential approach.

There are also mathematical techniques that attempt to determine the contribution of two or more overlapping fluorophores to a complex spectrum. These techniques, often referred to as spectral analysis or spectral unmixing, are discussed in Chap. 9 and are an advancement found in many laser scanning confocal microscopes produced after 2005. They allow for specific detection of fluorochromes with overlapping spectra. However, by far, the best and easiest approach to separating

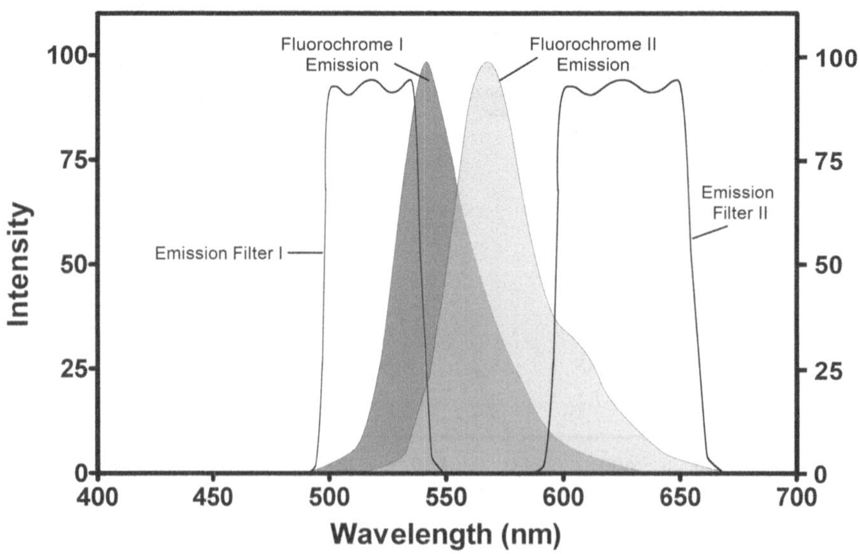

Fig. 3.22 Band-pass filters for separating non-overlapping regions of two fluorophores with overlapping spectra

fluorescent signals is to use fluorochromes with widely separated emission spectra whenever possible.

In all cases, it is necessary to confirm that selected filter combinations are working as expected. This can be done by exciting one fluorochrome at a time while monitoring the emission being transmitted through all the other filter combinations. There should be no signal above background through any of the filter sets other than the one for the fluorochrome of interest.

3.8 Optimizing the Fluorescent Signal

Because of a number of factors, the amount of light reaching the detector in a fluorescence microscope is relatively low. Even with very sensitive detectors, one still must insure that the signal strength is sufficiently high compared to background noise. Otherwise, you cannot be confident regarding the accuracy of what is detected. Throughout the remaining chapters of this book, we will discuss methods to increase the number of photons detected. This aspect of confocal microscopy cannot be overemphasized. "Photons are your friends, don't abuse them!" should become your mantra. In order to maximize signal, it is imperative to carefully review the excitation and emission spectra of the selected fluorophore(s) and choose filters based on these spectra. It is cheaper in the long run to buy additional filters, if needed, than to misinterpret critical data because of sloppy microscopy. Once the correct filters are in place, additional key elements for maximizing signal strength include (1) properly aligning all elements of the excitation and emission pathways; (2) removal of dirt and other obstacles that would impede the flow of photons to the detector; (3) matching, as close as possible, the refractive indices of all parts of the imaging path, including the specimen so that photons are not lost due to refractive index mismatch; (4) minimizing loss of energy to competing processes (e.g., photobleaching) which decrease the fluorescence signal; and (5) using a high numerical aperture lens to collect as much of the emitted light as possible.

Literature Cited

Borlinghaus R, Gugel H, Albertano P, Seyfried V (2006) Closing the spectral gap - the transition from fixed parameter fluorescence to tunable devices in confocal microscopy. Proc SPIE 6090:60900T60901-60906

Diaspro A, Federici F, Robello M (2002) Influence of refractive-index mismatch in high-resolution three-dimensional confocal microscopy. Appl Opt 41(4):685–690

Dunn, KW. And Wang, E. 2000. Optical aberrations and objective choice in multicolor confocal microscopy. BioTechniques 28(3): 542–550

Samoc A, Miniewicz A, Samoc M, Grote JG (2007) Refractive-index anisotropy and optical dispersion in films of deoxyribonucleic acid. J Appl Poly Sci 105:236–245

Voros J (2004) The density and refractive index of adsorbing protein layers. Biophys J 87:553–561

Chapter 4
Specimen Preparation

W. Gray (Jay) Jerome, John Fuseler, Caleb A. Padgett, and Robert L. Price

4.1 Introduction

In confocal microscopy, as with any other microscopy technique, there is no single best method of preparing a sample that can accommodate all types of samples, all possible methods of staining specific structures, and all possible modes of imaging. Optimum biological sample preparation for confocal microscopy is very dependent upon the cell or tissue type, the labeling technique, and type of data to be collected. In this chapter, however, we will review some basic concerns that need to be addressed for all types of samples and lay out some basic principles that help guide decisions about sample preparation.

A key difference between confocal microscopy and widefield microscopy is that, for confocal, the aim is to explore structure and structural relationships along the optical (Z) axis as well as in the X-Y plane. In other words, the investigation of spatial relationships is being evaluated in at least three dimensions. Often this involves the collection of a series of planar images along the Z-axis and reconstruction of the data into an image that depicts all three dimensions at once (see Chap. 10). However, even when the interest is solely in a single X-Y plane, the goal is to

W. G. (. Jerome (✉)
Department of Pathology, Microbiology and Immunology, Vanderbilt University School of Medicine, Nashville, TN, USA
e-mail: Jay.Jerome@Vanderbilt.edu

J. Fuseler
Department of Pathology, Microbiology and Immunology, University of South Carolina School of Medicine, Columbia, SC, USA
e-mail: John.Fuseler@uscmed.sc.edu

C. A. Padgett · R. L. Price
Department of Cell Biology and Anatomy, University of South Carolina School of Medicine, Columbia, SC, USA
e-mail: Caleb.Padgett@uscmed.sc.edu; Bob.Price@uscmed.sc.edu

W. G. Jerome, R. L. Price (eds.), *Basic Confocal Microscopy*,
https://doi.org/10.1007/978-3-319-97454-5_4

separate as thin a plane of information as possible from the planes directly above and below. When using confocal microscopy at medium to high magnifications, the Z-axis is usually less than 1 micrometer in thickness. Obviously, in order to acquire and analyze 3-D structural information at these resolutions, the 3-D structural relationships must be preserved during the preparation of the sample. In contrast, in standard widefield microscopy, information from 6 to 8 microns along the Z-axis is frequently imaged as a single two-dimensional image, thus compressing the Z-axis information. Because of the added need to preserve 3-D structure for confocal images, methods suitable for preparing samples for widefield microscopy must often be altered to accommodate the additional demand of preserving relationships in the third dimension.

In the past few years, considerable effort has been made to increase the depth of imaging, and hence increased 3-D reconstruction depth, that can be achieved with confocal microscopy. Many of these techniques involve clearing the tissue so the refractive index of the sample more closely matches that of the optical system and mounting media used in sample preparation. As discussed in Chap. 3, reducing photon scatter due to refractive index differences allows information to be collected from greater depths. Clearing of tissue typically involves the removal of the lipids from the sample rendering it optically transparent (Fig. 4.1). There are currently several clearing techniques in popular use. A few of these, BABB (benzyl alcohol/benzyl benzoate), a group of DISCO (three-dimensional imaging of solvent-cleared organ) techniques with slight variations in protocols, and X-CLARITY™, are discussed in more detail in Sect. 4.4.2.

4.2 Preserved Samples Versus Live Imaging

After defining the goals to be obtained for a particular confocal imaging application, the next decision is whether it is most appropriate to view living or preserved (fixed) samples. Key questions to guide this decision include (1) will rapidly changing events be analyzed, (2) will native fluorescent molecules or exogenous

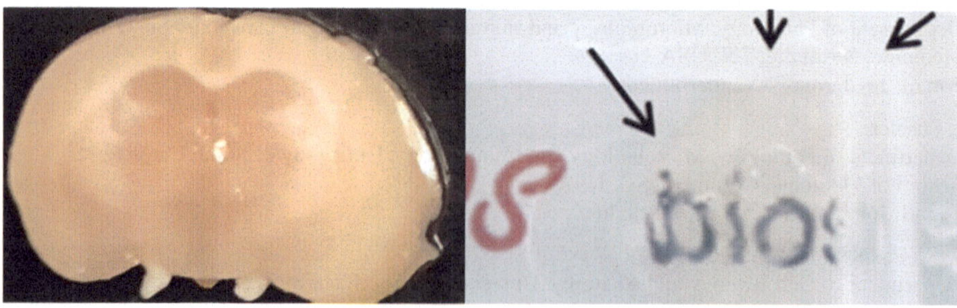

Fig. 4.1 Uncleared brain slice (left) showing opaque nature of the tissue and cleared brain slice (right) showing transparent nature of cleared tissues

fluorochromes be imaged, and (3) how bright is the fluorochrome? Live imaging is a powerful technique that allows the analysis of dynamic events and avoids some of the artifacts that can be introduced by specimen preservation and processing. Although it is possible to image thin tissue slices in organ culture, live cell imaging usually involves continuous viewing of cells in culture. Specimen preparation for live cell imaging is more straightforward than for imaging fixed material, but this often comes at a price. Since the specimen is not preserved, the microscopist must insure that the act of viewing the sample does not introduce artifacts. Fairly elaborate systems for keeping the cells healthy during imaging are required. In addition, the excitation (laser) energy generally must be kept at a low level to minimize photodamage and cell death. This reduces the signal that can be generated from fluorochromes and so increases the need to make sure that signal collection is maximal. If a fluorophore is already a weak emitter, it might be better to consider imaging it in fixed material where higher intensity excitation energy can be used because cell death is not a consideration. There are also fewer fluorophores available for live cell imaging, although this is slowly improving with the development of several varieties of fluorescent proteins such as GFP, RFP, mcherry, and others that have been designed for live cell work.

Finally, one needs to consider the rate at which the event of interest occurs. The number of frame captures per second in laser scanning confocal can be varied depending upon conditions but will generally fall between 0.1 and 1 frame per second in most point scanning confocal microscopes. This is adequate for following some processes, but many reactions such as calcium fluxes in cells happen at a faster rate than this. If the event of interest occurs at a faster rate, one should consider faster imaging methods, such as spinning disk or resonant scanning confocals optimized for live cell imaging, or abandon confocal altogether and use a widefield microscopy approach.

Despite its difficulties, viewing of live material in 3-D can provide dynamic information not obtainable by other techniques. If one is patient, is meticulous, and pays attention to correct microscope setup and specimen preparation, the rewards can justify the effort required.

4.3 Working with Fixed Samples

Working with fixed samples negates the ability to observe events in real time, although dynamic processes can sometimes be inferred from a series of static images taken over time. Fixed material, though, is easier and more forgiving to image. In fixed samples one does not have to worry about keeping the cells or tissue alive nor is specimen damage as much of a concern since fixed material will hold up to most confocal viewing modes. Fluorophore bleaching, though, remains a critical issue. Even though there is less concern about artifacts induced by the photon beam's interaction with the specimen when viewing fixed material, one must be very concerned about the effects of fixation and subsequent specimen preparation on the sample architecture. The choice of available fluorophores and techniques to employ fluorophores is also much greater when working with fixed material. For this reason, the majority of confocal experiments use fixed samples. However, the trade-off for the simplification of microscopy is the fact that more processing of the sample is required prior to microscopy and one must make sure that this processing does not induce unrecognized artifacts. There are few absolutes regarding the best method of specimen preparation. However, if one understands the general principles and has clearly defined the experimental goals, a reasonable set of decisions can be made regarding the most appropriate procedure.

4.3.1 Fixation

The first parameter to consider when working with fixed samples is which fixative to use. A perfect fixative should maintain all cellular structures in their native state, eliminate destructive autolytic processes, and protect the material against damage during subsequent processing and microscopy. In addition, if the sample will be immunostained, the ideal fixative should also preserve the ability of subcellular molecules to bind with their antibodies. As with most things in microscopy, the perfect fixative does not exist, and so the microscopist must weigh the relative benefits and drawbacks of each fixative protocol relative to the experimental goals.

Fixatives can be divided into two general types: precipitating fixatives and those that cross-link proteins. Precipitating fixatives are generally organic solvents. The most common ones in use for microscopy are alcohols and acetone. Precipitating fixatives work by dehydrating and precipitating some components, usually proteins, within the cell or tissue. Because many of the changes brought about by these fixatives can be reversed by rehydration during staining, they are often employed in immunocytochemistry experiments because the reversal sometimes increases antibody binding to the antigen of interest. An additional advantage is that they remove the lipids and small soluble cellular molecules. This permeabilizes the membranes and enhances the access of antibodies to cellular antigens. However, they should not be the first choice for confocal experiments because cell shrinkage can be as high as 70%, particularly along the Z-axis as cells are dehydrated and lose their contents. Although this is useful in widefield fluorescent experiments because the inherent flattening of the sample lessens the amount of out of focus material that blurs the image, this shrinkage can severely alter the 3-D relationship of intracellular structures. It is our experience that, often, structures not generally adjacent in real life become co-localized in space after ethanol fixation or dehydration.

An alternative to precipitating fixatives is cross-linking fixatives. Aldehydes are the most often used cross-linking fixatives with solutions of formaldehyde or glutaraldehyde being the most common. These work by forming intermolecular bridges between components of the cell. The fixation generally involves binding of the aldehydes to free amino groups to create a network linking the cellular constituents together and holding them in place. Cross-linkers do a better job of maintaining structural integrity than do fixatives that act by dehydration and precipitation. However, they often reduce the antigenicity of molecules more severely. In addition, because they do a better job of maintaining membrane structure, intracellular staining with antibodies generally requires sectioning, permeabilization, or both to allow stains to access intracellular structures.

Glutaraldehyde in a concentration range of 2–5% in buffer does a better job of preserving fine structural detail than does formaldehyde. However, glutaraldehyde often significantly alters antigenic sites to the point that primary antibodies no longer recognize a specific epitope resulting in significantly reduced labeling intensity. Glutaraldehyde is also highly autofluorescent, in part due to the presence of free amines after fixation. The autofluorescence can be partially quenched by incubation with 1% sodium borohydride, but, in general, it is best to avoid glutaraldehyde for confocal microscopy experiments unless you are using a very bright fluorophore which can be easily detected above the autofluorescence. If it is essential that glutaraldehyde be used, it may be possible to avoid the typical green-yellow autofluorescence of aldehydes by selecting a fluorochrome which emits in the far red range allowing you to filter out the green-yellow aldehyde fluorescence.

Dilute solutions of formaldehyde (1–5%) in buffer are the most common fixatives used for confocal microscopy. It is important to use an appropriate concentration for the type of sample being preserved. Over-fixed samples often exhibit lower labeling intensity due to reduced antigenicity, and under-fixed samples lose structural detail if stored for an extended period of time (a few weeks or more). In our laboratories we typically use 2% freshly prepared solutions for fixation of cell monolayers and 4% solution for fixation of larger tissue blocks. It is also important to use adequate quantities of fixative, so sufficient aldehyde groups are present for cross-linking of tissue components. Cell monolayers are generally not a problem to fix, but we often encounter poorly fixed tissue samples due to inadequate volumes of fixative being used to preserve tissues. A good rule of thumb is that the fixative to tissue volume ratio should be at least 20:1. Ratios less than this often result in poorly preserved tissue, especially in the central areas of the tissue because insufficient aldehyde groups are available for fixation.

An additional concern when using formaldehyde in solution is that it can quickly decompose to formic acid. This contaminant reduces the fixation efficiency of formaldehyde solutions. For this reason, working strength solutions of formaldehyde should be made fresh from paraformaldehyde powder or diluted from higher strength solutions of formaldehyde which have been stored in an inert environment. Sealed vials containing 10–25% formaldehyde in nitrogen gas can be purchased and used to make up working strength formaldehyde solutions. The lack of oxygen in the vial prevents formation of contaminants. Alternatively, formaldehyde, which is a gas, can be liberated into solution by heating paraformaldehyde powder in an alkaline solution. Paraformaldehyde is a crystalline condensate of 8–100 formaldehyde molecules. Regardless of how the working strength formaldehyde solution is made, it should be refrigerated and used within a few days to avoid the formation of contaminants. The purity of the formaldehyde solution can be tested by comparing the absorbance of a 0.5% solution at 235 nm and at 280 nm. If the ratio of these values is less than 2, it indicates minimal contamination, and the solution can be used. Formaldehyde fixatives having a ratio of higher than 2 should not be used as a fixative for confocal microscopy. A standard protocol for making a formaldehyde solution from paraformaldehyde is provided in Fig. 4.2.

Aldehyde fixation protocols vary depending upon the application and the method of staining. However, a few concerns are reasonably consistent across applications. Aldehyde fixatives should always be made fresh, kept cold, and be used within a few days. Although it is best to keep fixatives refrigerated until used, the actual fixation should take place at room temperature or warmer. Cold temperatures disrupt cytoskeletal elements and so can alter cellular size, shape, and the interrelationships of subcellular organelles. Both the sample and fixative should be at room temperature or warmer when fixation is initiated. After an hour of fixation, however, fixation can be continued in the cold. The length of time the sample is fixed and the concentration of fixative will vary depending upon the staining technique that will be employed. In general, however, a concentration of 4% formaldehyde for 1–2 h is generally suitable for samples that are thinner than 500 micrometers along at least one axis. An overnight fixation, with the first hour being at room temperature and the remaining time being in the refrigerator, is a general fixation protocol that is often employed. Shorter periods or more dilute solutions may be required for some antibodies used as stains. Empirical testing is the only sure way to maximize fixation protocols for your unique needs. It is wise to always review the literature when designing an experiment to determine if others have used a specific combination of

Making a Buffered Formaldehyde Solution from Paraformaldehyde.

To make 100 mL of a 10% solutions of formaldehyde in 0.1M Cacodylate buffer (this is sometimes referred to as a 10% paraformaldehyde solution):

1. In a 250 mL flask, add 20 grams of paraformaldehyde to 80mL of distilled water
2. Heat gently but do not allow to boil
3. At 60 C, water will start to condense on the inside of the flask and the paraformaldehyde will begin to dissolve.
4. When the solution has reached 60 C, add two drops of 2N sodium hydroxide.
5. Maintain the solution at ~ 60-65 C and do not allow it to boil. The solutions should slowly begin to clear as the paraformaldehyde crystals dissolve and the released formaldehyde goes into solution.
6. If the solution does not clear within 10-15 minutes, add an additional 2-3 drops 2N NaOH.
7. When clear cool the solution and add distilled water to 100 mL. (This replaces any water that evaporated).
8. Add 100 mL of 0.2M Cacodylate buffer.
9. Keep solution refrigerated and use within 3 days.

The working strength fixative solution is produced by diluting the 10% solution just prior to use. Some researchers include 1% calcium chloride to help preserve membrane structure.

Fig. 4.2 Protocol for preparing buffered formaldehyde from paraformaldehyde concentration, time, temperature, and antibody that has worked well and to use this as a starting point for an experiment or empirical testing.

It is always best to process, stain, and image samples immediately after fixation is complete. However, this is not always possible. Fixed samples can be kept refrigerated in fixative for weeks, but additional cross-linking may occur which can further alter epitopes leading to additional reduction in staining affinity. Glutaraldehyde-fixed samples are generally stable and can be rinsed and placed in a buffer solution after fixation for storage up to a few weeks. On the other hand, the cross-links formed by formaldehyde are less stable. In some cases, prolonged time in buffer solutions (or water) can reverse the effects of fixation. This can prove advantageous in some instances, since antigenicity may be restored and staining improved. However, for most applications, prolonged storage of formaldehyde-fixed samples in buffer should be avoided. It may lead to a reversal of fixation, increased autolysis, and loss of the 3-D architecture. In extreme cases loss of all sample structure may occur. If prolonged storage of formaldehyde-fixed samples is necessary, it may be advisable to store in buffer with 0.5–1% formaldehyde added. This solution should be changed weekly as formic acid and other contaminants may form in the stored sample.

4.4 Working with Very Thick Samples

As noted previously the goal of many confocal microscopy studies is the creation of 3-D volumes for examination of structural relationships in the Z-axis. The actual depth of a sample that can be imaged is dependent upon several variables as discussed in Sect. 3.4. In most tissues, even if the optical elements in the light path are optimized, less than 100 μm of tissue can be imaged reliably. The effective depth of viewing is dependent on the content of the sample. Beyond this depth, photon scattering and absorption degrade the signal to undetectable levels. To image deeper regions of a sample, two basic approaches have been employed. First, one can cut the tissue into sections (100–500 μm) to expose deeper areas and second tissue clearing protocols may be used to minimize refractive index mismatch.

4.4.1 Tissue Sectioning

In most cases, tissue sectioning is done by mechanically slicing the sample into segments with the aid of specialized microtomes. If the sample is to be stained with exogenously applied fluorochromes, the slicing is usually done prior to staining because sectioning the tissue facilitates penetration of the stains. The most common microtomes for slicing in conjunction with fluorescence staining are tissue slicers and vibrating blade microtomes. Tissue slicers consist of a blade attached to an arm which is mechanically raised and lowered, while the underlying sample is shifted slightly after each stroke to slice the sample like a loaf of bread. In contrast, a vibrating microtome, such as a Vibratome (Vibratome Inc., St. Louis, MO), shaves sections off the top surface of a specimen using a rapidly vibrating blade. The material is kept cool in a fluid bath (generally water for fixed tissue or buffer for live material). The cool temperature makes the tissue more rigid to enhance slicing, the buffer helps maintain tissue structure, and the presence of a fluid helps lubricate the cutting operation. Sections are then retrieved from the bath onto a glass slide or other substrate for staining.

Some fixed tissue, such as the liver, may be sufficiently rigid to allow sectioning without further stabilization. In most cases, though, it is preferable to add additional support to the sample before sectioning. This can be done by encasing the sample in a material that can be liquefied at low temperature for application, but at room temperature or when slightly cooled (but not frozen) becomes rigid and provides support to protect the tissue from being distorted during slicing. A particularly useful material for encapsulating the sample is a 3–5% solution of agarose or a mixture of agarose (1%) and acrylamide (3%) (Germroth et al. 2005). Compared to paraffin wax embedment or other methods of embedding samples for microtomy, agarose encasement is simple, minimizes specimen manipulation that can produce artifacts, and does not require harsh solvents such as xylene or toluene or the use of high heat. Figure 4.3 describes a standard procedure for preparing a sample for slicing.

Other solutions to sectioning of very thick material exist. One can employ standard sectioning methods such as paraffin embedding and microtome sectioning or cyromicrotome sectioning of frozen material. In addition, if specific layers from a multilayered tissue are to be imaged, gentle protease digestion can sometimes allow the desired layer to be separated from the rest of the tissue mass. None of these, however, are ideal in that they require additional tissue processing that has the potential to disrupt 3-D cellular and tissue architecture.

Preparation of Samples for Vibratome or Tissue Slicing

1. Melt Sigma type IA low gelling agarose in PBS in tube in a 55 C water bath to make a 5% agarose solution
2. Wash fixed tissue in same buffer as used for fixation.
3. Pour agarose solution into a small plastic weigh-boat. Immediately submerge tissue in agarose. Work quickly; the agarose will begin to set in about 5 minutes as it cools.
4. Chill sample in refrigerator to solidify agarose.
5. Remove the block from the weigh-boat.
6. Trim the block so that only a small amount of agarose surrounds the sample.
7. Blot the base of the block to remove excess fluid.
8. Superglue the agarose block to a Vibratome planchette or tissue slicer platform.
9. Use Vibratome or tissue slicer to slice into sections of desired thickness (often 50 to 100 μm) following manufacturer's directions for the equipment.

Fig. 4.3 Protocol for preparing tissue for slicing prior to staining

4.4.2 *Tissue Clearing*

Depending on sample density and other factors, even when sectioned at 100 μm, it may not be possible to image through the entire section volume due to refractive index mismatch that reduces laser penetration and detection of photons emitted from deep within the tissue slice. In these samples tissue clearing protocols may improve data collection from larger volumes of tissue.

There have been several recent reviews on tissue clearing (Richardson and Lichtman 2015; Azaripour et al. 2016; Silvestri et al. 2016; Tainaka et al. 2016), and a brief discussion of some common techniques is presented here. While there are a host of other clearing methods, this review will focus on the most effective and commonly cited protocols. Tissue clearing protocols are typically divided into two common categories: solvent-based and hydrogel embedding.

Solvent-based clearing methods such as benzyl alcohol with benzyl benzoate (BABB) and 3-D imaging of solvent-cleared organs (DISCO) dehydrate the tissue with a lipid solvent and then extract the lipids with additional solvation. This solvation can typically be accomplished by compounds such as tetrahydrofurane (THF), dichloromethane (DCM), or BABB. These compounds require minimal shaking and will clear lipids passively without an additional catalyst. While solvent-based methods are effective in optimizing the refractive index, marked sample shrinkage occurs due to the intensity of the dehydration steps. In addition, the toxicity of many of the solvents is a concern for the safety of the researcher (Table 4.1).

Hydrogel embedding techniques such as X-CLARITY™ (Clear Lipid-exchanged Acrylamide-hybridized Rigid Imaging/Immunostaining/In situ hybridization-compatible Tissue-hYdrogel) seek to address the problem of clearing large tissue samples without the removal of key proteins. As part of the processing protocol for very large pieces of tissue, fixation should be performed by perfusion with 4% paraformaldehyde. In some cases with dense tissue, such as the heart, it is also recommended that following paraformaldehyde fixation, the hydrogel solution also be perfused followed by passive incubation in the hydrogel solution. Samples are then cleared with 4% sodium dodecyl sulfate (SDS) or actively cleared using electrophoresis. While passive clearing may take days or even weeks, electrophoretic clearing can be accomplished in several hours, depending on the density of the tissue (Table 4.1). Dense samples such as heart tissue tend to require longer clearing times, while samples such as colon tissue require little time at all. After clearing, samples

Table 4.1 Characteristics of common tissue clearing methods

	BABB	DISCO	CLARITY
Method	Solvent-based	Solvent-based	Hydrogel electrophoresis
Toxicity	Moderate	Minimal	None
Cost	Minimal	Moderate	Higher start-up
Replicability	Minimal	Moderate	Maximum
Fluorescent protein preservation	None	Moderate	Maximum
Clearing time	Hours to days	Days	Hours to days

should be mounted in a histodenz or glycerol medium, which matches the refractive index of the cleared tissue. Hydrogel methods are typically faster and safer than solvent-based methods but are often more expensive to complete.

Cleared samples should be imaged using an objective appropriate for the desired mounting medium. For best results, samples should be mounted in a medium that matches the refractive index of the cleared tissue. This is typically accomplished by mounting in a small droplet of the clearing agent, such as BABB, THF, or DCM. X-CLARITY™-cleared tissues can be mounted in the proprietary X-CLARITY mounting medium or in 85% glycerol. When imaging, it is imperative to keep samples submerged in their respective media, as cleared tissue may rapidly oxidize.

4.4.2.1 BABB

BABB was the first solvent-based clearing technique (Zucker et al. 1998; Miller et al. 2005) developed and gained popularity for being a quick and simple clearing protocol with a low cost of operation. Studies by Miller et al. (2005) demonstrated that tissues cleared with BABB could be imaged at depths of up to 2 mm, suggesting that the limiting factors for imaging depth were objective working distance and degree of antibody penetration.

In the BABB procedure, specimens are fixed, typically in 4% formaldehyde, vibratome sectioned to a thickness optimized for the region of interest, and immunolabeled using the protocol of choice. Samples are then rinsed with phosphate-buffered saline (PBS) and gradually equilibrated into methanol by gently shaking for 15 min each in increasing methanol concentrations of 20%, 40%, 60%, 80%, 100%, and an additional 100%. Samples are then transferred into the BABB solution, which is a 1:2 mixture of benzyl alcohol and benzyl benzoate. Clearing of tissue sections of approximately 100 μm can be completed in 24 h, but some dense tissues such as the heart or brain may require 2 days or more. Trimming the tissue to isolate the region of interest can decrease the amount of time needed to obtain a cleared area for imaging. After the specimen has become clear, it must be mounted within a droplet of BABB before imaging. Figure 4.4 shows a BABB cleared chicken embryo.

While BABB is a quick and efficient method for clearing, great care must be taken to avoid contact with plastics. BABB is corrosive and can damage many parts of the microscope and other surfaces. In addition, tissues cleared with this method tend to exhibit significant shrinkage from methanol dehydration, which may lead to potential issues when studying volume and morphology. BABB can drastically reduce GFP fluorescence and moderately reduce RFP and YFP fluorescence, making it impractical for clearing tissues with an inherently low expression of fluorescent proteins (Table 4.1). Despite these drawbacks, BABB continues to be the cheapest and most widely utilized method in tissue clearing.

Fig. 4.4 Mouse embryo cleared using the BABB technique. Often large pieces of tissue or embryos such as this remain slightly yellow in color following clearing

4.4.2.2 DISCO

DISCO is another example of a solvent-based clearing technique. Originally developed as 3DISCO by Dr. Ali Erturk (Ertürk and Bradke 2013), DISCO sought to improve upon the clearing power of BABB while protecting fluorescent protein signal (Dodt et al. 2008). Renier et al. (2014) developed alternative DISCO techniques to optimize immunolabeling reproducibility and to eliminate the use of harmful solvents such as tetrahydrofuran (THF) in the clearing protocol. While not as common as BABB, DISCO techniques can produce the same final refractive index with less toxicity than that associated with BABB. The primary concern is the volatility of dichloromethane (DCM) that may result in an inhalation hazard causing dizziness, fatigue, and nausea if not handled properly. However, DCM is used in food processing techniques such as caffeine removal from coffee so in general is not considered as hazardous as the chemistry associated with BABB.

Following fixation, samples are pretreated by equilibration into methanol using hourly incremental concentrations of 20%, 40%, 60%, 80%, 100%, and an additional 100%. Following dehydration, samples are incubated in a 2:1 mixture of DCM and methanol at 4 °C overnight. Samples must then be rehydrated using an hourly methanol gradient of 100%, 80%, 60%, 40%, 20%, and ending in PBS.

If antibodies found incompatible with methanol are needed for immunolabeling, specimens may be pretreated by incubation overnight at 37 °C in a 0.2% Triton/0.1%

Tween/0.1% deoxycholate/0.1% NP40/20% dimethyl sulfoxide (DMSO) solution in order to solubilize membrane proteins and increase the penetration of the label. Following pretreatment, tissue samples must be permeabilized by incubating in 0.2% Triton/20% DMSO/2.5% glycine solution at 37 °C for 48 h (Renier et al. 2014). Specimens are then immunolabeled using a protocol of choice, with antibody solutions that contain 5–10% DMSO, which increases the mobility of the antibody into the DMSO-incubated tissue. After immunolabeling is complete, tissues are again dehydrated using a methanol gradient of 20%, 40%, 60%, 80%, 100%, and 100%. Samples are incubated for 3 h in 2:1 DCM:methanol and then for 30 min in 100% DCM. Specimens are stored and mounted in dibenzyl ether (DBE) to prevent oxidation.

DISCO tends to be more time-consuming than BABB but circumvents the use of toxic benzyl compounds. DISCO-cleared tissues exhibit minimal shrinkage and preserve the signal from endogenous fluorescent proteins such as RFP for a longer amount of time (Table 4.1). However, DISCO is markedly more expensive and time-consuming, and precise clearing results have proven difficult to reproduce. Similar to BABB, DBE will corrode plastics and microscope parts and should be handled with care. Additionally, samples in DBE must be completely sealed with no exposure to air, as they will oxidize rapidly. It is recommended that a slide chamber built with a DBE-resistant polymer be utilized for imaging (Fig. 4.5).

4.4.2.3 X-CLARITY

The CLARITY method for tissue clearing was invented by Chung (2013) and Chung and Deisseroth (2013) in order to better understand neural mapping of the brain. This method embeds cellular components onto a hydrogel scaffold in order to clear away lipids without compromising the position of key biological constituents. The technique can be performed manually, but for consistency between

Fig. 4.5 Slide chamber used for containment of tissue clearing fluids

runs, Logos Biosystems (Annandale, VA) has recently commercialized a complete X-CLARITY™ system (Fig. 4.6) for vacuum preparation and polymerization of the hydrogel in embedded tissue. The X-CLARITY™ protocols provide deeper penetration of reagents into three-dimensional tissues than other clearing techniques while preventing loss of structural integrity resulting from solvent-based clearing protocols.

Following perfusion fixation to insure complete preservation of large segments of tissue, samples are incubated overnight at 4 °C in a 1:100 mixture of polymerization initiator/hydrogel solution. This incubation insures that the hydrogel solution penetrates the entire segment of tissue before polymerization, which is performed in a vacuum system at −90 kPa and 37 °C for 3 h. Following polymerization, samples are gently shaken for 15 min in PBS to rinse excess hydrogel from the tissue.

Once samples are properly polymerized, they are cleared by electrophoresis. Depending on the density of the tissue, samples may be cleared at a current between 1.0 and 1.5 A at 37 °C. Clearing time may vary from 1 h for tissues that are relatively thin and not very dense (colon) to 6 h for brain tissue and 24 h for dense cardiac tissue. Monitoring of the samples is recommended every few hours to ensure proper clearing. Once the tissue is clear, it must be washed in PBS overnight to remove any residual sodium dodecyl sulfate (SDS) from the electrophoresis, which may interfere with immunolabeling. Following labeling with a protocol of choice, samples are washed three times in 5-min increments in distilled water to remove phosphates, which can precipitate during mounting. Samples are then incubated in 85% glycerol for 1 h and then mounted in fresh 85% glycerol for imaging. Occasionally, samples may remain opaque prior to mounting and may require additional time in glycerol to completely clear. Examples of X-CLARITY™-cleared tissue are shown in Fig. 4.1 above and in Fig. 3.15 where a section of nearly 1 mm of basolateral amygdala from the brain was imaged.

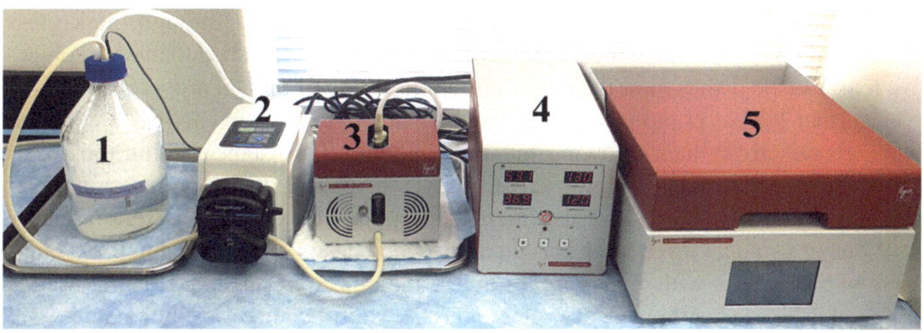

Fig. 4.6 Components of the Logos Biosystems X-CLARITY™ tissue clearing system. (1) Chemical reservoir for clearing fluid exchange, (2) pump to transfer clearing fluid from the chemical reservoir to the electrophoretic clearing (ETC) chamber, (3) ETC clearing chamber, (4) control unit for setting time and temperature of the ETC unit during clearing, (5) hydrogel embedding and polymerization unit

Due to the highly automated process of electrophoretic clearing with the Logos Biosystems, X-CLARITY™ system results tend to be more replicable and experiments easier to carry out. The benefits of more complete and reliable clearing, minimally altered tissue morphology, and reduced time for clearing tend to outweigh drawbacks such as the higher start-up cost and use of proprietary reagent kits. Aside from the slightly higher cost, CLARITY provides the lowest final refractive index of the other methods discussed, contains primarily nontoxic compounds, and preserves endogenous fluorescent protein signal to a greater extent than other clearing protocols (Table 4.1).

In addition to the above advantages, Leica has also developed a CLARITY objective (Fig. 3.14) for their microscopes that optimally match the refractive index of CLARITY-prepared tissues. While expensive (in the range of $30 K), this objective optimizes refractive index in a large percentage of the optic path for maximum depth and resolution of imaging in cleared tissues.

4.5 Mounting Specimens

Immunostaining of fixed specimens is covered in Chap. 5. After staining, the prepared samples can be viewed with an inverted microscope without further processing. Culture chambers attached to thin cover glasses or tissue culture plates with part of the bottom of the plate replaced with a cover glass allow cells to be grown, processed, and observed directly with an inverted confocal microscope. These tissue culture/viewing chambers can be made in the laboratory or purchased from suppliers of microscope accessories. Figure 4.7 provides a few examples of the many alternatives that are commercially available. Chambers that are affixed to glass microscope slides or plastic dishes can also be used to view samples directly. However, using standard glass slides or culture dishes results in part of the working distance of the objective lens being used to image within the glass, and this reduces how much of the specimen depth can be probed.

The more important reason, though, for avoiding imaging through thick glass slides or glass dishes is that their optical properties are not matched to modern objective lens. Plastic dishes have the same limitation, and their optical properties result in even further loss of photons and decreased signal-to-noise ratio (SNR). In contrast, cover glasses are thinner allowing more usable working distance within the specimen. Chambers mounted on thin cover glasses or plastic dishes with a thin cover glass window for use with inverted microscopes are readily available commercially.

Whether using an inverted or upright microscope design, it is critical to use a number 1 ½ cover glass with your samples unless the objective lens that is being used is designed to be used without a cover glass present. Almost all high-resolution lenses are specifically designed to work with a number 1½ cover glass. This represents a thickness of about 170 μm. Since the refractive index of a 1 ½ cover glass is considered in the design of the microscope lens, cover glasses of other thicknesses can produce a mismatch of refractive indices leading to a loss of photons collected from the image. The deleterious effects of refractive index mismatch were discussed further in Chap. 3.

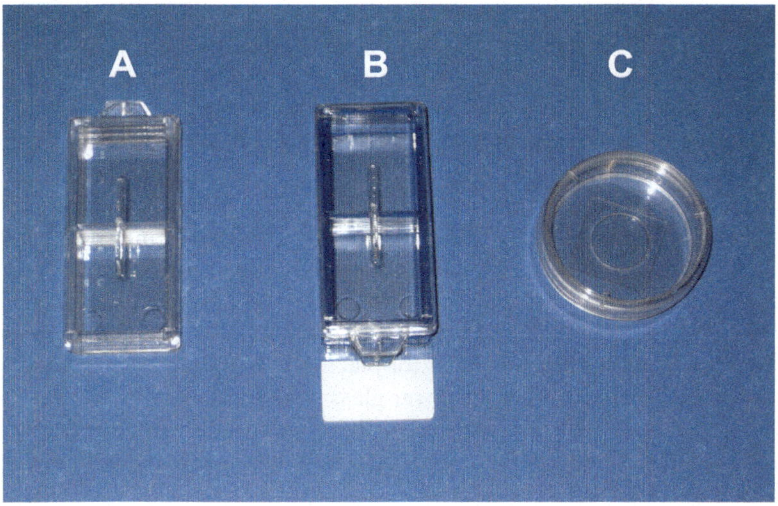

Fig. 4.7 Commercially available culture chambers adapted for microscopy include culture chambers on cover glasses such as this Lab-Tek chamber cover glass (**a**) from Thermo Fisher Scientific (Rochester, NY), culture chambers on glass microscope slides such as this Lab-Tek chamber slide (**b**) from Thermo Fisher Scientific, and plastic culture dishes with cover glass over a central viewing hole in the plastic such as this MatTek glass bottom dish (**c**) from MatTek Corporation (Ashland, MA)

The closeness of the tolerance permitted in number 1½ cover glasses varies among suppliers, but no group of 1½ cover glasses will all be exactly 170 μm thick. Subtle differences in cover glass thickness will influence the image quality, and this will be particularly apparent with high numerical aperture lenses. For this reason, the best of these lenses are supplied with correction collars that allow the microscopist to adjust the tube length to compensate for the actual cover glass thickness as discussed in Chap. 3. Suffice it to say that using the proper cover glass thickness is important for achieving maximum image quality.

For widefield microscopy, one additional benefit of having a cover glass present is the fact that it flattens the viewed surface of the specimen. This minimizes variations in light scattering which can occur from a rough surface. With most samples, the cover glass flattens more than just the surface; it can greatly compress the entire thickness of the sample. In widefield microscopy, the decrease in specimen thickness decreases the amount of out of focus information from the sample and, thus, increases specimen contrast and improves visibility. Of course, this is at the expense of preserving the 3-D architecture. For this reason, in confocal imaging where the goal is to analyze the 3-D architecture, it is best to avoid this kind of artifactual flattening. The best way to do this is to provide some type of riser to maintain the cover glass at a suitable height above the specimen. However, the cover glass should not sit too high above the specimen because this would result in too much of the working distance of the lens being used to image the fluid above the specimen, thus limiting the depth within the specimen that can be imaged.

Risers can be fashioned out of any material. Many microscopists place shards of broken cover glasses at the four corners of a cover glass to hold the cover glass at a height of 170 μm. Risers can also be fashioned out of VaLaP (Molé-Bajer 1968). This is a nontoxic mixture composed of equal parts (by weight) of Vaseline petroleum jelly, lanolin, and low melting point (56 °C) paraffin. All the components are available from most scientific supply houses, and Vaseline and lanolin are also available at most pharmacies. VaLaP is a compressible solid at 37° C but melts to a liquid at low heat (45–50 °C). This makes it easy to fashion into desired shapes within a temperature range which will not damage cells.

VaLaP risers are made by applying a small drop of liquid VaLaP on each corner of a cover glass and letting it cool to hardness (Fig. 4.8). A wooden applicator stick or fine point watercolor brush is ideal for placing VaLaP drops on the cover glass. When the VaLaP has hardened, the cover glass is inverted and placed over the wet sample on a glass slide, extruding excess fluid but not allowing the cover slip to smash the sample. Gentle pressure will compress the VaLaP risers to the desired

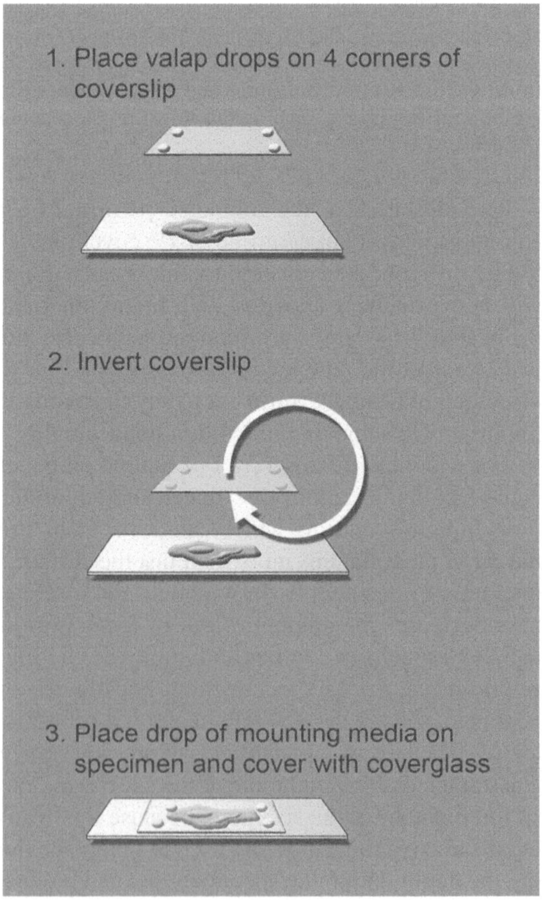

Fig. 4.8 The use of VaLaP to prevent cover glass from overly compressing mounted samples

height and adhere the risers and cover glass to the underlying slide. Liquid VaLaP can then be painted around the edges of the mount. When hardened, this will form a watertight seal that will prevent the mounting fluid from evaporating and drying out the sample. Brief exposure of cells to the 45–50 °C temperature of warm VaLaP is well tolerated by most cells.

An additional benefit to using VaLaP as a sealant is that if some of the VaLaP should accidentally be transferred to an objective lens by contact, it can be easily removed. This is not true for some of the commercially available sealants or the ubiquitously used fingernail polish. Many an expensive objective lens has been destroyed by contact with not quite dry fingernail polish. The VaLaP seal, however, can break down over time and so is not suitable for long-term storage of slides. If long-term storage is required, after initial viewing, the VaLaP sealant (but not the risers) can be peeled away and replaced with a more permanent sealant. Just make sure that ample time is allowed for the sealant to completely dry before the preparation is viewed under the microscope. A wait of at least 2 days before viewing is a good rule of thumb.

Samples for confocal microscopy are usually wet when mounted. The use of a suitable mounting media is encouraged. Commercially available mounting media, such as DABCO (1,4-diazabicyclo[2,2,2]octane (Sigma-Aldrich), are available for this use. They are applied to the sample just prior to cover slipping the sample. These mountants match the refractive index of most biological samples and contain antioxidants that can retard photobleaching. However, equally good mounting media can be formulated inexpensively in the laboratory. Figure 4.9 provides a recipe for a very useful "lab-made" mounting media.

Mounting media that are applied as liquids and then harden to a solid that encases the sample are wonderful for long-term preservation of fluorescent samples. ProLong® (Invitrogen, Carlsbad CA) is an example of this type of mountant. Unfortunately, although these mounting media are excellent for preserving samples, they contain polyvinyl alcohols and so are generally not suitable for most confocal experiments. Polyvinyl alcohol can produce severe tissue shrinkage that will disrupt the three-dimensional architecture of the sample. Polyvinyl alcohol is also a powerful lipid solvent, so it is unsuitable for experiments where lipids must be preserved.

Although most objective lenses are designed for use with cover glasses, one class of lenses, known as dipping lenses, is made to be used without a cover glass (Fig. 4.10). These lenses are waterproof and can be lowered directly into a fluid environment, such as tissue culture medium, in order to directly image a specimen. A cover glass or mounting medium is not required with these lenses. This lessens the number of refractive indices that need to be matched. Dipping lenses generally also have longer working distances (up to about 2.5 mm). However, they have slightly lower numerical apertures (about 1–1.1) compared to similar water immersion lenses that are designed for use with cover glasses (usually NA of 1.3). They do, however, allow direct viewing of samples using upright microscopes without the worries of correctly cover slipping the sample. Further information on the use of various dipping and immersion lenses was presented in Chap. 3.

Photobleach Retarding Mounting Media.

Phosphate buffered saline........................100 ml
Glycerol..35 g
Para-phenylenediamine..........................0.25 g

Mix the ingredients and aliquot into 1ml units. Wrap in foil to protect from light. Store at -80 C.

This solution will keep for several years at -80 C.

Before use, the solution can be diluted with PBS to more closely match the refractive index of the specimen.

For most specimens 1 part PBS to 3 parts mounting media works well.

Para-phenylenediamine is an antioxidant which reduces photobleaching. It is light sensitive so avoid prolonged exposure to light.

Fig. 4.9 Recipe for a mounting media with suitable refractive index and anti-fade properties

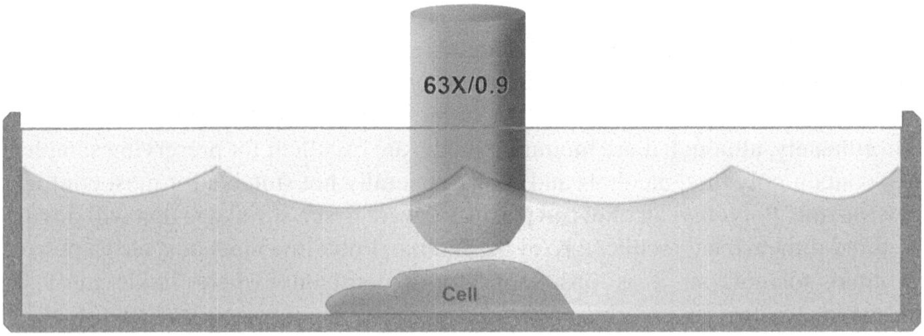

Fig. 4.10 Use of a dipping lens for viewing cells or tissue without a cover glass

4.6 Working with Live Cells or Live Tissue

Although tissue slices or thin organs (such as isolated blood vessels) can be imaged, the bulk of live imaging involves viewing cells. The principal benefit of live imaging is the ability to observe events in cells or tissue as they happen. However, to image samples in the live state requires that the cells or tissue remain viable and without artifactual structural changes. There are real challenges to accomplishing this goal and insufficient space in this text to address all of the concerns. However, a few guiding principles are worth mentioning. For a more complete discussion of the topic of live imaging, the reader is referred to a review by JC Waters (Waters 2007).

4.6.1 Live Cell Imaging Instrument Configuration

With respect to instrumentation, live imaging can be carried out with standard laser scanning confocal microscopes. However, for most applications, spinning disk confocals and multiphoton excitation imaging (Chap. 8) on an inverted microscope platform provide better alternatives because their imaging methodologies are not as likely to generate some of the negative influences on cell viability discussed below. In addition to the confocal capabilities, the microscope should ideally be equipped with phase-contrast or differential interference-contrast and polarized light objectives. These modes of optics for live cell imaging are most useful in studies involving events associated with cell migration, motility, and cell division (karyokinesis and cytokinesis).

The primary concern associated with imaging of live cells and their responses to experimental conditions is being able to maintain the cells alive while under the stressful conditions of imaging. A key to cell viability is keeping the cell environment in the necessary range with respect to temperature, CO_2 concentration, and humidity. In addition, these parameters need to be stable throughout the analysis period which can be minutes to days or even weeks. For short-term experiments, some simple solutions can be devised, but longer experiments require special systems to maintain a suitable and stable environment. Most biological properties of living mammalian cells must be recorded at 37 °C. This requires that heat be put into the system to maintain the cells. Numerous solutions have been applied in trying to resolve this problem. One early approach was to heat the entire microscope by putting the instrument into a warm room with atmospheric control. This was impractical because the microscope could not be used for anything else and water contamination was a severe problem leading to degradation of the microscope and optics.

Several commercially available stage heaters have been developed to heat the stage, optics, and various size chambers that may enclose the entire microscope or only a small platform on which the cells are grown. An innovative approach has been to construct a miniature version of an environmental control chamber (tissue culture chamber) which is mounted directly on a heated microscope stage. Various types of culture chambers can be used including both closed and open systems. The simplest and least expensive of these culture systems are specialized tissue culture dishes (MatTek) that have a cover slip over an opening in the bottom of the dish on which the cells are grown. Cells grown in MatTek culture plates maintained in an environmental chamber have been continuously imaged for up to 72 h. These dishes are an open system which has the advantages of being inexpensive and easy to use. The main disadvantages of these culture dishes are that media exchange is difficult when performing washin-washout and laminar flow experiments.

For live cell imaging, plastic tissue culture dishes are usually unsuitable for several reasons. The thickness of the plastic culture dish is usually >2 mm which is much greater than the depth of focus of both medium (20×) and high (40–100×) NA objectives meaning the specimen cannot be brought into focus. Additionally, the plastic dish is not optically flat or homogeneous over the surface on which the cells are grown. This results in degradation of the image of the specimen using any objective. In addition, the plastic used in the dish may be fluorescent which can add high background fluorescence. Most plastics used in the culture dishes are also birefringent and cannot be used with DIC or polarized light optics.

4.6.2 Live Cell Imaging Modes

Once the problem of keeping cells viable is solved, the next problem is deciding which imaging mode to use. Many biological and cellular processes have been studied in living unlabeled cells using specialized optical methods. The processes studied cover both intracellular events (chromosome movements, vacuole formation kinetics, etc.) and cellular events (cellular migration and motility, cilia movements, cell division, etc.). The most commonly used optical system for imaging live cells involves phase-contrast optics (for review, see Slayer EM (1976)). However, with phase contrast no optical sectioning is possible since phase-contrast imaging is accomplished in the aperture plane of the microscope objective. Phase-contrast imaging has limited use when combined with fluorescence imaging, but the fidelity of the fluorescent image or phase-contrast-fluorescent overlay is dependent on absorptive characteristics of the phase plate and annulus which may adversely affect the fluorescent signal.

The application of DIC optics to a live cell imaging system, like phase-contrast optics, produces an image in which differences in the optical path through the specimen are made to appear as differences in intensities. The optical section which comprises the focused DIC image is remarkably shallow. The DIC image is free of specimen edge halos and disturbances from out of focus objects. This property makes the DIC image neatly sliced and isolated from a complex three-dimensional phase object. Since DIC image formation is accomplished in the image plane of the objective, optical sectioning of the specimen becomes possible. Additionally, DIC images are completely compatible with fluorescence signals from the same specimen making overlays of confocal fluorescence and DIC images useful.

4.6.3 Selection of Fluorescence Imaging Mode

It is possible to image endogenous molecules, such as NADPH (Piston and Knobel 1999; Williams et al. 1994), but more often the fluorophore is delivered exogenously. There are far fewer possible fluorophores that can be adapted to live imaging compared to those available for labeling fixed material. However, as discussed in Chap. 2, the advent of genetically encoded fluorescent molecules, such as green fluorescent protein (GFP), the spectral variants of GFP, and red fluorescent protein (RFP), has expanded the repertoire of possible live imaging probes. Application of these selective fluorescent probes offers several distinct advantages for live cell imaging. Included in these advantages are a high degree of sensitivity as in many cases the fluorescence signal allows visualization of single molecules. These can have a high degree of selectivity for detecting and visualizing precise locations of molecules to very specific regions or moieties of the molecules of interest. Under appropriate imaging conditions, the intensity of the fluorescence can also be used to quantify the number of molecules or binding sites of interest since, under normal conditions, fluorescence emission is linear with respect to the number of probe molecules bound to the molecule or site of interest.

There are several strategies available for loading living cells with a specific fluorescent probe. The easiest and most straightforward are those probes that are readily taken up by cells and include vital dyes and the AM-ester compounds. Vital dyes (e.g., acridine orange) are minimal or nontoxic at low concentration and readily penetrate cell membranes where they associate with the appropriate intracellular organelles. In general, vital dyes change emission spectra and/or intensity of emission when the intracellular environment is altered. For example, the AM-ester compounds are cell permeant as a result of the presence of the AM-ester moiety. Once the AM-ester compound enters the cell, the AM-ester moiety is removed by nonspecific cytoplasmic esterases from the probe rendering it incapable of leaking through the membrane and back out of the cell. This approach allows for a precise amount of the fluorescent probe compound to be introduced into the cell and the emission quantified. These compounds are typified by the AM-ester probes used for determination of intracellular calcium concentrations.

Fluorescent probes can also be taken up by cells by making use of the natural processes of pinocytosis or phagocytosis. In general, fluorescent molecules can be attached to substrates such as dextran which the cell perceives as a usable sugar molecule. Dextran is not degraded by intracellular enzymes and the dextran: probe conjugate is stable in the intracellular environment. This process can be enhanced by using additional agents to induce elevated rates of pinocytosis to increase the influx of the fluorescent reagent into the cell.

Introducing other types of probes, such as labeled ligands or antibodies, into cells and assuring they reach their proper destination and do not perturb cell viability or structural integrity remain problematic, and there is no single solution to this quandary. Gentle and transient permeabilization of cell membranes, microinjection of probes, and crafting probes that are freely permeable to membranes are just some of the approaches that have been used. Other more complicated approaches for introducing probes into cells including physical microinjection and electroporation have also been used with varying degrees of success (Beckerle 1984; Berglund and Starkey 1991).

4.6.4 Selection of a Fluorescent Probe

An efficient fluorescent probe for live cell imaging should have a high absorbance, a high quantum yield, a large Stokes shift, and a narrow emission spectrum. Additionally, the probe should have high specificity for the target molecule since non-specific binding of the probe will increase background fluorescence and result in degraded contrast of the image. In designing experiments it is advisable to select fluorescent probes which use excitation/emission wavelengths close to those filter sets that are most commonly available. Most commonly used filter sets for single and multiple labeling include DAPI (360/460), Cy2, FITC (490/515), Cy3, TRITC (555/570), and Cy5 (635/670).

4.6.5 Imaging the Living Cells

In collecting live cell images, the interaction of the excitation beam with the specimen is an important consideration. Major adverse conditions may arise as light interacts with the probe causing the generation of toxic compounds (phototoxicity) or photobleaching. Phototoxicity may also be induced by bright field imaging, but it is more commonly associated with fluorescence imaging. In experiments where time-lapse image collection of a fluorescent probe is extended for several hours, and in particular when image collection is carried out over several days, phototoxicity becomes a real issue. Excitation of fluorescent molecules in the presence of oxygen leads to fluorochrome bleaching and generation of free radicals. These free radicals are predominantly derived from oxygen and induce potent and toxic hydroxyl radicals, singlet oxygen, and superoxide radicals which react with water leading to the production of hydrogen peroxide. These free radicals and hydrogen peroxide can rapidly kill the cells.

Additionally, the interaction of bright field or fluorescent light with cells leads to the production of heat. Elevation of heat within the cell rapidly alters cellular function and results in cell death. To maintain the cells under normal homeostatic conditions, steps must be taken to limit the effects of chemical and heat phototoxicity. The time that the cells are exposed to the incident illumination must be carefully limited. This can be accomplished by limiting exposure of the cells to periods of light only during image collection. Under these conditions the cells are exposed to light for the optimal exposure time as determined by the experimental setup and camera used for image collection. Maximization of light collection which results in shorter exposure times can be accomplished by use of high NA and immersion objectives. In general the choice of optics and objectives will depend on the time duration of the experiment and how frequently images must be acquired to generate the data set.

Another problem frequently encountered in real-time or time-lapse imaging of cells is loss of image quality due to image and/or focus drift. In severe cases the image of the specimen may be completely lost during short time periods of time-lapse image collection. There are two principle causes of image drift, mechanical and thermal. The major component of mechanical drift occurs as the microscope stage is moved to focus the specimen. For live cell time-lapse imaging, it is best to use a fixed stage microscope where focusing of the specimen is carried out by movement of the objectives. Loss of image quality can also be induced by focus drift of the objectives. Most live cell imaging systems have motorized focusing in the Z-axis to focus the objectives. The most common system in microscopes to move the objectives is a friction drive to control the level of fine focus of the specimen. The use of friction drives to control the fine focus is thus indirect and subject to slippage and focus drift. Drift is always a problem in these microscopes. Ideal control of the focal plane of the specimen is best acquired by use of stepper motors directly mounted to gears controlling the fine focus. The use of gears and stepper motors constitutes a direct drive system with no possibility for focus drift.

Loss of specimen image quality due to thermal drift is the result of expansion and contraction of elements of the microscope. Since most biological properties must be recorded at 37 °C for mammalian cells, this temperature must be maintained for the live cell specimen and elements of the microscope directly involved in viewing the specimen. Maintenance of constant temperature serves to minimize the effects of thermal expansion and contraction on focus. In all microscopes, inverted or upright, the stage and especially the objectives act as heat sinks and are the main source of thermal-induced focus shifts. The easiest solution to minimizing thermal drift is to heat just the stage and the cells or specimen. This can be accomplished by the application of an environmental chamber to enclose the stage, cells, and objective being used. This type of system requires heat, CO_2, and humidity controllers and is relatively expensive but easy to control and use. An alternative to these elaborate systems is a simple warm box mounted on the stage which encloses the specimen and objective being used. This simple arrangement has a significant drawback in that it is difficult to accurately control the CO_2 and humidity but has the advantage of being inexpensive and easy to use.

4.6.6 *Summary of Live Cell Imaging*

As can be seen from the above discussion, imaging of living cells and tissues is not a trivial issue. It is imperative to limit specimen exposure. This involves working with low levels of excitation light, limiting the area probed, restricting the time the excitation beam dwells on a single microvolume, and scanning an area as few times as possible. However, all of these factors can lead to minimal interaction of the excitation photons with the sample and result in greatly reduced signal-to-noise ratios. It is always a good idea to optimize the system to collect as much fluorescence signal as possible from your sample, but in live cell imaging, this is imperative. To accomplish this, the refractive indices along the optical path must be accounted for and accommodated to make sure as much of the signal as possible remains focused on the confocal pinhole and thus reaches the detector. In some cases, it might be necessary to increase the size of the pinhole. This reduces the resolution along the optical axis (Z-axis) but allows more light to reach the detector. The use of high NA lenses and high quantum efficiency detectors are also important factors in the collection of sufficient signal. Chapter 5 includes a discussion of recently developed high sensitivity detectors that have improved the ability to image live cells. Finally, the use of scrupulously clean optical elements is important because even a small amount of dirt on a lens surface can deflect sufficient photons to noticeably degrade the image.

Literature Cited

Azaripour A, Lagerweij T, Scharfbillig C, Jadczak AE, Willershausen B, Noorden CJ (2016) A survey of clearing techniques for 3D imaging of tissues with special reference to connective tissue. Prog Histochem Cytochem 51(2):9–23. https://doi.org/10.1016/j.proghi.2016.04.001

Beckerle M (1984) Microinjected fluorescent polystyrene beads exhibit saltatory motion in tissue culture cells. J Cell Biol 98:2126–2132

Berglund D, Starkey J (1991) Introduction of antibody into viable cells using electroporation. Cytometry 12:64–67

Chung K (2013) Structural and molecular interrogation of intact biological systems. Nature 497:332–337. https://doi.org/10.1038/nature12107

Chung K, Deisseroth K (2013) CLARITY for mapping the nervous system. Nat Methods 10(10):1035–1035. https://doi.org/10.1038/nmeth1013-1035a

Dodt H, Jährling N, Becker K (2008) The glass brain: visualization of neuronal networks in the whole mouse brain by Ultramicroscopy. Biomedical Optics. https://doi.org/10.1364/biomed.2008.btua4

Ertürk A, Bradke F (2013) High-resolution imaging of entire organs by 3-dimensional imaging of solvent cleared organs (3DISCO). Exp Neurol 242:57–64. https://doi.org/10.1016/j.expneurol.2012.10.018

Germroth PG, Gourdie RG, Thompson RP (2005) Confocal microscopy of thick sections from acrylamide gel embedded embryos. Microscopy Res Tech 30(6):513–520

Miller CE, Thompson RP, Bigelow MR, Gittinger G, Trusk TC, Sedmera D (2005) Confocal imaging of the embryonic heart: how deep? Microsc Microanal 11(03):216–223. https://doi.org/10.1017/s1431927605050464

Molé-Bajer (1968) Studies of selected endosperm cells with the light and electron microscope. The technique. Cellule 67:257–265

Piston D, Knobel S (1999) Real-time analysis of glucose metabolism by microscopy. Trends Endocrinol Metab 10:413–417

Renier N, Wu Z, Simon DJ, Yang J, Ariel P, Tessier-Lavigne M (2014) IDISCO: a simple, rapid method to immunolabel large tissue samples for volume imaging. Cell 159(4):896–910. https://doi.org/10.3410/f.725229558.793505797

Richardson DS, Lichtman JW (2015) Clarifying tissue clearing. Cell 162(2):246–257. https://doi.org/10.1016/j.cell.2015.06.067

Slayer EM (1976) Optical methods in biology. Robert Krieger Publishing Co, Huntington, NY, pp 288–302

Silvestri L, Costantini I, Sacconi L, Pavone FS (2016) Clearing of fixed tissue: a review from a microscopist's perspective. J Biomedical Optics 21(8). 081205-1-081205-8
Tainaka K, Kuno A, Kubota AI, Murakami T, Ueda HR (2016) Chemical principles in tissue clearing and staining protocols for whole-body cell profiling. Ann Rev Cell Dev Biol 32:713–741. https://doi.org/10.1146/annurev-cellbio-111315-125001
Waters J (2007) Live-cell fluorescence imaging. Methods Cell Biol 81:115–140
Williams R, Piston D, Webb W (1994) Two-photon molecular excitation provides intrinsic 3-dimensional resolution for laser-based microscopy and microphotochemistry. FASEB J 8:804–813
Zucker RM, Hunter S, Rogers JM (1998) Confocal scanning microscopy of apoptosis in organogenesis-stage mouse embryos. Cytometry 33:348–354

Chapter 5
Labeling Considerations for Confocal Microscopy

R. M. Albrecht and J. A. Oliver

5.1 Introduction

Some components of biological systems can be readily identified solely by their unique structure or other intrinsic physical properties which are evident when visualized in brightfield or various types of interference-based light microscopy (LM). However, for the unambiguous identification and localization of most biological molecular or macromolecular elements within a structural framework, some type of staining/labeling must be employed. This is important for a variety of applications including identification of particular tissues, identification of cells and subcellular components/structures, tracking of cells or subcellular components, as well as co-localization of cells and cell components on or within tissues or cells. Labeling is also used to provide quantitative comparisons of epitope density, cell numbers, organelle numbers or volume, and a variety of other types of quantitative data. However, considerable caution must be taken when attempting quantitative or even semiquantitative analyses. Efficiencies of labeling for different epitopes and antibodies or antibody mixtures are variable as highlighted throughout this book. The exact relationship of color density, particle numbers, or fluorescence intensity (which can fade during observation) to the actual numbers of labeled sites is critical and often not known. These factors often make quantitative estimations or comparisons very risky.

R. M. Albrecht (✉)
Department of Animal Sciences, Pediatrics. and Pharmaceutical Sciences, University of Wisconsin – Madison, Madison, WI, USA
e-mail: albrecht@ansci.wisc.edu

J. A. Oliver
Department of Biological Sciences, University of Wisconsin – Milwaukee, Milwaukee, WI, USA
e-mail: joliver@uwm.edu

W. G. Jerome, R. L. Price (eds.), *Basic Confocal Microscopy*,
https://doi.org/10.1007/978-3-319-97454-5_5

5.2 Types of Labels

A number of opaque, chromatic, and fluorescent-based labels have been developed for staining/labeling purposes. A wide variety of histo(cyto)logical stains such as the well-known Wright's stain or H&E stain and many others are commonly used. More recently the conjugation of antibody to various metal nanoparticles including gold, silver, palladium, iron, and others has facilitated direct observation by interference LM methodology. The particles can be metal enhanced following labeling to render them visible in brightfield or white light confocal LM.

Histo(cyto)chemical staining techniques also make substantial use of chromatic and opaque methodologies. Various opaque or colored precipitates and dyes are available. This includes stains for enzymes such as peroxidase, phosphatase, or esterase. Certain enzymes can be conjugated to antibody labels, and the specific cytochemical reaction is subsequently used to localize the antibody. A typical example is the peroxidase stain where a precipitate forms at the site of the enzyme-antibody complex. This is identified in LM by its dark brownish color and can be stained with osmium. The resulting black precipitate can be visualized in LM. It is also electron dense and visible in the electron microscope. The fluorescence-based labels also have broadly based applications particularly in the research arena. The advent of various nanoparticle-based labeling technologies has been particularly valuable for correlative light-, electron-, and force-based microscopies. While certain of the nanoparticles are fluorescent, others rely on their chromatic or opaque nature for identification. Currently, the prevailing use of confocal microscopy is largely in the area of fluorescence-based imaging; however it should be noted that white light confocal microscopy or simultaneous confocal and wide field imaging can be used to image and reconstruct images enhanced by chromatic or opaque labeling techniques. The development of high-sensitivity electron multiplying charge-coupled device (EMCCD) cameras (discussed in Chap. 7) used in conjunction with spinning disc confocal systems has facilitated the use of chromatic and opaque labeling along with the more common fluorescent labeling technology.

The use of small organic dyes or quantum particles which can be excited at one wavelength of light and then emit light at a different wavelength provides an extremely versatile labeling technology. The fluorescent dyes or particles can be conjugated directly to specific molecules of interest or to highly specific "identifier" species such as antibodies or ligands which bind selectively to the species of interest. When illuminated by the appropriate excitation wavelength, the resulting emission can be viewed. This provides an accurate localization for the dye molecule or particle and hence the conjugate as well. Fluorescent species having different, non-overlapping, excitation, and/or emission properties can be employed simultaneously to facilitate comparisons of the location of different molecular species and to examine co-localization of the different species.

Unfortunately, other than in specialized systems such as multiphoton excitation or light sheet systems, any fluorescent molecules in the pathway of the illuminating excitation beam, whether in front or in back of the plane of focus, will also be excited and fluoresce. Due to the minimal depth of focus in photon-based imaging, this results in generation of substantial out-of-focus fluorescent light. The confocal microscope provides a mechanism of illumination in which out-of-focus information is minimized providing a clear view of fluorescence from a single in-focus plane. Individual planes can be stored and subsequently combined to reconstruct a totally in-focus 3D image of the label localization. Such images can be superimposed on wide field or interference base images in order to put the labeled species more in the context of the biological structure.

5.3 Practical Considerations in Labeling

There are a number of practical considerations which have to be addressed relative to the particular question the investigator wishes to address. Suitable dyes have to be selected and matched with excitation sources of appropriate strength and wavelength. Historically, fluorescent dyes were developed which matched the emission peaks of mercury vapor lamps. Xenon lamps with a more broad emission peak permitted the use of dyes with excitation peaks not covered by mercury. A number of laser-scanning types of confocal light microscopes have been developed to provide appropriate illumination for confocal fluorescent labeling. This has resulted in the development of newer fluorescent dyes whose excitation matches that of the available, reasonably priced, laser excitation wavelengths. The spinning disc type of confocal microscopes, in addition to providing real-time imaging, can use either the traditional vapor lamp source of photons or laser sources in addition to newer diode illumination. This generally provides a wider range of excitation wavelengths.

The actual labeling process is not necessarily complex, but considerable attention to detail is required. Perhaps the first question that needs to be addressed relates to the information that the investigator wishes to obtain. This volume deals primarily with confocal microscopy, and, while confocal microscopy is an extremely useful tool, it is wise to consider alternative methodologies relative to the particular question at hand. Wide field imaging, often in conjunction with image deconvolution methodology, will often be the method of choice for thin specimens, particularly if living material is being observed. Images of live or otherwise dynamic specimens can be readily obtained. Substantially less light is required and lasers are unnecessary. As with spinning disc confocal imaging, a wide variety of excitation/emission combinations are available, and imaging is in real time. Rapid and direct comparisons of fluorescence images with brightfield or various interference-based imaging modes can be made. Simultaneous interference-based imaging with wavelengths of light in, for example, the IR or far red region can be used to follow living material without damage, while simultaneous fluorescent imaging or fluorescent ratio imaging can be performed. A variety of fluorescent and nonfluorescent labels, opaque and chromatic, can be used simultaneously.

Identification of structures or areas of interest is absolutely dependent upon distinguishing those areas from everything else. This requires a specific label. A wide variety of labeling methods can be employed, but almost all methods rely on creating a visual difference between the area of interest and the surrounding tissue.

5.4 Fluorescence

As discussed in Chap. 2, in single-photon confocal microscopy, a fluorochrome absorbs a photon at one wavelength, enters an excited molecular state, and then emits a photon at a longer, less energetic wavelength when the fluorochrome returns to its resting state. Distinguishing this emitted light from the incident light is the basis of image formation in fluorescence microscopy (Chap. 3).

Fluorescent markers are especially well-suited for confocal microscopy. They can provide high quantum yield and, under optimized labeling and specimen preparation conditions, provide excellent signal-to-noise ratio (SNR).

5.4.1 Photobleaching

A significant disadvantage of fluorescence detection methods is the occurrence of photobleaching or fading. Bombarding fluorochrome molecules with high-energy illumination at their optimal excitation wavelength (i.e., maximum energy absorbance) can damage their chemical structure. In addition, fluorochromes are highly reactive in their excited state and can form covalent associations with other molecules, decreasing their fluorescence capacity. Fading is a particular obstacle when imaging weakly labeled specimens, as the extended illumination times required to obtain a detectable signal can lead to extensive photobleaching. In addition, in single-photon confocal microscopy, the entire depth of the specimen is illuminated throughout the scanning period, even though signal is collected from only one focal plane at a time. Fluorochromes that are highly susceptible to photobleaching, like fluorescein, can have a significant proportion of the available molecules bleached by the time the last step of the Z-series is imaged.

A number of approaches can be employed to address photobleaching. First, fluorochromes are differentially susceptible to fading. Replacing the susceptible molecule with a comparatively resistant fluorochrome possessing a similar excitation and emission profile leads to improved imaging. For example, substituting fluorescein (excitation peak 494 nm, emission peak 521 nm) with the more photostable Alexa Fluor 488 (excitation peak 495 nm, emission peak 519 nm) allows for higher-intensity illumination and higher signal yield or prolonged illumination at a lower intensity.

A second approach to diminishing photobleaching is to use anti-fade reagents during imaging. Most photo damage occurs when excited fluorochromes interact with molecular oxygen to produce singlet oxygen, a reactive oxygen species. Depleting oxygen in the specimen environment can attenuate photo damage; of course, this is not practical in many live samples. Including reagents that scavenge reactive species and free radicals can markedly prolong the photostability of the fluorochrome. Examples of common anti-fade reagents are *p*-phenylenediamine (PPD), *n*-propyl gallate (NPG), and 1,4-diazabicyclo[2.2.2]octane (DABCO). A number of commercially available anti-fade mounting media such as ProLong are also available (Vector Labs, Molecular Probes). It is important to note that not every anti-fade reagent is compatible with every fluorochrome, and incompatibility can result in nearly complete quenching of the fluorescent signal.

Finally, particles such as quantum dots or other doped metal oxide particles such as europium-doped yttrium vanadium oxide (YVO) particles can be conjugated to antibodies or antibody fragments. Fluorescent yields are good and there is little to no fading with these particles.

5.4.2 Autofluorescence

The fluorescent signal in samples arises from a number of sources. Endogenous compounds, such as flavins and nicotinamide adenine dinucleotide (NAD/NADH) that function in metabolic oxidation-reduction reactions, absorb and emit light at wavelengths very similar to those used for imaging, producing what is known as autofluorescence. Lipofuscins, collagen, and elastin are also autofluorescent compounds. In plants, chlorophyll, carotene, and xanthophyll all absorb and emit light resulting in autofluorescence. Pollen grains produce such a strong autofluorescent signal that they are often used as demonstration samples for traditional fluorescent and confocal microscopes.

In addition to naturally occurring fluorescence, aldehyde fixatives also produce autofluorescent signal. Formaldehyde gives moderate fluorescence, but glutaraldehyde gives strong fluorescence. Free aldehyde groups are thought to be responsible for the signal detected with these fixatives; however, it has been proposed that some signal results from the interaction of the aldehydes with amines. Paraformaldehyde solution, which contains formaldehyde monomers freshly released from the solid paraformaldehyde polymer, gives the lowest autofluorescence levels among the aldehyde fixatives.

5.4.3 Sources of Fluorescence

The source of fluorescence most commonly used for imaging is organic dyes. Fluorescein and rhodamine, which give green and red emission spectra, respectively, are probably the most broadly used fluorochromes. Most dyes in current use were developed because they have absorption spectra that roughly match the spectral lines, or output peaks, produced by high illumination intensity mercury vapor lamps. High-intensity xenon vapor lamps give more uniform signal output across the visible spectrum than do mercury sources and work well with the same dyes developed for mercury lamps. In confocal microscopy, intense monochromatic laser light sources are usually employed, and the fluorochromes used have at least one absorption peak that matches the laser emission. Many of the dyes most frequently used for confocal microscopy and flow cytometry are excited with a 488 nm blue laser. Examples include fluorescein, Alexa Fluor 488, PerCP, R-phycoerythrin, and R-phycoerythrin coupled to cy5, cy5.5, cy7, or Texas Red.

R-phycoerythrin and PerCP (peridinin-chlorophyll protein) are naturally occurring fluorochromes that are derived from red algae and photosynthetic dinoflagellates, respectively, and function to transfer energy to chlorophyll for food production. Another naturally occurring fluorochrome is GFP, green fluorescent protein, which was originally isolated from the bioluminescent jellyfish *Aequorea victoria*. GFP is a small (27 kD) protein that has been adapted for use as a fluorescent reporter of protein expression (Chalfie et al. 1994).

Mutations that have been introduced include a single point mutation that shifted the absorption maximum to 488 nm (Heim et al. 1995). Additional modifications have resulted in yellow (YFP), cyan (CFP), and blue fluorescent reporters (BFP) (Heim and Tsien 1996). BFP can be used as the donor in fluorescence resonance energy transfer (FRET) studies (Fig. 2.4), with GFP serving as the fluorescence acceptor (Heim and Tsien 1996). However, optimal results are obtained using GFP or CYP as the donor, with YFP as the fluorescence acceptor, as it allows for excitation at a visible wavelength rather than in the UV range. GFP and its variants provide excellent photostability. In addition, when employed as recombinant fusion proteins, they provide unequivocal detection of the recombinant protein, in that every molecule expressed carries the marker. Depending on the number of GFP gene copies per cell and the strength of the promoter driving its expression, GFP reporters can provide extremely high signal. More recently a wide range of red-shifted fluorescent proteins (RFPs) that are also useful for FRET and live cell imaging experiments have been developed (Piatkevich and Verkhusha 2011).

5.5 Desirable Features of Molecules Used as Markers

As discussed above, localizing a fluorochrome marker to a molecule or structure of interest is the primary challenge of labeling. In the absence of genetically engineered fluorescent fusion proteins or autofluorescence, an exogenously applied fluorescent identifier is required to label cells and tissues for confocal microscopy. A variety of molecule types can be used in identifiers, but good markers must all have a number of features in common.

5.5.1 Affinity

First, they must bind to the labeling target with high affinity. Once bound to the target, a high affinity label will remain in place, even through multiple wash steps in which the concentration of the label in the surrounding medium becomes essentially zero. For live labeling studies, the affinity should be so high (K_d values in the low nanomolar, or even picomolar, range) that for practical purposes, it is irreversible. If the label is to be fixed in place, a lower affinity can be tolerated, but perhaps at the risk of giving higher background staining as a result of less vigorous washing.

Binding of antibodies to their antigenic targets is among the highest affinity interactions employed in labeling protocols. Antibody affinities can range from micromolar on the weak end of the spectrum to high picomolar at the strong end. Most monoclonal antibodies available for investigational use are at the mid-to-high end of the range. An even higher affinity interaction is seen between the small vitamin biotin and the large tetrameric protein avidin. The avidin-biotin interaction has the highest affinity described, with a K_d in the femtomolar range (Green 1963). Together with the amplification made possible through the four biotin-binding sites found in avidin, this interaction has been put to very successful use in labeling protocols.

5.5.2 *Avidity*

Secondly, avidity also influences labeling efficiency. Avidity is the number of binding sites for the target molecule that a label possesses. A label with moderate affinity but high avidity can give labeling efficiency equivalent or superior to a high affinity label with lower avidity.

5.5.3 *Cross-Reactivity*

A third feature of a good marker is low nonspecific binding capacity. If the label can bind to the cell or tissue through cross-reactivity (binding to molecules similar to, but distinct from, the target) or is undesirable, albeit specific binding, its usefulness may be limited. When using antibodies for labeling, a serious concern is whether the antibody has cross-reactivity with antigens other than the target. Generally, this is a bigger problem with polyclonal antibodies than with monoclonals (see Sect. 5.6) because the source animal serum will contain at least some antibodies to all the antigens to which the animal has ever been exposed, intentionally or not. Polyclonal reagents can be affinity purified on immobilized antigen, thus removing reactivity to all but the target antigen. While monoclonal antibodies are less likely to have cross-reactivity than polyclonals, some closely related proteins have highly conserved sequences that can lead to significant cross-reactive staining. However, the incidence of antibody cross-reactivity in intact cells or tissues is relatively infrequent. This is because intact epitopes are three-dimensional. They rarely involve contiguous amino acid sequences and are heavily influenced by the environment provided by the neighboring amino acid sequence, also in three dimensions (Horsfall et al. 1991). Antibodies that recognize linear sequences, such as those that can be used for Western blotting, often show significant cross-reactivity in confocal labeling protocols.

Antibody epitopes are also influenced by posttranslational modifications, such as phosphorylation, acetylation, sulfation, or amidation. This has the potential to be problematic if one requires a reagent that labels a target under all circumstances. However, a level of antibody specificity that distinguishes between modified and unmodified molecules can make a very powerful research tool. For example, antibodies that recognize only the phosphorylated version of signal transduction molecules have been the cornerstone of cell signaling studies. On the other hand, laboratory manipulations can destroy the ability of an antibody to recognize its antigen. For example, fixation, especially with cross-linking aldehydes, destroys many monoclonal antibody-binding sites and a significant portion of polyclonal sites as well. The best retention of antigenicity is obtained with live or frozen specimens.

5.5.4 Stability

Stability is also an important feature of a good marker. The molecule and its interaction with the target need to hold up through the entire specimen preparation and imaging procedures. Likewise, stability of the target is an important consideration. If the marker recognizes the target after specimen fixation, the high quality of the target preservation can result in optimal imaging. For any labeling application, good access of the marker to the target is critical. If the markers are small and flexible, binding will be maximized, even if the target is partially obscured. Target access can be increased by permeabilization (mild detergent), proteolysis (enzymatic digestion), or unmasking (heating at low or high pH or with chelators) techniques commonly referred to as antigen retrieval techniques (Sect. 5.10.1).

5.5.5 Identification

Finally, identification of the marker in the microscope is crucial to its function. Multiple techniques can be used to identify the marker. It may have some inherent characteristic, such as a readily identifiable shape. Alternatively, it may have some characteristic, such as antigenicity, that makes it identifiable with a secondary marker, such as a fluorescent- or particle-conjugated "second" antibody. It may also be directly tagged with a fluorochrome or enzyme that identifies the location of the marker with higher resolution than is generally achieved with a secondary marker.

A number of types of molecules can be used as identifiable markers for labeling the feature of interest in specimens. Most common is the use of antibodies, glycoproteins produced in an animal as a specific, defensive response to a challenge with a foreign protein. However, ligands that bind specifically to cellular receptors can also be used as markers. These can be the natural ligand that binds as part of a normal physiological function, or it can be a molecule that serves as an agonist or antagonist at the receptor. Likewise, molecules that bind to some particular component of a cell organelle can be used as an identifying marker. For example, fluorescent dyes that interact with nucleic acids can be used to stain nuclei (DAPI, propidium iodide) or even mitochondrial DNA (ethidium bromide). Reagents that sequester preferentially in one type of organelle by virtue of charge or pH (e.g., LysoTracker) or become trapped in a cellular compartment following an enzymatic cleavage (e.g., acetoxymethyl esters of fluorochromes) or an oxidation reaction (e.g., MitoTracker) are routinely used for identifying structures in specimens.

Another approach that can be exploited in the identification of structural elements is the binding of toxins to their cellular target. An excellent example of this method is labeling with fluorochrome-conjugated phalloidin, a toxin isolated from the poisonous mushroom *Amanita phalloides*. Phalloidin binds selectively to filamentous actin and stabilizes the filaments, helping to preserve their structure throughout the specimen preparation procedure.

Yet another labeling approach is the use of lectin conjugates. Intracellular structures enriched in particular carbohydrate moieties can be identified through the selective binding of lectins to the sugar or glycoprotein. For example, the interaction of wheat germ agglutinin (WGA) to N-acetylglucosamine and N-acetylneuraminic acid (sialic acid) residues concentrates WGA binding within the Golgi apparatus. The interaction of concanavalin A with alpha-mannopyranosyl and alpha-glucopyranosyl residues can be used to identify smooth endoplasmic reticulum. The utility of using lectins as markers varies according to whether the carbohydrate to which the molecule binds is broadly distributed through cells or tissues. Affinity of the lectin-carbohydrate interaction is also a significant factor. Some interactions are of such low affinity that washing after the lectin incubation is sometimes eliminated from the labeling protocol in order to better preserve labeling.

5.6 Antibody Generation

Antibodies are immunoglobulins that recognize foreign proteins, known as antigens, at specific regions, known as epitopes. Generation of antibodies is part of the normal adaptive immune response of vertebrates and provides for an especially diverse immune response in mammals, where the theoretic capacity to generate 10^{10} distinct antigen recognition sequences is obtained via an elegant genetic recombination mechanism (French et al. 1989). Antibodies are produced by B lymphocytes in response to an antigen binding a B cell receptor that specifically recognizes it. B cell proliferation and maturation occur upon cross-linking of B cell receptors, together with a second signal provided by antigen-specific T helper lymphocytes in secondary lymphoid tissues (e.g., lymph nodes, spleen, tonsils), or cross-linking by a T-independent antigen. A B lymphocyte that is terminally differentiated to secrete large amounts of antibody is known as a plasma cell.

Antibodies can be described as polyclonal or monoclonal, depending on how many epitopes of the immunizing antigen they recognize. Polyclonal antibodies are made by immunizing animals and bleeding them to obtain antiserum. The serum can be used directly, or the immunoglobulin fraction can be purified from it. Polyclonal reagents contain a collection of antibodies produced by multiple clones of antibody-secreting plasma cells. These cells are located in lymphoid tissue such as lymph nodes or spleen. Each individual clone produces antibody of only one specific type and epitope specificity. The different species of antibody in a polyclonal will include high and low affinities, specificities toward linear and nonlinear epitopes, and multiple immunoglobulin classes (see below). Because they contain multiple antibodies, polyclonal reagents can be thought of as "hearty." That is, if target antigenicity is compromised during specimen preparation, it is likely that one or more species of antibody in the polyclonal will still be able to bind antigen. However, the more generalized binding characteristics associated with polyclonals may lead to undesirable cross-reactive binding (see below).

Monoclonal antibodies are of a single binding specificity. The first step in generating monoclonals is also immunizing an animal (usually a mouse), followed by fusing B cells from secondary lymphoid tissue such as the spleen with an immortal myeloma cell line. The resulting hybridoma cells are cloned by limiting dilution so that each colony that grows originates from a single fused cell; i.e., the cells are monoclonal. The antibodies secreted by the hybridoma cells selected through screening will be of a single antigen specificity that is the same as that of the B cell from which it arose. Antibodies of relatively high affinity are usually selected for further study during the screening process, although antibodies with lower affinities will be selected if a clone with a very useful or rare specificity is found. Because the antibodies are of a single specificity, cross-reactivity is seen less often with monoclonals than with polyclonal reagents.

5.7 Immunoglobulin Classes and Structure

Immunoglobulin class G (IgG) is found at the highest concentration in mammalian plasma, but IgM is also present at significant levels. IgA is present in plasma at lower concentrations than IgG and IgM but is the immunoglobulin at highest concentration in secretions such as mucus, saliva, and tears. The structures of IgG, IgM, and IgA are shown in Fig. 5.1. Free IgD and IgE are found only at low levels in plasma because these molecules are generally sequestered on cell surfaces. Most monoclonal antibodies used in labeling applications are IgGs; some IgMs with unique epitope recognition are also commonly used. IgMs are much larger than IgGs and are generally of lower affinity, making them less desirable than IgGs for many labeling procedures. IgGs and the functional fragments made from them have predictable sizes and structures that give consistent labeling efficiencies that vary only according to the individual antibody affinity and target antigen accessibility. The average diameter of correctly folded IgG is 12.9 nm; $F(ab')_2$ fragments (see below) of IgG have an average diameter of 9.8 nm; and Fab fragments average 5.3 nm.

IgG is a large glycoprotein of 150,000 kD that is made up of four subunits – two 50,000 kD heavy chains and two 25 kD light chains (Fig. 5.2). The heavy chain distinguishes the Ig classes from one another: IgG has a heavy chain of isotype gamma (γ), IgM has isotype mu (μ), IgA has isotype alpha (α), IgD has isotype delta (δ), and IgE has isotype epsilon (ε). In birds, reptiles, and amphibians, the dominant circulating immunoglobulin is IgY, named because it has a heavy chain isotype

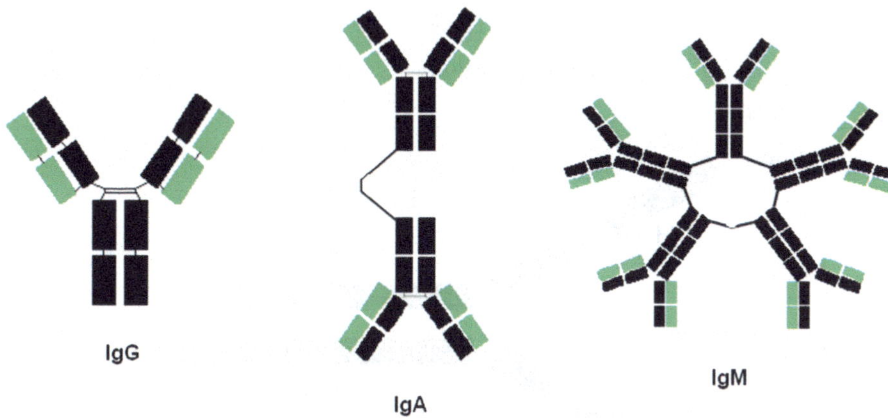

Fig. 5.1 Basic organization of the major antibody classes. IgG is the dominant species in the blood. It is monomeric and divalent. IgM exists as a pentomere with ten antigen-binding sites. IgA is generally present in dimeric form and is found primarily in secretions

found on antibodies that can be isolated from egg yolk. Immunoglobulins found in bony fish include IgM, IgD, and IgT. IgT carries a unique tau (τ) heavy chain restricted to teleost, or bony, fish (Hansen et al. 2005).

Immunoglobulin light chains in mammals are of two types, kappa (κ) and lambda (λ). Birds lack kappa light chains. A sigma (σ) light chain was first found in *Xenopus* and has since been described in all cold-blooded vertebrates, including teleost fish. Another primordial light chain that originated before bony fish diverged from cartilaginous fish (σ-cart) has been described (Criscitiello and Flajnik 2007) and is restricted to elasmobranchs. Antibodies of mammalian and avian (specifically, chicken) origin are routinely used in labeling applications.

5.7.1 *Immunoglobulin Subclasses*

There are four well-described subclasses of IgG in both humans and mice, each carrying a subclass of the γ heavy chain. In humans, the subclasses are IgG1, IgG2, IgG3, and IgG4; in mice they are IgG1, IgG2a, IgG2b, and IgG3 (Fig. 5.3). The molecular formulas of the different mammalian immunoglobulin classes differ. As mentioned above, IgG consists of two heavy (γ) chains and two light chains (κ or λ), giving formulas of $\gamma 1_2\kappa_2$ or $\gamma 2b_2\lambda_2$, for example. By contrast, circulating IgM is pentameric, with a molecular formula of $(\mu_2\kappa_2)_5$ or $(\mu_2\lambda_2)_5$. In addition, IgM contains an 18 kD joining, or J, chain (Fig. 5.4a). IgA consists of two subclasses and is generally dimeric, giving example molecular formulas such as $(\alpha 1_2\kappa_2)_2$ or $(\alpha 2_2\lambda_2)_2$. However, IgA can also be tetrameric. Like IgM, IgA contains a J chain that holds the immunoglobulin monomers together (Fig. 5.4b). Additionally, secretory IgA is associated with an aptly named secretory component. This is a remnant of a poly-Ig

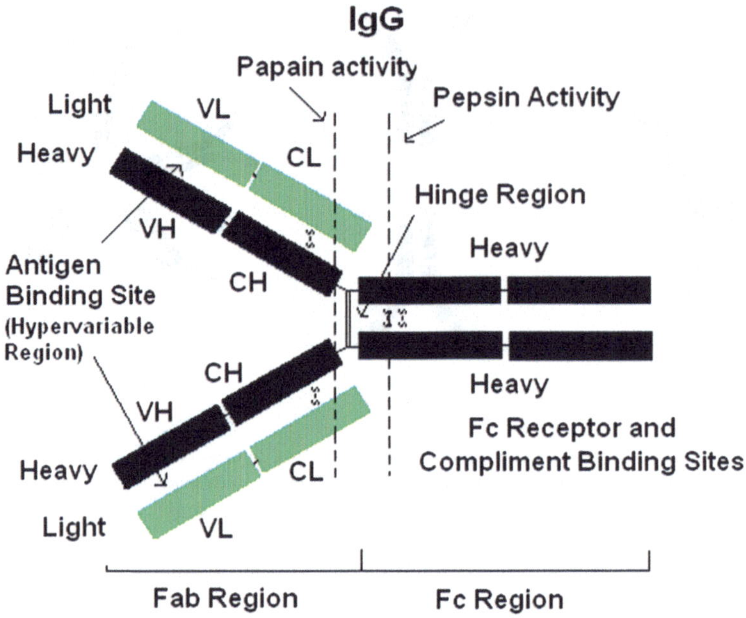

Fig. 5.2 A typical IgG antibody. The molecule is composed of two heavy and two light chains (either kappa or lambda) which are connected by disulfide bridges. The tail or Fc region is composed of heavy chain constant regions. The Fc portion of the molecule contains Fc receptor-binding sites and complement binding sites. The Fab portion of the molecule is comprised of constant, "CH" and "CL," and variable, "V_H" and "V_L," regions of the heavy and light chains, respectively. A hypervariable region within the variable region is the site of epitope binding. The IgG molecule can be enzymatically cleaved at various places resulting in specific antibody fragments. Pepsin cleavage results in separation of the two Fab fragments from the Fc portion. The Fab fragments remain connected resulting in an $F(ab')_2$ fragment and an Fc fragment. Cleavage by papain results in two separate Fab fragments and an Fc fragment. Separation of the active antigen-binding site from the remainder of the antibody molecule produces an single-chain variable fragment, scFv (Fig. 5.5). The fragments from the light and heavy chains must be reconnected via a short peptide bridge. scFv can be genetically engineered via phage display technology and produced in bacteria

receptor responsible for uptake of IgA on the basolateral side of mucosal epithelial cells and its incorporation into endocytic vesicles. The vesicles are transported to the luminal surface of the epithelial cells, where enzymatic cleavage of the receptor allows subsequent release of IgA and the remaining associated receptor fragment into the lumen (Kindt et al. 2007). IgD and IgE are monomers.

5.7.2 Antibody Structure and Fragments

The shape of antibody molecules is similar to the letter "Y." If portions of the molecule are described in terms of function, the "arms" of the "Y" are called the variable, or antigen-binding regions, and the base of the "Y" is called the constant

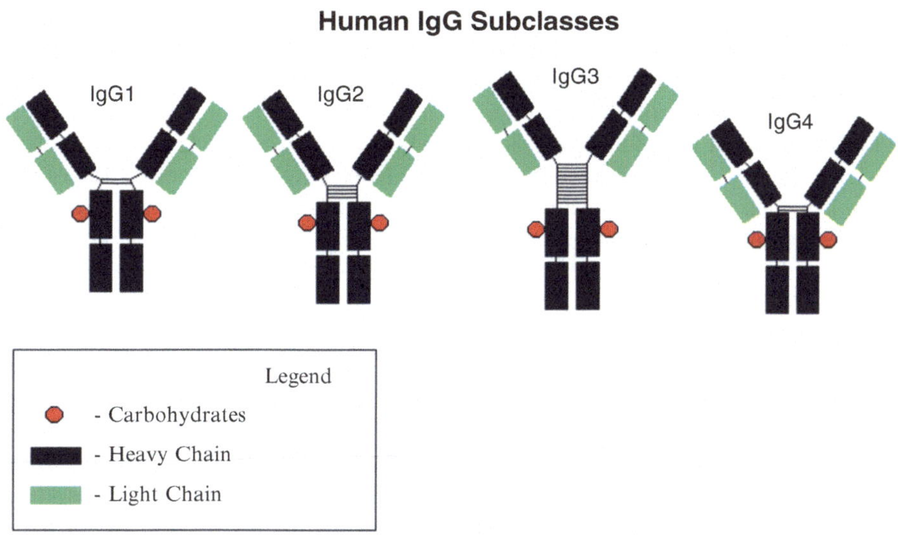

Fig. 5.3 Examples of human IgG subclasses. Variation is seen in the hinge regions of the different subclasses

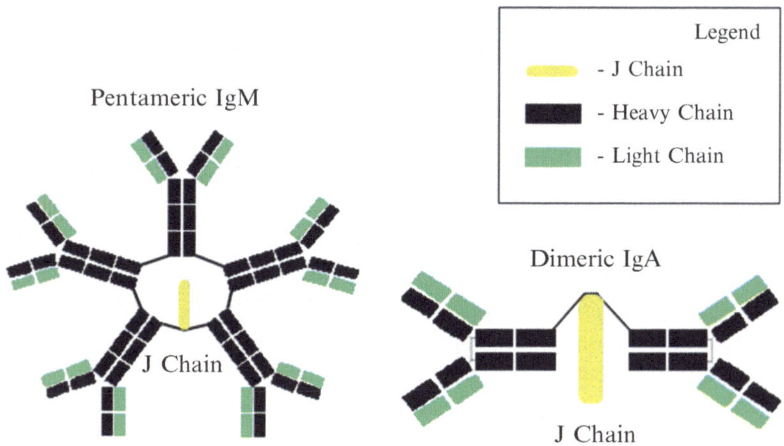

Fig. 5.4 Structure of an IgM molecule and of an IgA molecule. The IgM is comprised of five individual immunoglobulin molecules that are cross-linked to each other by disulfide bonds as well as connected via a joining or "J" chain. The IgA is usually in a dimeric form also connected by a joining chain

region. The variable, or Fab, region is responsible for antigen recognition. The constant, or Fc, region is responsible for biological effector functions of antibodies, such as binding complement to induce lysis of antibody-coated cells such as invading bacteria, or binding antibody-coated bacteria to specific Fc receptors on phagocytic cells that are capable of engulfing and killing pathogens (Fig. 5.2).

Disulfide bonding is critical to the structure of antibodies. Intrachain disulfide bonds create the Ig loops that characterize members of the immunoglobulin superfamily, which contains a wide variety of proteins of diverse functions. Some members of the Ig superfamily have roles in normal immune function, but many do not. IgG, IgD, and IgA have four Ig loops in their heavy chains, while IgM and IgE each have five Ig loops in their heavy chains. Light chains have two Ig loops. Each of the loops is named according to whether it is on the heavy or light chain, and whether it is in the variable or constant region of the molecule. Beginning at the variable portion of an IgG (the "arm" of the "Y"), the structure of the heavy chain is V_H, C_H1, C_H2, and C_H3. The light chain is made up of V_L and C_L (Fig. 5.2).

Interchain disulfide bonds are responsible for the association of the two heavy chains with one another and for the association of a light chain with each of the heavy chains. In IgG, two or more disulfide bonds between heavy chains at the "neck" of the "Y" are found in what is called the hinge region. This is a flexible area of the antibody structure that allows for the arms to move relatively freely with respect to one another during antigen binding. The heavy-light interchain disulfide bonds are found near the hinge region, slightly closer to the variable region (Fig. 5.2).

Functional antibody fragments can be generated by enzymatic digestion of whole antibody molecules. Digestion with pepsin cleaves the heavy chains just below the interchain disulfide bonds. The result is a bivalent product capable of cross-linking antigen through its two antigen-binding sites but missing its Fc portion. These fragments are called $F(ab')_2$ (Fig. 5.5), and they are very useful in labeling applications where Fc-mediated antibody binding gives high background staining. Digestion of whole IgG with papain cleaves the molecule at a site between the heavy-heavy chain disulfide bonds and the heavy-light chain disulfide bonds. The product is two, monovalent fragments with antigen-binding capacity and one Fc fragment. The antibody-binding portions are called Fab fragments (Fig. 5.5), and they are useful in labeling applications where both Fc binding and antigen cross-linking are undesirable. Both $F(ab')_2$ and Fab fragments are smaller than whole IgG molecules (110 and 50 kD, respectively, as compared to the 150 kD whole molecule) and can therefore provide better specimen penetration and potentially more accurate localization of antigens. The increased penetration is particularly important in the case of confocal microscopy where imaging can extend tens of microns and well into the specimen.

5.7.3 Variable Antibody Domains

The smallest functional antibody fragment is a combination of the V_H and V_L domains. These fragments can be generated through protein engineering rather than enzymatic digestion of whole molecules. The product is called a single-chain variable fragment, or scFv (Fig. 5.5) (Huston et al. 1988). The variable regions of the heavy and light chain do not have a naturally occurring connection such as an interchain disulfide bond, so a bridge between the two sequences is engineered into the construct used to produce the single-chain product. This bridge generally spans

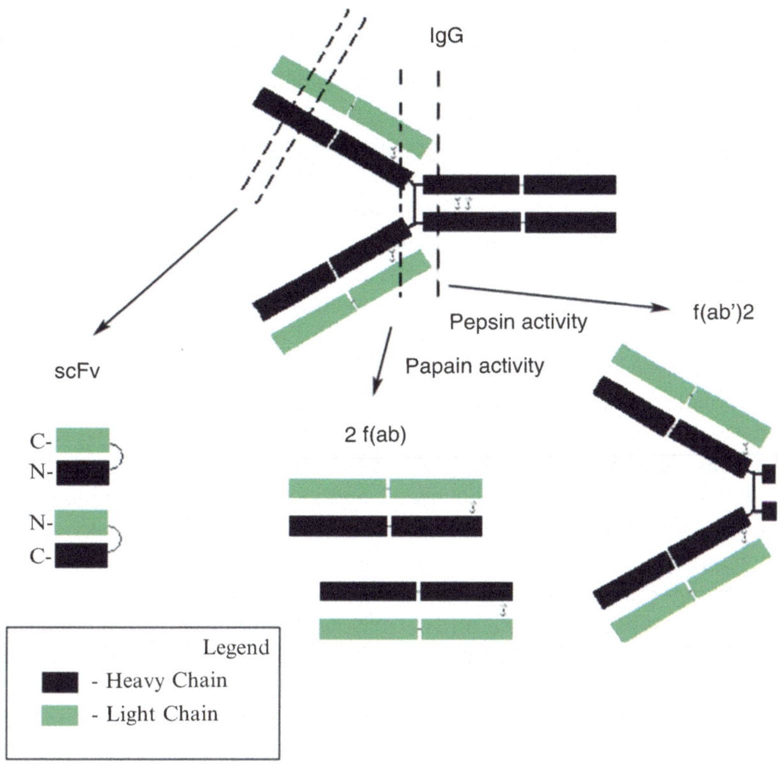

Fig. 5.5 Active antibody fragments, F(ab')$_2$, Fab, and scFv in order of decreasing size, that can be used to reduce overall label size and eliminate Fc receptor or complement binding. The F(ab')$_2$ is divalent, while the Fab and scFv are monovalent with respect to the number of specific antibody-binding sites

3.5–4.0 nm. It has been reported that introducing disulfides to stabilize the structure without altering the antigen-binding sequence can increase affinity and half-life of the molecules Reiter et al. 1994). Affinity purification methods can be used with scFv, similar to those used with whole molecule antibodies. However, the bacterial proteins A and G (originally derived from *Staphylococcus aureus* or group G streptococci, respectively), which bind to the Fc fragment for purifying whole antibodies, will not work with fragments. Instead, immobilized bacterial protein L (originally derived from *Peptostreptococcus magnus*) that binds to $V_L\kappa$ can be used to purify Fab and scFv (Nilson et al. 1993). This method will not work for antibody fragments with λ light chains, but 60% of human and 95% of mouse antibodies have κ light chains (Kindt et al. 2007).

Because the structures of scFv are relatively simple and do not require posttranslational modifications, they can be produced in *E. coli*. The DNA sequence of the variable region can be cloned from cDNA of hybridoma cells producing monoclonal antibodies of known antigenic specificity and affinity. Producing the recombinant protein in bacteria allows for high-volume production more quickly and affordably than can generally be achieved with whole antibodies grown from hybridoma cells in mouse ascitic fluid or in mammalian cell culture.

Within the variable regions of antibodies, there are specific amino acid positions that vary more than the others; these variations account for a great deal of antigen specificity. Clusters of these hypervariable amino acids found in V_H and V_L are referred to as the complementarity-determining regions, or CDRs. The CDRs are found on the outside of fully folded immunoglobulin molecules, where they have maximal access to antigenic epitopes. In the construction of scFv molecules, screening antibody phage display libraries for antigen binding can identify CDR DNA sequences that will produce the desired specificity (McCafferty et al. 1990). CDR sequence modules can be introduced into scFv constructs for high-throughput generation of panels of recombinant antibody fragments.

Another, even smaller variation of antigen-binding regions is seen with nanobodies. These can be derived from antibodies made in camelids – dromedary camels, Bactrian camels, llamas, alpacas, guanacos, and vicuñas as well as some shark species. Animals from this family make antibodies consisting of two heavy chains and no light chains (Fig. 5.6), in addition to the more typical two heavy chain + two light chain molecules (Hamers-Casterman et al. 1993). Naturally occurring antibodies devoid of light chains are unusual for two reasons. First, the antigen recognition site has contributions from both V_H and V_L, and both are expected to be necessary to achieve a complete antigen-binding repertoire. However, the heavy chain-only antibodies found in camelids have broad antigen-binding capacity. Second, heavy chains are dependent on light chains for good solubility. Isolated heavy chains from other mammals tend to agglutinate, but non-aggregated camelid forms are readily

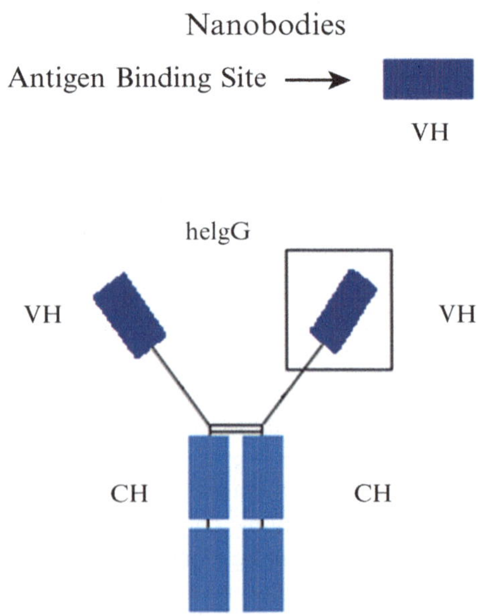

Fig. 5.6 Nanobodies are heavy chain variable fragments from *Camelidae* sp. which exhibit specific binding without the need for a variable light chain and are robust with respect to variation in temperature, pH, and resistance to enzymatic degradation. The variable heavy chain region, "VH," is isolated from the heavy chain immunoglobulin molecule, "hcIgG"

found in circulation. These characteristics allow for the production, through protein engineering, of fully functional V_H segments that are one-tenth the size of whole IgGs. The smaller size gives nanobodies greater access to antigens than larger antibodies or fragments have. Nanobody conjugates or fusion proteins used for detection in labeling protocols will also be significantly smaller than the corresponding whole molecule or fragment, providing more accurate localization of target molecules or structures. The camelid-derived nanobodies are significantly more resistant to heat, pH, and enzymatic digestion than other antibodies or fragments from other species. This provides flexibility in labeling protocols that could potentially extend to the oral administration of labels to live animals.

5.8 Labeling Considerations

5.8.1 Review of the Literature

There are literally thousands of protocols developed for a wide variety of labeling applications. Perhaps the best rule of thumb is to clearly define what information is needed and then proceed with a thorough review of the literature looking for previous studies designed to obtain similar information. This often avoids a substantial amount of trial and error. Sources of antibody as well as types or subtypes, specific fluorescent dyes, use of chemical or physical fixation, and even buffers and conditions of labeling can all be reasonably estimated from prior labeling protocols used by others. Obviously modifications may have to be made, and in some instances totally new protocols will have to be developed. Once the information required is defined, it is also advisable to attempt the least complex protocols first. There are no extra points for overly complex protocols when the desired information can be reliably obtained by a relatively simple direct approach. Try to avoid "overkill" which may be impressive technologically and which may be able to provide additional information. However, if the additional information is irrelevant to the study at hand, much additional time and expense may be saved by a less impressive but more direct methodology.

5.8.2 Use a Hierarchy of Applications

There is often a hierarchy of technologies with each designed to better maintain epitope antigenicity. This often involves starting with a classical buffered aldehyde fixation. The aldehydes interact chemically with a number of functional groups in biological specimens. Formaldehyde is thought to cause less cross-linking and hence preserve antigenicity better than the dialdehyde, glutaraldehyde; however, this may not always be the case. Mixtures of formaldehyde and glutaraldehyde may be preferable to preserve structure for interference-based LM structural studies or correlative

EM studies. If a classic fixation adversely affects antigenicity of one or more of the epitopes of interest, a brief fixation in dilute, 0.5% or less formaldehyde followed by labeling and then a more concentrated, structure maintaining, fixative may be used. For preservation of even less robust epitopes, fixation in the cold may be sufficient to inhibit epitope inactivation. More complicated procedures such as fixation in the cold followed by freezing and labeling of frozen sections or freeze substitution procedures with fixation may be needed. Finally a physical fixation approach may be required. This can involve ultra-rapid freezing followed by labeling of cryosections without any fixation. As a last resort, it may be necessary to label living specimens and follow labeling with fixation or freezing steps. However labeling parameters for living cells must be designed to account for the effects of the formation of antibody-antigen or ligand-receptor complexes in or on living cells and any subsequent changes in cell structure, cell activity, or movement of the antigen-antibody complexes.

5.8.3 Know the Antibody

With regard to the antibody or antibody fragments used in the labeling process, the primary antibody should be well characterized as to species, type, and subtype if possible. It is important to know if a polyclonal antibody (cross-absorbed) or monoclonal antibody is being employed. It is also important to know if whole antibody or active antibody fragments are being used. The affinity and avidity of the antibodies is also useful information. This is particularly the case with monoclonal antibodies where specificity may be high but the affinity may be lower. Monoclonal antibodies may be extremely specific, with minimal cross-reactivity, but they react with only one epitope. This may present difficulty if the preparation of the sample masks or alters the single epitope. Polyclonal antibody may have the advantage in such cases because it often contains higher avidity antibody and antibodies specific to several or all of the epitopes. Thus inactivation or blocking of one epitope will not necessarily compromise identification of the entire molecule.

However, as noted above, this may also lead to higher levels of cross-reactivity. Cross-reactivity may be substantially reduced by cross-absorption, while antibody species with high affinity can be selected by specific absorption against the antigen or even the specific epitope of interest.

Other considerations in the labeling process include the isoelectric point of the antibody. Antibodies are proteins and behave as such. Labeling at pH values substantially divergent from the isoelectric point may compromise affinity and/or avidity. This is not usually a problem with whole antibody where the isoelectric point will be in the range of commonly used physiologically relevant buffers. However active antibody fragments, described above, may have a wider range of isoelectric points necessitating modification of labeling buffer pH. The use of particular ions to maintain epitope structure may also affect antibody affinity and avidity.

5.8.4 Antibody Concentration

Antibody concentration is also of obvious importance. A reasonable estimate of the concentration is essential. Often antibodies are supplied or purified to be in the range of 1 mg/ml or roughly $1 \times 10^{15-16}$ antibody molecules/ml. Standard instructions often recommend dilutions of 1:10, 1:100, or even 1:1000 prior to use in labeling protocols.

When concentrations reach the 1×10^{12} range, 1 ug/m/ or below, particular attention needs to be paid to the time of labeling since 100% epitope coverage can be attained in several minutes at 1×10^{13}, while it may take hours to days at 1×10^{11}. The former may be useful for live cell labeling in cases where maximal labeling needs to be accomplished rapidly. The lower concentrations may be useful to minimize background where labeling is performed overnight in the cold. The antibody concentration in commercial antibody preparations often depends on whether the antibody is polyclonal or monoclonal.

Another factor affecting antibody concentration is related to freezing and thawing. On thawing, a precipitate may form in frozen antibody preparations. The usual procedure is to centrifuge or filter the sample in order to remove the precipitate. However it must be kept in mind that the precipitate generally contains a portion of the antibody and the concentration of soluble antibody may be significantly reduced in the remaining soluble fraction. Antibody should always be divided into aliquots and frozen. Aliquots are thawed as needed, so the antibody is frozen and thawed only once. Multiple freezing and thawing will promote precipitate formation and a substantial reduction in antibody concentration or even the total elimination of antibody.

5.8.5 Antibody Conjugation and Level of Fluorescence

The level of fluorescence is determined not only by the number of antibody molecules or number of active antibody fragments but also by the number of fluorescent molecules per antibody molecule. In general each antibody molecule can be conjugated with four to five fluorescent molecules with little effect on parameters of binding. As noted previously other factors specific to the dye also play a role. These include wavelength of excitation and emission, quantum efficiency, extinction coefficient, and quenching rate.

Direct conjugation of the primary antibody to fluorescent or other markers has several advantages (Fig. 5.7a). There is less chance of cross-reactivity when the specimen is incubated with only one antibody. Cross-reactivity and nonspecific binding of the second antibody are eliminated. Considerations relative to the binding avidity and affinity of the second antibody to the primary antibody are also eliminated. Finally spatial resolution is maximized, and measurements of semi-quantitative aspects are more reliable.

However, primary antibody is often available in small quantities and, if purchased commercially, is generally expensive. Fluorochrome-conjugated primary antibody is often not available commercially. Conjugation to a fluorochrome can significantly decrease affinity or avidity, therefore, in house conjugation may

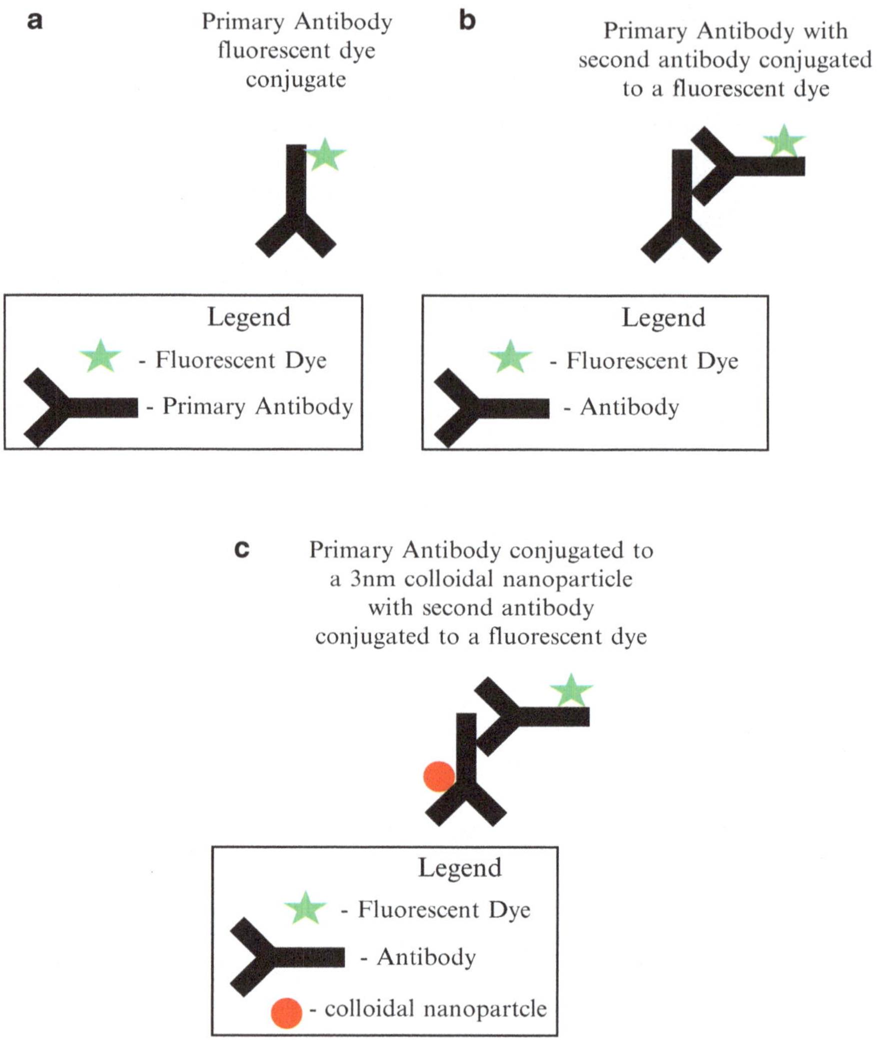

Fig. 5.7 (**a**) Primary antibody-fluorescent dye conjugate. Generally three or four individual dye molecules are linked to each IgG molecule. (**b**) Primary antibody with second antibody conjugated to a fluorescent dye. (**c**) Primary antibody conjugated to a 3 nm colloidal nanoparticle with second antibody conjugated to a fluorescent dye

require testing and the availability of relatively larger amounts of specific antibody. In order to avoid conjugation of primary antibody to a fluorescent dye and the potential loss of antibody and/or activity, a second antibody-fluorescent dye conjugate can be used in what is referred to as a sandwich technique; antigen bound to primary antibody bound to fluorochrome-conjugated secondary antibody (that recognizes primary antibody). The second antibody reacts specifically with the primary antibody and can be more general in nature (Fig. 5.7b). Second antibodies can be conjugated to a variety of fluorescent dyes so that there are a number of possible excitation/emission combinations available.

If the primary antibody is mouse, then a second antibody from any of several other species such as rabbit, chicken, goat, and so forth can serve as the second antibody. In many cases, especially when working with mouse tissues, only mouse primary monoclonal antibodies for a specific antigen may be available. In these cases, some companies such as Vector Labs (Burlingame, CA) sell mouse on mouse (MOM) labeling kits. While extensive blocking and controls must be performed to insure no nonspecific mouse primary monoclonal antibody is sticking to random mouse tissue, these kits do provide a mechanism for labeling of mouse targets with mouse antibodies.

Generally multiple second antibody molecules will bind to each primary antibody molecule. Each of the second antibody molecules will be conjugated to four or five fluorescent molecules, and hence the signal will be substantially enhanced. The source of the fluorescence will come from a slightly larger area around the specific labeled epitope, but at LM levels of resolution, this is not a problem. With molecular-resolution LM or high-resolution EM studies, the closer the indicator is to the actual location of the epitope of interest the better.

A combination of small nanoparticles conjugated to the primary antibody to provide maximal spatial resolution for EM in combination with fluorochrome-conjugated second antibody for LM and confocal imaging can be employed in these cases (Fig. 5.7c) (Kandela et al. 2007, 2008; Albrecht and Meyer 2008). This also avoids quenching of fluorescent dye which occurs when it is conjugated directly to the same antibody molecule as the colloidal metal particle (Kandela and Albrecht 2007). It is also important to realize that primary conjugates used for cell sorting purposes may not be applicable to microscopic imaging with standard CCD cameras. This is due to differences in the quantum efficiency of sorting, as opposed to imaging instrumentation. High sensitivity EMCCD and hybrid cameras (discussed in Chap. 7) may be more useful in the detection of lower signal levels from primary conjugates.

5.8.6 Use of Second and Third Antibodies

All the considerations relative to the primary antibody must also be observed in the case of the second (or third) antibody. If an antibody fragment is used as the primary antibody to facilitate penetration throughout the sample or for facilitating semiquantitation of epitopes, it is important to also use an active antibody fragment for the second antibody. Whole second antibody will defeat the purpose of the small primary probe. The concentration, affinity, and avidity of the second antibody should be known. Second antibodies are often better characterized than primary antibodies because of their general use. The cost also tends to be substantially less. The use of a second antibody that is more specific for the primary antibody is preferred. This facilitates labeling of more specific determinants rather than the more general shared determinants. For example, if the primary antibody is a mouse IgG2b, it is preferable to use an anti-mouse IgG2b rather than simply using a polyclonal antibody which labels mouse whole H + L chains. This reduces nonspecificity and can enable simultaneous multiple labeling for other mouse primary antibodies such as an IgG2a.

5.9 Antibody Sources

There are a variety of sources for polyclonal as well as monoclonal antibodies. Offerings from commercial sources fill volumes and can be searched on the Internet. A useful source for information on antibodies and related products is Linscott's Directory of Immunological and Biological Reagents which currently has over 480,000 different listings in their antibody section (Linscott's Directory of Immunological and Biological Reagents 2018). While an antibody may have specificity for an antigen or epitope of interest, it is important to know the context of that binding. Antibody may be used for labeling of in vivo or in vitro systems, ELISA, Western blots, cell sorting applications, or various binding assays or functional inhibition tests. A specific antibody may be effective in a variety of these labeling applications or restricted to several or even one application. Binding to an antigen in one context does not guarantee binding in another.

5.9.1 Polyclonal Whole Molecule

Various levels of purity for polyclonal preparations are available. Whole serum from immunized animals; purified immunoglobulin fractions; fractions of particular immunoglobulin subtypes such as IgG, IgM, IgE, or IgA; and highly cross-absorbed and affinity-purified antibodies are all available. The cost is generally directly proportional to the level of purification and absorption. Some more commonly used primary antibodies can be purchased as fluorescent conjugates, but generally this is not the case. Primary antibody, as above, may be directly conjugated in house to an appropriate fluorescent dye, or, more commonly, they are used with a fluorochrome-conjugated second antibody. The commercially available antibody preparations are generally well characterized. Antibody preparation, concentration, known cross-reactivity, preservatives and preservative concentrations, as well as references to papers describing the initial preparation and subsequent uses are generally included or available online from the supplier.

Antibodies may sometimes be obtained directly from other investigators who have generated the antibody and have sufficient amounts to share either as gifts or at relatively low cost. This may be the case where one group has developed an antibody with a specific activity and the antibody is not yet available commercially. Gift antibodies often are less well defined than those available from commercial sources, but they have generally been tested and work well for the specific needs of their laboratory of origin. They may be in various stages of purification, from whole serum to affinity-absorbed specific antibody, depending on the needs of the original investigator. When obtaining such antibodies, it is important that there be a clear understanding of how these antibodies will be used, author requirements for publications, patents, and other public disclosures, as well as limitations with respect to the use of antibodies by individuals other than the specific investigator receiving the gift.

If no other sources exist, antibodies can be prepared in house. Often there are biotechnology facilities that will guarantee specific antibody to a particular antigen which you supply in a pure (as is reasonably possible) state and at a concentration required by the antibody production facility. For in-house preparation, there are requirements for animals and animal care, antigen preparation and purification, adjuvant selection, and an immunization protocol of which there are many depending on the animal, the antigen, and the type of antibody required. Most antibodies must be separated from the blood serum although IgY from chickens is purified from egg yolks. Antibody is then purified and absorbed to increase specificity and reduce cross-reactivity. The entire process is involved and may be costly, but necessary, if you are working with a unique antigen or epitope. It also facilitates studies where large amounts of antibody are required or where direct conjugation of primary antibody to fluorescent dye or other indicators is desirable. There are also a number of commercial concerns who will work with the investigator to provide specialty antibodies. Cost of antibody production will vary and may be an issue if funds are limited.

5.9.2 Monoclonal Whole Antibody

Commercial groups are the primary source for monoclonal antibodies. Literally thousands of monoclonal antibodies are available commercially. The antibody is generally well characterized as to antibody type and subtype. The concentration is usually provided, but affinity and avidity are often listed as "high" or "low." While specificity of monoclonal antibodies may be high, it may take a certain amount of good fortune to find a monoclonal antibody with exceptionally high affinity, so care must be exercised in establishing the affinity and avidity of any monoclonal antibody.

It is helpful if the specific epitope on the antigen recognized by the antibody is also known, but this may not be available. Often references are provided which discuss the original preparation and use of the particular antibody. This may provide more information on subtypes, epitope specificity, cross-reactivity, and avidity/affinity information. Most commercially available monoclonal antibodies are sold in small volumes and are relatively expensive.

Monoclonal antibodies may also be obtained as gifts. As is the case with polyclonal antibodies, monoclonal antibodies directly from colleagues or other investigators can be the only source when antibodies to specific epitopes are otherwise unavailable and the laboratory originally producing the antibody has sufficient stocks to provide antibody to other investigators. Monoclonal antibodies obtained as gifts may come well characterized and with publications describing their preparation and use; however, information provided often is less than provided with commercial preparations. One major advantage is direct advice that may be provided via the principal investigator or other investigators in the laboratory producing the particular monoclonal. Also, the investigators may have produced several clones with antibody specific to epitopes in your antigen of interest. It is sometimes helpful to test antibodies from multiple clones to find the antibody best suited to provide the information you require. In some instances investigators will provide cells from the specific hybridoma clone. These antibody-producing cells can be propagated in vivo or in vitro, thus insuring a substantial supply of antibody. Agreements for the use of monoclonal antibodies or for cell clones producing monoclonal antibodies have to be carefully prepared to avoid any subsequent issues related to commercialization or patents.

Monoclonal antibodies can also be produced in house. The process is involved and costly. Many investigators have access to a biotechnology laboratory that will produce a guaranteed number of hybridoma clones each of which produces antibody specific for the antigen you provide. The affinity of the monoclonal antibody from the clones is often a problem. However if monoclonal antibodies which are specific for a unique antigen or epitope of interest do not exist elsewhere, this is the only option. If studies require large amounts of antibody, having a hybridoma clone which produces a functional monoclonal antibody insures a sufficient supply of antibody. This is important when primary antibody is directly conjugated to fluorescent dyes or colloidal nanoparticle markers since a substantial amount of antibody may be used in refining the conjugation process. It is also important when active antibody fragments are prepared by fractionation of whole antibody molecules.

5.9.3 Antibody Fragments

Active fragments such as Fab, $F(ab')_2$, or scFv, single-chain variable fragment, for some antibodies are available from commercial sources. However the demand for fragments from specific antibodies is general such that it is not profitable to produce fragments which are held in stock. Often commercial or in-house biotechnology services can prepare various fragments on a fee-for-service basis. Kits are also available from various manufacturers which can be used to prepare Fab or $F(ab')_2$ fragments from whole antibody preparations or cloned in bacteria. For in-lab or service facility preparation, sufficient antibody must be available as loss of antibody during the fragment preparation is inevitable. The scFv fragment is a fusion protein of the heavy and light chain variable regions. The label can be made by converting monoclonal antibody into the scFv form. Other methodology includes the use of a phage display and with expression of the scFv in one of several bacterial systems.

5.9.4 Second Antibody

Second antibodies are generally available commercially. The expense is usually less than primary antibody since substantial amounts of the same second antibody are used to identify a variety of primary antibodies from a specific animal source. For example, goat anti-mouse can be used to label a wide variety of primary mouse antibodies. Second antibodies generally are well characterized and used routinely. They can be produced from a number of sources including chicken (egg yolk) or serum from rabbits, goats, donkeys, mice, and other species. Various preparations are available commercially which include a hierarchy of specificities ranging from specificity for any immunoglobulin from a given species through individual immunoglobulin subtypes from a given species, to antibody fragments from a given species, and so on.

Second antibody can be obtained as a gift through colleagues, and it can be generated in house using the primary antibody or antibody fragment as antigen and highly cross-absorbing the specific antibody. It can be affinity purified if necessary.

5.9.5 Control Antibody

Control antibodies can be obtained commercially, as gifts, or generated in house. They are often less costly than primary specific polyclonal or monoclonal antibodies. There are several levels of sophistication. The use of pre-immune serum as a control simply provides a mixture of blood proteins and antibody none of which, unless there is cross-reactivity, binds to the specific epitope. Pre-immune serum permits delineation of general nonspecific binding.

The use of purified immunoglobulin fractions as a control narrows the specificity to nonspecific binding of immunoglobulins, while the use of a purified immunoglobulin subtype identical to the primary antibody provides an even more precise control.

Perhaps the most specific control involves the use of an antibody which is specific for the inappropriate haplotype of the antigen. These controls can be identical in every respect to the primary antibody with the exception that they bind to an alternative antigenic haplotype. Since they are generally monoclonals to an inappropriate haplotype, there is no cost saving when compared to the primary antibody which does bind to the appropriate haplotype.

5.10 Conditions of Labeling

It is not the intent here to review any of the thousands of labeling protocols or even to try to provide a "one size fits all" generic protocol. Hopefully the discussion here simply serves as a reminder that there are some basic conditions which must be met for labeling to be successful.

5.10.1 Maintain Antigenicity

First the conditions prior to and during labeling must maintain antigenicity of the antigen or epitope. With polyclonal antibodies this means at least one and preferably as many as possible of multiple epitopes on a given antigen need to remain recognizable by a specific antibody. For monoclonal antibodies the specific epitope recognized by the antibody has to retain antigenicity. As noted previously there is often a hierarchy of technical approaches which can be employed to protect the antigenicity of epitopes. This usually begins with a standard chemical, usually aldehyde, fixation, as an initial protocol, and may ultimately necessitate labeling of physically fixed, i.e., frozen without chemical, fixation or even labeling of living material to prevent epitope inactivation. Other factors important in retaining antigenicity include ions and ion concentration, pH, buffer type, temperature and time of fixation and labeling, and the use of agents, such as the detergent Triton X 100, to increase permeability and penetration.

5.10.2 Maintain Optimal Affinity and Avidity

The affinity and avidity of the labeling molecule, whether antibody or ligand, must also be optimized. This requires close attention to buffer composition, osmolarity, and temperature. Often 0.1 molar phosphate buffer at pH 7.2 or 0.15 molar phosphate-buffered saline with up to 1.0% bovine serum albumin are used in washes and in the specific labeling step. However, particularly for active antibody fragments, ligands or ligand fragment, or scFv constructs, the buffer type and pH may vary.

5.10.3 Minimize Nonspecific Binding

It is also critical to minimize nonspecific binding. Blocking agents such as albumin, pre-immune serum, skimmed milk, or fish gelatin can be used to mask “sticky sites” where proteins bind nonspecifically. Blocking agents vary in their ability to mask nonspecific labeling in a particular labeling situation. If one type of blocking agent does not provide the level of activity needed, then it is important to try others either individually or in combination. Specific anti-Fc receptor antibodies, aggregated IgG, or Fc fragments can be used to block Fc receptor binding if pre-immune serum proves unsatisfactory. Salt solutions such as sodium chloride or various surfactants such as Tween 20 can be used to minimize surface charge and limit electrostatic charge-charge interactions. The use of free aldehyde blocking agents such as 0.5–1.0% sodium borohydride or 0.01–0.04 M glycine should be included in rinse buffers to eliminate any free aldehyde groups generated during the fixation process. Such groups can nonspecifically cross-link proteins, including antibodies, to the specimen.

5.10.4 Preserve the Specimen Morphology

Generally there is also need to preserve morphology at some level of resolution such that the labeling can be viewed in the context of biological structure. Again there is a hierarchy of methodology. Structural preservation for imaging with photon-based methodology may not be as stringent as for electron- or force-based imaging. Preservation of structure is compatible with a number of labeling protocols although modifications may be necessary particularly where high-resolution structural information is required.

5.11 Considerations in Labeling with Ligands

Ligands, particularly those with small dissociation constants, can be useful in order to label specific receptors or binding sites (Fig. 5.8). They often can be used to determine if binding sites are in their active state. Specific antibody may label epitopes within components of receptor sites without regard as to the activity of the receptor. Specific ligands bind only to the active receptor site. Thus antibody labeling can be used to determine the location and density of receptor molecules, while ligand labeling will identify only the active receptor sites. As with specific antigens or epitopes for antibody binding, fixation can often alter receptor structure and reduce or eliminate ligand binding.

There are a number of options for ligand labeling. Natural ligand can be directly fluorochrome conjugated or conjugated to fluorescent, chromatic, or opaque nanoparticles. Ligand can also be identified with second, fluorescent- or particle-labeled, anti-ligand antibody. For some receptors a number of pseudo-ligands, some of which are drugs, may exist. Attachment of ligands to receptors or binding sites can proceed by various mechanisms. Bond types and strengths vary; either non-covalent or semi-covalent binding can occur depending on the particular ligand-receptor combination. Noncompetitive agonists or antagonists are preferred because of their strong irreversible binding; however, competitive agonists or antagonists and even partial agonists can be used and concentrations adjusted to provide a high efficiency of labeling.

As with antibody, active fragments of ligand can be derived from whole ligand or synthesized directly to provide smaller labels to promote penetration or to better provide semiquantitative information with respect to receptor or binding site density (Fig. 5.9).

Ligand labeling is particularly effective in live cell microscopy where tracking of ligand-receptor complexes is desirable or the effects of ligand-receptor interaction on subsequent cell activity and structure is needed. Just as for whole IgG antibody molecules, some ligands are divalent and can link two or more receptors on cell surfaces leading to various types of receptor ligand-induced activity. Ligands can be obtained from a variety of sources including commercial concerns, as gifts, or purified/synthesized in house. Few commercial preparations of fluorochrome-conjugated ligands, pseudo-ligands, or active ligand fragments are available.

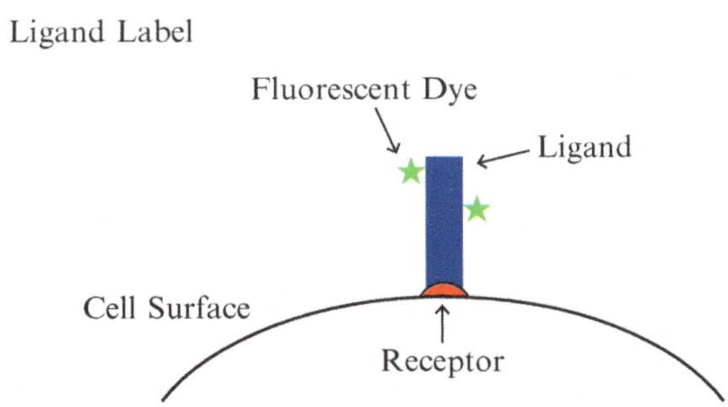

Fig. 5.8 Ligand conjugated to a fluorescent dye. Ligand conjugates can be effective for labeling active receptor sites particularly for live cell applications

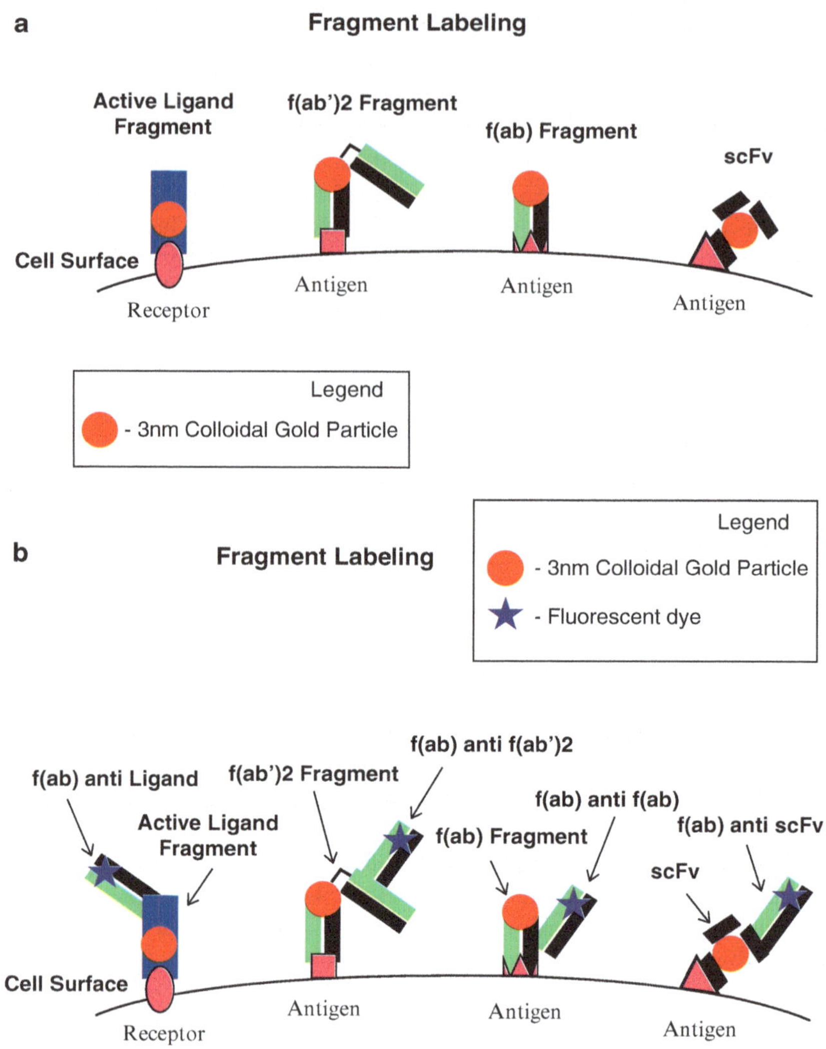

Fig. 5.9 (**a**) Labeling receptor sites or antigenic sites with ligand or antibody fragments with particles. Colloidal metal conjugates can be used for EM applications and can be further enhanced for LM as well. Conjugation of fluorescent dye to fragments may also provide LM signal if sufficient dye and conjugates are present and/or if a highly sensitive CCD camera is employed. (**b**) The primary fragments can also be identified using fluorescinated anti-ligand antibody or fluorescinated anti-fragment second antibody. If penetration is a problem due to the nature or thickness of the specimen, the fluorescinated second antibody may also have to be an antibody fragment as shown. Whole antibody, with more dye molecules, is preferred if signal strength is a limiting factor. The use of a fluorescent third, anti-second antibody may also be considered to further increase fluorescence levels

5.12 Live Cell Labeling

Live cell labeling may be employed where any type of chemical or physical fixation that alters the antigenicity of epitopes or modifies ligand-receptor interactions must be avoided. Live cell labeling may also be important to follow movement or endocytosis of antibody-labeled and cross-linked antigens on or in cells or to follow movement of receptor-ligand complexes. When cross-linking is not desirable, monovalent Fab or scFv labels can be used as can active ligand fragments. Cells may be cooled prior to labeling provided the cooling process does not compromise the presence or availability of the epitopes of interest. Where possible labeling should be performed on washed cells in simple buffer or in cell culture media with no or minimal levels of added serum protein. In some instances low levels, 2% or less, of serum may be required in the washing and/or labeling buffer to block nonspecific sites and to reduce Fc binding if this is an issue. However it should be kept in mind that serum contains a mixture of proteins, some of which may also bind to the living cells and influence cell structure or function. Serum-free media preparations can be useful if this becomes an issue. After labeling, cells can be returned to standard culture media if necessary.

High concentrations of antibody or ligand should be employed to facilitate rapid binding of available antigenic sites and a high efficiency of labeling. It is often necessary that primary antibody be directly conjugated to fluorescent dyes or particles. There may be insufficient time to employ a second antibody to label primary antibody or specific ligand. The second antibody may also produce extensive cross-linking of the primary antibody-antigen complexes.

Penetration of antibody, ligand or active fragments of antibodies, or ligands can be problematic. Various permeabilization techniques can be used. These range from the use of electric currents as in electroporation, ballistic penetration, use of specific endocytosis activators, or detergents which produce a reversible temporary increase in cell permeability.

5.13 Labeling with Particles

Colloidal nanoparticles have been used for a number of years as labels in electron microscopy and for correlative LM/EM, as well as in force-based labeling (Albrecht and Meyer 2008; Eppell et al. 1995). Conjugation of particles to primary or secondary antibody provides markers visible in TEM and SEM. Bulk concentrations of particles provide an opaque image in LM, or, in the case of gold, this can be black or a red to orange color. Individual particles as small as 10 nm can be identified and tracked by their inflated diffraction image in interference-based LM imaging. For correlative confocal LM/EM, imaging particles as small as 3 nm can be conjugated to the primary antibody or antibody fragment. This provides spatial resolution for epitope identification and localization in the submolecular range for EM.

A fluorochrome-conjugated second antibody can be used for LM/confocal localization. While the fluorescent signal is coming from a larger area around the epitope, it is still well within the resolution range of the LM. Having a fluorescent dye on the second antibody also substantially reduces fluorescent quenching which is often greater than 99% when metal particles and fluorescent dyes are on the same antibody molecule. Recently colloidal particles of uniform small size (to 3 nm) but having different elemental composition such as gold, palladium, platinum, silver, and others have been conjugated to different antibody labels so that simultaneous multiple labeling at EM levels of resolution can be performed. This is analogous to the use of multiple fluorescent dyes on different antibodies to provide simultaneous labeling of antigens at the LM levels of resolution. Particles can be differentiated by EDX or EELS in an aberration corrected STEM/TEM or by ESI/EELS imaging in an EFTEM, although EELS identification requires particularly thin specimens. Core-shell configurations further expand the number of labels that can be differentiated and hence used simultaneously. Particles of similar size but having different shapes have also been developed. For particles of different composition or of different shapes, the correlative confocal LM proceeds as above. Each particle-antibody combination is labeled with a second antibody with a particular fluorescent dye of unique excitation/emission properties. This allows multiple antigens or epitopes to be differentiated both in confocal/LM and subsequently at much higher resolution in EM (Kandela et al. 2007).

Other approaches to correlative fluorescent LM/EM have employed quantum dot particles. The quantum cores of these particles consist of cadmium-selenium or tellurium (Bruchez et al. 1998; Chan and Nie 1998). The cores fluoresce when they are illuminated with light of sufficient energy. The wavelength of illumination is not critical. The size of the quantum core determines the emission wavelength. They also exhibit minimal photobleaching. This makes them particularly desirable for confocal imaging over long time periods. The core is generally surrounded by a shell of zinc sulfide. The particle is further stabilized by an amphiphilic polymer with a functionalized, often PEG, surface. The final size is substantially greater than the 5–10 nm quantum core and can be as large as 20–30 nm for the particle without conjugated antibody. This can reduce their effectiveness for high-resolution EM labeling. The density of the quantum core is also not great, and identification by EM is possible but somewhat more difficult than the heavy metal nanoparticles. Ultraviolet light can destroy the cores and release toxic cadmium, but the stabilized particles appear to be generally stable in the visible spectrum.

Correlative confocal labeling with other fluorescent nanoparticles that are stable and nontoxic is under development. Oxides, such as YVO, doped with rare earth elements, for example, europium, fluoresce at specific wavelengths and can be identified in EM, but additional studies are necessary to determine the ultimate utility of these particles for confocal labeling or in correlative confocal LM/EM studies.

5.14 Sequential Versus Simultaneous Labeling

For simultaneous multiple labeling, the question of sequential vs. simultaneous labeling of the multiple antigens/epitopes or receptors/binding sites is a consideration. For live cell imaging, usually simultaneous labeling with high concentrations of each label is necessary as sequential labeling may require too much time. The order of labeling may also affect cellular responses particularly if antibody or ligand cross-linking is involved. For fixed material, if there is minimal cross-reactivity and a blocking regime that reduces nonspecific binding for all the antibodies can be employed, simultaneous binding can be used. However concentrations of each label may have to be adjusted relative to the specific antigen-antibody interactions.

The affinities of the different labels may vary, and the overall conditions should allow for the maximal efficiency of labeling to be attained for each label whether a ligand or an antibody. In some cases a sequential approach may be desirable, particularly where optimal labeling for each antibody requires significantly different pH or specific ion concentrations. Brief fixation steps may be required between application of each label, but care must be exercised not to reduce or eliminate antigenicity of unlabeled epitopes prior to their labeling. Sequential labeling can involve labeling of a living system followed by fixation and subsequent labeling of additional epitopes depending on the information required.

5.15 Troubleshooting

There are three basic areas where problems arise.

5.15.1 "Operator Error"

The first, and most common, is the "operator error" category. Often the symptom is no labeling or substantially reduced labeling. This may involve improper storage or too lengthy storage of fixatives which can result in polymerization and inactivation of fixative.

One of the more common operator errors involves composition and dilution of reagents. This may result in buffers or fixatives lacking critical components or concentrations of components that can be an order or more in magnitude too weak or too strong. Primary, second, and third antibodies are all generally diluted in buffer prior to use. Any error in antibody concentration or in the composition of the dilution buffer can reduce the efficiency of labeling. Improper dilutions resulting in higher than required antibody concentrations can result in an increase in nonspecific background labeling.

Errors in buffer composition can lead to labeling at inappropriate pH or osmolarity which can eliminate or substantially reduce labeling. Length of time of labeling relative to antibody concentrations is also important as is temperature of incubation. Even differences in prevailing "room temperature" may alter labeling. A shortened or lengthened time of incubation with primary or secondary antibody or with blocking reagent can also adversely affect the efficiency of labeling.

The labeling process is multistep, and each step must be carefully carried out. In instances where a particular protocol is unfamiliar to the investigator, it is not uncommon that certain steps may be accidentally or purposely omitted. Generally each step is important to the final result, and missing or skipping steps can have unintended results. All labeling, washing, and blocking steps need to be completed. Once good labeling is obtained via an existing protocol, the investigator can modify the protocol to eliminate possible unnecessary steps or streamline the process.

Care must be taken not to repeatedly freeze and thaw antibody as this can result in aggregation and/or denaturation of antibody. Long-term storage at temperatures just above freezing will often lead to a gradual loss of antibody activity. Each antibody preparation behaves somewhat differently. Some antibodies stored in the refrigerator may maintain their activity over several months or longer, while others show marked deterioration in days or weeks.

5.15.2 Background Labeling

A high level of background labeling is a second problem commonly encountered. This often results in substantial levels of labeling throughout the sample and/or significant amounts of labeling seen in controls. This can result from several sources. Fc receptors on cells in the sample may bind to Fc portions of the whole antibody molecule. This can occur with primary or second antibody or both. There is substantial cross species Fc-Fc receptor binding. For example, murine Fc, particularly IgG2a and IgG3, will bind strongly to human Fc receptors, while other murine antibody subtypes bind but at somewhat lower affinity (Lubeck et al. 1985). Use of Fab or $F(ab')_2$ antibody fragments which lack the Fc is one approach to eliminating Fc binding. Blocking of Fc receptors via the use of pre-immune serum can also be used.

Non-epitope-specific binding of an antibody via complement components can also produce background which can be eliminated via the use of Fab or $F(ab')_2$, which also lack the complement binding sites. Specific blocking of complement is also possible.

Nonspecific binding of antibody by hydrophobic bonding or by charge-charge bonding is also a possibility. At its isoelectric point, an antibody molecule is maximally hydrophobic and can bind to hydrophobic regions on other molecules. In addition the antibody molecules, even at their isoelectric point, may have charged regions within the molecule facilitating charge-charge interactions with oppositely charged regions in other protein molecules. Usually these nonspecific antibody-protein interactions can be reduced or eliminated with an increase in blocking agent concentration or via the application of additional blocking agents.

High background may also be a result of excessive concentrations of primary or second antibody. This may be remedied by simply decreasing antibody concentrations. If some, or all, of the specific antigen is diffusible, the antigen or even antigen-antibody complexes may redistribute throughout the sample. This may also result in diffuse labeling which appears nonspecific. This type of artifact can be reduced through the use of more aggressive fixation or a more thorough washing.

Autofluorescence or aldehyde autofluorescence can also produce a diffuse background or a specific pattern of nonspecific labeling. Autofluorescence is difficult to eliminate and may require the use of fluorescent labels with excitation/emission wavelengths that will not be confused with autofluorescence. Aldehyde or aldehyde-induced fluorescence can be eliminated by labeling specimens prior to fixation or after a very light fixation with formaldehyde. The use of cryo-preparative methodology which does not use chemical fixatives also may be employed. Coagulating fixatives such as ethanol may prove useful to eliminate fixative fluorescence, but the alcohols tend to extract or dissolve specimens. However for some antigens, alcohol fixation is required. The use of counterstains to mask background is also possible. Stains such as toluidine blue or pontamine sky blue have been recommended as potential counterstains to mask nonspecific background due to autofluorescence or aldehyde fixatives.

Some nondiffuse but nonspecific labeling can result from several sources. Fixation may produce or expose normally hidden cross-reactive antigenic sites resulting in labeling which may appear specific. The use of monoclonal primary antibody against selective epitopes may be used to reduce this type of nonspecificity. As above, labeling prior to fixation, labeling after a very light formaldehyde fixation, or cryo-preparative methodology which does not employ fixatives may reduce aldehyde-induced cross-reactive antigens.

Sequestered or trapped antibody may also appear similar to specific labeling. Antibodies may diffuse into specimens but subsequently have difficulty diffusing out of specific locations. Additional washing and attention to the specific location relative to cell and tissue structure may help to avoid misinterpretation of these labeling patterns. Phagocytosed antigen may appear in phagosomes or phagolysosomes of phagocytic cells if at least some of the antigen epitopes are resistant to degradation in the phagolysosome or if labeling is sufficiently soon after antigen ingestion so as to avoid significant destruction of antigen epitopes.

5.15.3 Specificity and Affinity

A third significant area where problems can arise relates to levels of specificity and affinity. The intensity of labeling may appear substantially lower than expected based on other measurements of antibody-antigen binding. This unexpectedly reduced binding may occur as a result of several factors. Antibody avidity and specificity can be checked via Western blots, ELISA, or both. Variations between techniques can be due to technique-dependent structural variation in the antigen or the antibody. Epitopes available when an antigen is in a gel or in solution may not be available in vivo. Thus antigens in ELISA or Western blots may label extremely well, but labeling in vivo is still substantially less than expected.

Specific epitopes may be masked or unavailable in the context of cells or tissues. This is more of a problem with monoclonal species since only one epitope is labeled. If that particular epitope is partially masked or unavailable in vivo, labeling can be reduced or absent, and conclusions relative to the presence or absence of a particular molecular species may be in error. Where molecules expected to be present are not labeled by a specific monoclonal antibody, it is useful to employ a polyclonal antibody as a check. In this case multiple epitopes on the molecule can be recognized by the polyclonal antibody. Since there is less of a chance that all epitopes within the molecule of interest are masked in vivo, there is a much greater probability that labeling will occur. However, as discussed previously, there is an increased chance of cross-reactivity and nonspecific labeling with polyclonal antibody preparations. Absorbing polyclonal antibody against the antigen and eluting the specifically reactive antibodies will eliminate antibody molecules which are not specific for the antigen of interest. This may reduce nonspecific binding and also increase overall specific binding with a concomitant increase in signal. Specificity can be checked prior to labeling using a Western blot with the specific antigen and with closely related antigens.

Finally for post-embedding labeling of sectioned samples, masking by the embedding media material is a consideration. With the exception of dried, unembedded frozen sections or sections where the embedding media is dissolved away, antibody access to epitopes is limited to the surface of the section. This severely restricts the number of exposed epitopes that can be labeled. Furthermore specific antigens or epitopes may interact with the embedding media such that they are masked. This further restricts labeling. Solutions include the use of different embedding polymers with different properties or the use of etching agents to dissolve a very thin layer of polymer on the surface of the section. This latter approach increases the amount of embedded material on the surface, and hence additional epitopes, if they resist the etching process, are available to the antibody labels.

5.16 Masked Epitopes and Antigen Retrieval

There are also a variety of antigen "retrieval" protocols which can be used to restore antigenicity of fixed and processed antigen. In some cases, especially when tissue has been extensively processed via aldehyde fixation followed by paraffin embedding tissues, antigens may become masked. In cases where embedding with paraffin interferes with labeling, it may be possible to simply fix the tissue and vibratome section (Chap. 4) with no further embedding. This also eliminates much of the harsh tissue processing with ethanol, clearing reagents, and heat that may affect antigenicity.

A large number of antigen retrieval protocols have been developed and typically involve treatment with combinations of heat, low pH buffers, and/or enzymes (IHC World: http://www.ihcworld.com/ epitope_retrieval.htm_). Antigen retrieval is usually performed on tissue sections but can also be applied to cell preparations (such as cytocentrifuge specimens) or tissue blocks. With all methods, the user should be aware that re-titering of antibodies is likely necessary, as the optimal dilution before and after antigen retrieval may be very different.

Heat is a major factor in breaking the aldehyde bonds and unfolding the protein antigens in the heat-induced epitope retrieval (HIER) methods, while the buffer conditions selected serve to maintain the unfolded conformation. The exact buffer and temperature conditions required for unmasking needs to be optimized for every antigen-antibody combination, as does the time of unmasking treatment. Some sources describe "universal" reagents that work well with a variety of antigens in commonly encountered samples (abcam product #ab208572; Namimatsu et al. 2005). Generally, these reagents can be considered to work well with specific antibodies (or panels of antibodies) but must be tested for their effectiveness with new antigens. Sample heating can be performed by boiling slides in a microwave or on a hot plate, either in a beaker or a steamer. Alternatively, they can be heated under pressure in a pressure cooker or autoclave. The length of exposure and number of cycles vary with the type of sample and stability of the antigens. Vigorous boiling can potentially damage tissue sections on slides. The buffers used in HIER vary from low to high pH, sometimes with the inclusion of calcium chelators such as citrate or EDTA. Citraconic anhydride has also been used.

Some retrieval methods can be performed at room temperature (RTER) and make use of acid treatment. Formic acid retrieval has been reported to work well with amyloid plaque deposits associated with Alzheimer's disease (https://www.protocolsonline.com/histology/immunohistochemistry-histology/formic-acid-antigen-retrieval-protocol/). The exact mechanism of unmasking in these methods is unknown.

Another approach is proteolytic-induced antigen retrieval (PIER). These methods make use of treatment with trypsin, pepsin, proteinase K, pronase, or other enzymes. PIER is considered to be harsher and to have greater potential to negatively impact tissue morphology. However, some antigens may produce better results or similar results more quickly using protease treatment.

It is important to consider that the conditions under which some antigens can be retrieved have the potential to harm others. The extent of epitope recovery varies between antigens and may vary between antibodies for a given protein. Epitopes for monoclonal antibodies are especially susceptible to conformational changes. Interpretation of the level of protein expression may be compromised by a variable degree of antigen retrieval. Likewise, studying the co-localization of proteins could be confounded if the two epitopes are differentially retrieved.

The level of tissue fixation can influence antigen retrieval. Equilibrium penetration of fixative into a tissue block can occur in minutes to hours, while complete fixation can take 24 h to 7 days. Less complete fixation may make epitopes more easily retrieved.

While many of these procedures can be carried out by hand, there are also commercial systems available such as the Lab Vision™ PT Module (Thermo Fisher Scientific, Waltham, MA). All antigen retrieval protocols treat samples (proteins) harshly in an effort to unmask the antigenic sites. Use of commercial systems and standardized chemistry provides some consistency and reproducibility in the procedure which is essential when comparing fluorescent labeling in various samples. Always keep in mind that antigen retrieval methods might also expose epitopes on molecules other than your antigen of interest creating false-positive signal. Thus, proper controls are even more critical when using retrieval techniques. The necessity of proper controls when using these procedures cannot be underestimated.

5.17 Fixed Epitopes

It has been suggested that the generation of antibodies against fixed epitopes may be helpful for the labeling of specific epitopes in fixed specimens. Unfortunately fixatives generate a number of fixation-specific epitopes that are the same in all fixed tissues resulting in substantial cross-reactivity. Possibly preparation of monoclonal antibodies for specific fixed epitopes may be useful; however, to date, this has not been the case.

Literature Cited

Albrecht RM, Meyer DA (2008) Molecular labeling for correlative microscopy: LM, LVSEM, TEM, EF-TEM, and HVEM. In: Schatten H, Pawley J (eds) Ch 6, Low Voltage Scanning Electron Microscopy. Springer Science, New York, pp 171–196

Bruchez M Jr, Moronne M, Gin P, Weiss S, Alivisatos AP (1998) Semiconductor nanocrystals as fluorescent biological labels. Science 281:2013–2016

Chalfie M, tyu Y, Euskirchen G, Ward WW, Prasher DC (1994) Green fluorescent protein as a marker for gene expression. Science 263:802–805
Chan WCW, Nie S (1998) Quantum dot bioconjugates for ultrasensitive nonisotopic detection. Science 281:2016–2018
Criscitiello MF, Flajnik MF (2007) Four primordial immunoglobulin light chain isotypes, including lambda and kappa, identified in the most primitive living jawed vertebrates. Eur J Immunol 37(10):2683–2694
Eppell SJ, Simmons SR, Albrecht RM, Marchant RE (1995) Cell surface receptors and proteins on platelet membranes imaged by scanning force microscopy using immunogold contrast enhancement. Biophys J 68:671–680
French DL, Laskov R, Scharff MD (1989) The role of somatic hypermutation in the generation of antibody diversity. Science 244:1152–1157
Green NM (1963) Avidin 1. The use of [^{14}C]biotin for kinetic studies and for assay. Biochem J 89:585–591
Hamers-Casterman C, Atarhouch T, Muyldermans S, Robinson G, Hamers C, BajyanaSonga E, Bendahman N, Hamers R (1993) Naturally occurring antibodies devoid of light chains. Nature 363:446–448
Hansen JD, Landis EE, Phillips RB (2005) Discovery of a unique Ig heavy-chain isotype (IgT) in rainbow trout: Implications for a distinctive B cell developmental pathway in teleost fish. Proc Natl AcadSci U S A 102(19):6919–6924
Heim R, Tsien RY (1996) Engineering green fluorescent protein for improved brightness, longer wavelengths and fluorescence resonance energy transfer. Curr Biol 6:178–182
Heim R, Cubitt A, Tsien R (1995) Improved green fluorescence. Nature 373(6516):663–664
Horsfall AC, Hay FC, Soltys AJ, Jones MG (1991) Epitope mapping. Immunol Today 12:211–213
Huston JS, Levinson D, Mudgett-Hunter M, Tai M-S, Novotny J, Maroglies MN, Ridge RJ, bruccoleri RE, Haber E, Crea R, Oppermann H (1988) Protein engineering of antibody binding sites: Recovery of specific activity in an anti-digoxin single-chain Fv analogue produced in *Escherichia coli*. Proc Natl Acad Sci U S A 85:5879–5883
IHC World (2018) http://www.ihcworld.com/epitope_retrieval.htm
Kandela IK, Albrecht RM (2007) Fluorescence quenching by colloidal heavy metal nanoparticles: Implications for correlative fluorescence and electron microscopic studies. Scanning 29:152–161
Kandela IK, Bleher R, Albrecht RM (2007) Multiple correlative immunolabeling for light and electron microscopy using fluorophores and colloidal metal particles. J Histochem Cytochem 55(10):983–990
Kandela IK, Bleher R, Albrecht RM (2008) Correlative light and electron microscopy imunolabeling on ultrathin cryosections of skeletal muscle tissue. Microsc Microanal 14:159–165
Kindt TJ, Goldsby RA, Osborn BA (eds) (2007) Kuby Immunology, 6th edn. WH Freeman and Company, New York
Linscotts Directory of Immunological and Biological Reagents (2018) Linscotts USA. https://libraries.usc.edu/databases/linscotts-directory-immunological-and-biological-reagents https://www.linscottsdirectory.com
Lubeck MD, Steplewski Z, Baglia F, Klein MH, Dorrington KJ, Koprowski H (1985) The interaction of murine IgG subclass proteins with human monocyte Fc receptors. J Immunol 135:1299–1304
McCafferty J, Griffiths AD, Winter G, Chiswell DJ (1990) Phage antibodies: filamentous phage displaying antibody variable domains. Nature 348:552–554
Namimatsu S, Ghazizadeh M, Sugisaki Y (2005) Reversing the effects of formalin fixation with citraconic anhydride and heat: a universal antigen retrieval method. J Histochem Cytochem 53:3-11
Nilson BHK, Lögdberg L, Kastern W, Björck L, Åkerström B (1993) Purification of antibodies using protein L-binding framework structures in the light chain variable domain. J Immunol Methods 164:33–40
Piatkevich KD, Verkhusha VV (2011) Guide to red fluorescent proteins and biosensors for flow cytometry. Methods Cell Biology 102:431–461
Reiter Y, Brinkmann U, Jung S-H, Lee B, Kasprysyk PG, King CR, Pastan I (1994) Improved binding and antitumor activity of a recombinant anti-erbB2 immunotoxin by disulfide stabilization of the Fv fragment. J Biol Chem 269:18327–18331

Chapter 6
Digital Imaging

W. Gray (Jay) Jerome

6.1 Introduction

Image capture and rendering in confocal microscopy is a digital technique. To optimize confocal image acquisition, the operator needs to have a basic understanding of what constitutes a digital image and how digital imaging can be employed effectively to help visualize specific details. In this chapter, we will introduce the topic of digital images, discuss the basic components of the digital image, and provide some basics on how to optimize collection of these components with regard to confocal microscopy.

6.2 Analog Versus Digital Information

An analog signal is a continuous one. Its possible values can be $-\infty$ to $+\infty$ in infinitely small increments. This means that values such as 1, 2, 2.02, and 3.0000023 are all possible. In fact, the set of possible values includes all real numbers. In comparison, a digital signal is composed of discrete elements that have values of one unit or some whole multiple of one unit. Thus, you cannot have a value that is a fraction of one unit, such as 0.33, or a fraction of a whole multiple, such as 2.57. Of course, at this point we have not defined the basis of our unit. In fact it could be a micrometer, a yard, or 2.65 feet. However, once the unit is defined, then all elements of the set must be in whole multiples of that unit, and no element can be less than one unit in size. Thus, the value of 1 unit (1 quantum of information) defines the lower limit of resolution.

W. G. Jerome (✉)
Department of Pathology, Microbiology and Immunology,
Vanderbilt University School of Medicine, Nashville, TN, USA
e-mail: Jay.Jerome@Vanderbilt.edu

W. G. Jerome, R. L. Price (eds.), *Basic Confocal Microscopy*,
https://doi.org/10.1007/978-3-319-97454-5_6

Analog to digital conversion consists of converting the set of analog values to a set of digital values. Two steps are involved in this conversion. The first is determining the value of one unit. The second step involves deciding the means by which the analog values are assigned to one of the discrete digital values. This process is depicted in Fig. 6.1 for analog values varying with respect to an X and a Y parameter. The value of 1 quantum in the X and Y directions is indicated by the width of the tic marks. The Y value for a specific quantum on the X-axis is determined by taking all the analog values occurring within the range of a single quantum on the X-axis and converting them to the nearest quantum along the Y-axis. Often this is done by taking the average of all the values within one X quantum and then rounding to the nearest Y quantum value. In the low-resolution conversion of Fig. 6.1, a relatively high value has been chosen for 1 quantum of X and Y. In the high-resolution conversion, a smaller value has been chosen. Of course, this smaller value results in a closer approximation of the digital graph to the original analog information. Similarly, selecting a high-resolution value in the confocal microscope software will result in a better approximation of the analog information viewed

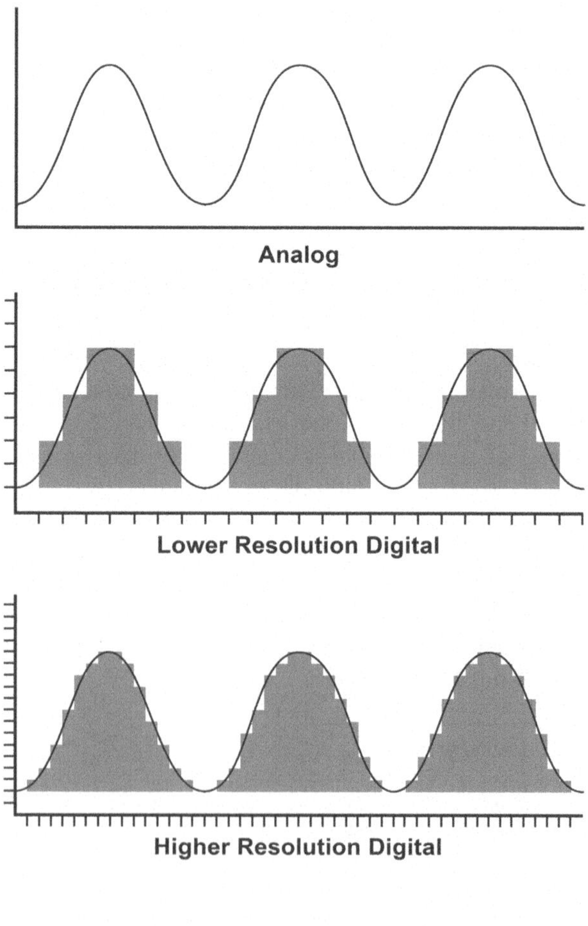

Fig. 6.1 Analog to digital conversion of a linear signal

through the microscope with our eyes. However, as discussed below and in other chapters of this book, there are guidelines in selecting the appropriate resolution values available in the software. It is not always the best practice to select the highest digital resolution available.

6.3 Pixels and Pixel Size (Spatial Resolution)

In the case of a two-dimensional micrograph, the spatial information in the image is divided up into a series of discrete two-dimensional picture elements (pixels). A pixel is one unit wide by one unit high. A pixel does not have to be square; in other words one quantum in width does not necessarily have to be the same as 1 quantum in height. However, in microscopy, square pixels are usually employed, and we will confine our discussion for the rest of this chapter to square pixels.

A two-dimensional digital image is made up of a mosaic of pixels. Besides the determination of the X and Y quanta, a third value must be determined, that of what value to paint in the pixel. Whatever the value, it will be uniform within the pixel. For a fluorescence image, we must convert the analog brightness information in the original sample to a digital representation. To do this, all of the values within a pixel area are averaged to a single value as depicted in Fig. 6.2. For a gray-scale image,

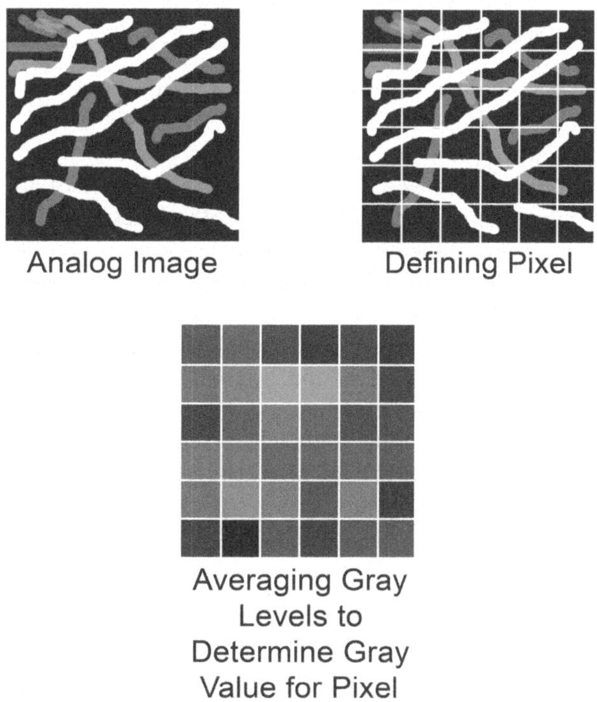

Fig. 6.2 Analog to digital pixel conversion

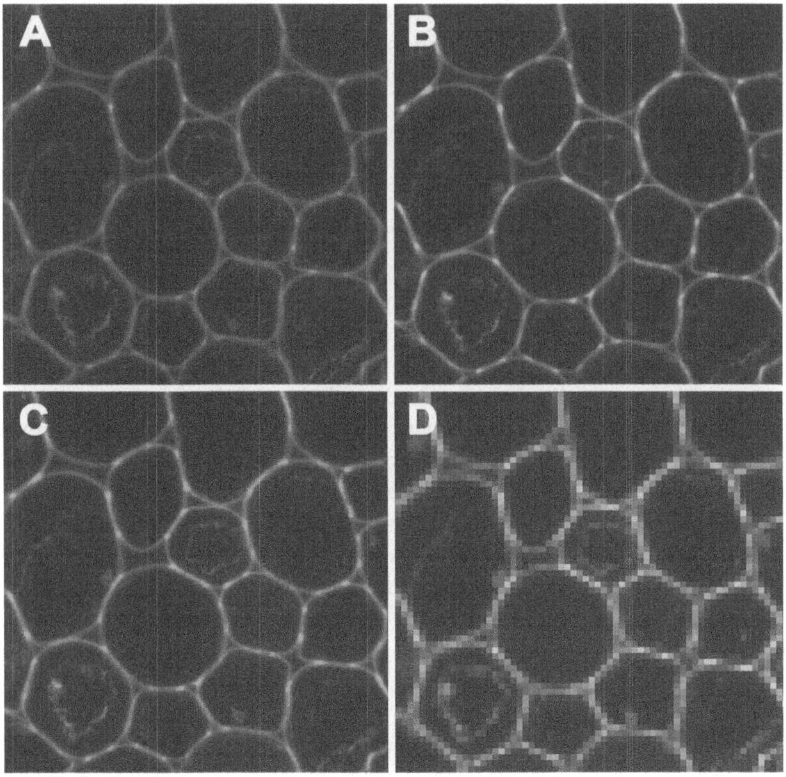

Fig. 6.3 Effects of different pixel sizes on image resolution

the value would be some gray value between black and white. In the simplest conversion, this would involve using the average of all the gray values in the pixel. For example, a pixel that contained 50% black and 50% white would be averaged to a gray value at the middle of the scale. Because all of the details within a pixel have been converted to a single value, the smallest detail that can be represented in the digital image is no smaller than one pixel.[1]

It follows, then, that the smaller the pixel size, the more faithfully will be the digital representation of the original analog information. Figure 6.3 shows the results of different pixel sizes on an image. The pixels in Fig. 6.3a are 0.05 mm in height by 0.05 mm in width. The pixels in Fig. 6.3b are 0.11 mm high × 0.11 mm wide. With the larger pixel size in Fig. 6.3b, some of the fine detail of the image is lost. This is even clearer in Fig. 6.3c where the pixels are 0.22 mm × 0.22 mm.

[1] In microscopy, the Nyquist-Shannon sampling theorem (Shannon 1998) tells us that we cannot accurately detect microscopic information smaller than 2 pixels by 2 pixels of the detector. This is related to the variability in signal collection and not specifically digital imaging, so we will ignore this signal processing issue until the discussion of detectors in Chap. 7.

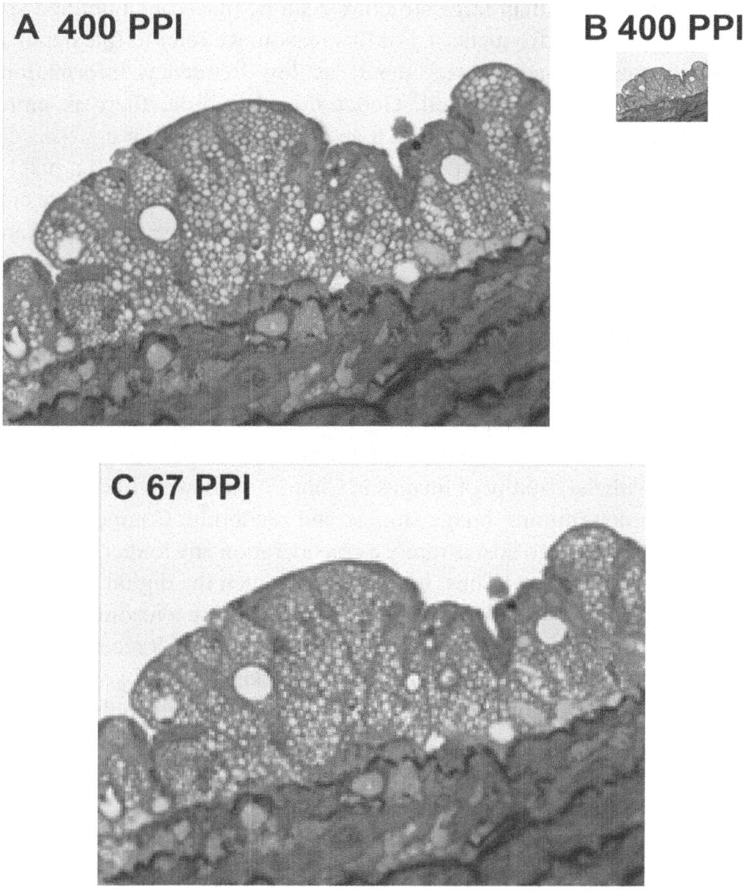

Fig. 6.4 Pixels per inch and image magnification

Finally, in Fig. 6.3d, the pixels are large enough (0.9 mm × 0.9 mm) to be visible. Notice that each pixel in Fig. 6.3d contains a single uniform gray tone.[2]

The number of pixels per inch (or per mm) is a useful value for comparing the amount of information contained in two images. However, in microscopy this term is deceptive. It is only an accurate comparison if the two images are magnified to the same extent. In Fig. 6.4, both images a and b have 400 pixels per inch (PPI). However, if we magnify Fig. 6.4b to the equivalent size as Fig. 6.4a (the result is displayed as Fig. 6.4c), the final rendering only has 67 PPI. Fine details present in Fig. 6.4a are not present in either Figs. 6.4b or c. Since, in microscopy, we are always dealing with magnification of images, pixels per inch is only a useful comparator of the amount of information contained in two images when the microscopy magnification of both images is equivalent.

[2] Most analog to digital conversions include some smoothing and blending functions, but we will ignore these for the purposes of this discussion.

More small structures than large structures can be placed in a defined area (i.e., they can occur with more frequency). For this reason, we refer to fine detail as high-frequency information and grosser detail as low-frequency information when describing the object being imaged. Under this definition, there is more high-frequency information contained in Fig. 6.4a than in Fig. 6.4b or c.

At first blush, it would seem that you would want to use as small a pixel as possible for imaging. However, the smaller the pixel size, the more pixels are contained in the image. This requires more computer memory for storage and more bandwidth for transferring the image information to the rendering hardware (printer, monitor). In addition, it requires more time to capture the information during the microscope session. All of these parameters need to be considered when determining the optimum pixel size for an application. It is inefficient to store an image at a resolution greater than the capability of the image capture method. The excess empty resolution costs both time and memory. For most applications, you should always attempt to match the resolution requirements with the final use of the image.

We will deal with the capture of images in Chap. 7. Here we will consider practical considerations in determining image storage and rendering. Computer memory has become extremely cheap, so cost is rarely a consideration any longer. However, bandwidth still remains a concern. Thus, it is useful to match the digital resolution to the output requirements. Most modern computer monitors have a resolution in the range of 60–130 PPI (most being between 90 and 100 PPI), although recently some monitors with >300 PPI resolution have come on the market. Thus, if you are only going to display an image on a monitor and the monitor being used is only capable of 100 PPI resolution, it is wasteful to store images at greater than this resolution. The pixels per inch displayed by computer projectors currently use very similar standards to computer monitors, and so the same principles apply. In microscopy, however, monitor display is usually not the only or most important endpoint for our images. We usually will also print our images, enlarge selected areas of our image, and often want to collect quantitative information from our captured images. All of these functions are capable of displaying or analyzing much higher resolution information.

Printers are calibrated in dots per inch. This is not the same as pixels per inch, and the two systems should not be confused. However, for digital printing, a stored image with a resolution of between 300 and 600 PPI is preferred. At this resolution, the human eye will generally not resolve the individual pixels in the print when the print is viewed at a normal viewing distance. Of course, if one gets very close to the image, the individual pixels can begin to be apparent. It is estimated that the human sight can resolve about 1,200 PPI at a very close viewing distance. For normal viewing distances, though, a good rule of thumb is to prepare images at 600 PPI for printing. This is also true of images for submission to journals unless specifically instructed otherwise by the journal's editorial policy.

A computer, of course, does not have the same resolution constraints as the human eye. It sees the image as a series of numbers and so can discriminate differences as small as one 1 unit. So, if the final use of the image will be to have the computer quantify physical characteristics, such as size, area, volume, or co-localization statistics, then one should store images at the maximum resolution feasible taking into consideration only the sensitivity of the image capture device and any constraints on the resolution imparted by the capture device. Resolution limits imparted by the microscope and image collection method are discussed in Chap. 7.

6.4 Pixel Density

For the analog to digital (A to D) conversion in Fig. 6.2, each pixel was filled with the average of the gray value contained within the pixel area. However, we did not define the range of gray levels that were possible. In computerized digital imaging, the range of possible values (gray levels) is based on binary mathematics, since computers are binary instruments. In binary space, the only values are 0 and 1. In other words, a one digit number in binary space can have a value of 0 or 1. This means that we can only depict two gray levels. If we want to depict a range from black to white, then that means only black or white is possible. By convention, black is usually designated as the lowest value and white as the highest. This situation is depicted in Fig. 6.5a. If, however, we divide our scale from black to white using a two-digit binary number, we have four possibilities: 00, 01, 10, or 11. Each digit is called a bit; Fig. 6.5b depicts the possibilities for a two-bit image. Increasing to two bits expands our range of possibilities so that we can now depict black, dark gray, light gray, and white. Three digits expand the range (bit depth) even further, so we can now represent even more gray levels (Fig. 6.5c). The range increases by 2 to the power of the number of bits. Thus:

1 digit = $2^1 = 2$ (1 bit of information)
2 digits = $2^2 = 4$ (2 bits of information)
3 digits = $2^3 = 8$ (3 bits of information)
4 digits = $2^4 = 16$ (4 bits of information)

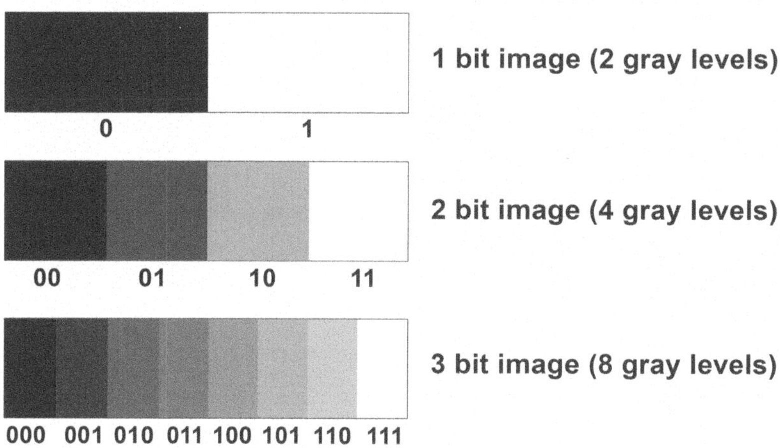

Fig. 6.5 Gray-scale values for various digital bit levels

As the number of possible values available to define the range from black to white increases, the range encompassed by each single quantum decreases. In fluorescence microscopy, smaller gray-scale quanta allow for the depiction of subtler differences in fluorescence intensity. An 8-bit image allows the depiction of 256 (2^8) gray levels. Based on the ability of the human eye to discriminate differences in contrast, this is considered the minimum required to accurately depict visual information. Of course, machines can discriminate more accurately, so, for quantitative work, even more gray levels are preferable. Many confocal systems are capable of accurately collecting 12-bit images (2^{12} or 4096 gray tones) or even higher bit depths. The advantages of 12-bit images will be discussed in subsequent chapters.

Most microscopists are not facile at working with binary numbers. For this reason, the binary representation of a pixel value is usually converted to a base ten number (Fig. 6.6 a). In the case of a 2-bit image (four gray levels), 0 is used to indicate black, and 3 indicates white. The values of 1 and 2 indicate gray levels based on dividing the range from black to white into four representative gray levels.

In modern computing, 8 bits is termed one byte. Thus, each pixel in an 8-bit images contains one byte of information. If you have an 8 inch by 10 inch gray tone image stored at 600 PPI, the image size would be 28,800,000 pixels. This is because the image will be 4,800 pixels wide × 6,000 pixels high. This equals 28,800,000 pixels (28.8 megapixels). Since each pixel contains one byte of information, it takes 28,800,000 bytes (28.8 megabytes) to store the gray-scale information of the image.

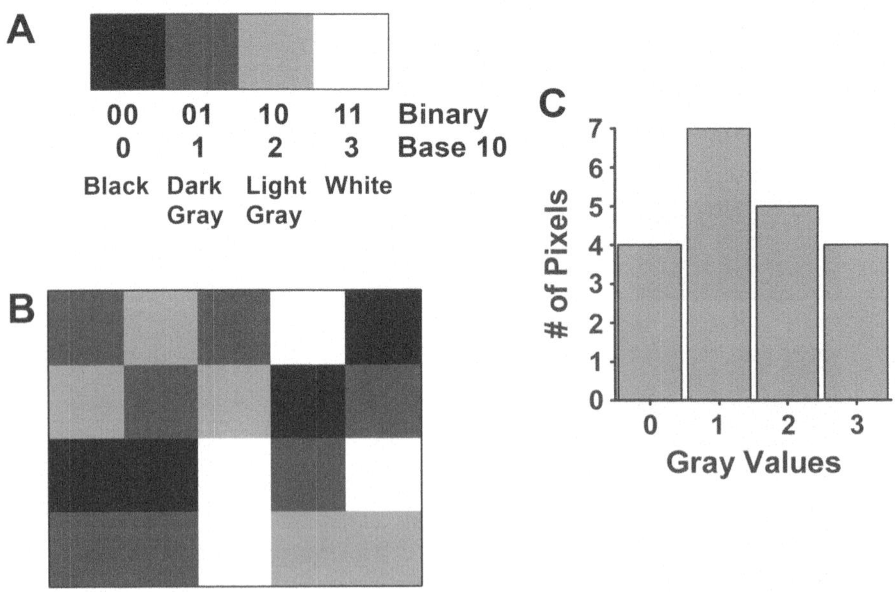

Fig. 6.6 Digital image histograms

6.5 Pixel Histogram Analysis

A useful way to analyze the gray levels in an image is by plotting a histogram of the number of pixels that have a specific gray level (Fig. 6.6b, c). Figure 6.6b shows a two-dimensional array of pixels from a two-bit image (four possible gray values). Figure 6.6c shows a histogram indicating the number of pixels in Fig. 6.6b which have values of 0, 1, 2, or 3. Note that the histogram does not provide any spatial information. We do not know the location of the four black pixels indicated by the bar height for the X value of 0 (black); we have only recorded that there are four black pixels somewhere in the image.

6.5.1 Histogram Stretch

Despite its lack of spatial information, the histogram can be very useful in analyzing and improving an image. Figure 6.7a shows an 8-bit (256 gray levels) image and the histogram from that image. The image has limited contrast with all of the gray level values bunched around the center (medium gray) values. Increasing the difference in density between two adjacent pixels (contrast) can be useful for increasing the visibility of structures. This can be done, based on the gray level (intensity) histogram, by stretching out the values to cover the full dynamic range of 256 gray levels (0 black to 255 white). Figure 6.7b illustrates the increased image contrast that results from spreading the gray values out to use the full dynamic range and also displays the new histogram after spreading.

Most imaging software has at least one routine for reassigning pixel values to stretch or compress the gray level histogram. An important point about histogram stretching for scientific imaging is that histogram stretch does not alter the relative relationship order of one pixel value to another; it only increases the difference in gray tone (contrast) between two pixel values and thus makes the difference more visible to the human eye. Thus, for simply discriminating the location of some structure or examining fine detail, the data contained in the image has not been altered in any significant way; individual structures have simply been made more visible. However, if you are using the absolute pixel gray value as some measure of concentration of one analyte compared to another, you have altered this data by performing a histogram stretch or any other method of reassigning pixel values. For example, in Fig. 6.7b, pixels that originally had a pixel value of 200 now are brighter with a pixel density value 227. Thus, even though procedures such as histogram stretch are relatively benign, they still need to be done with full knowledge of what values are being changed. In looking at the histogram of the stretched image in Fig. 6.7b, you will note that there are now gaps in the histogram. So, although structural information stored in the image file have been made more visible, there are structural details that were not captured. This information cannot be regained. Thus, it is preferable to capture the original image using the full available range of gray values rather than having to extensively stretch the image after collection. Figure 6.7c

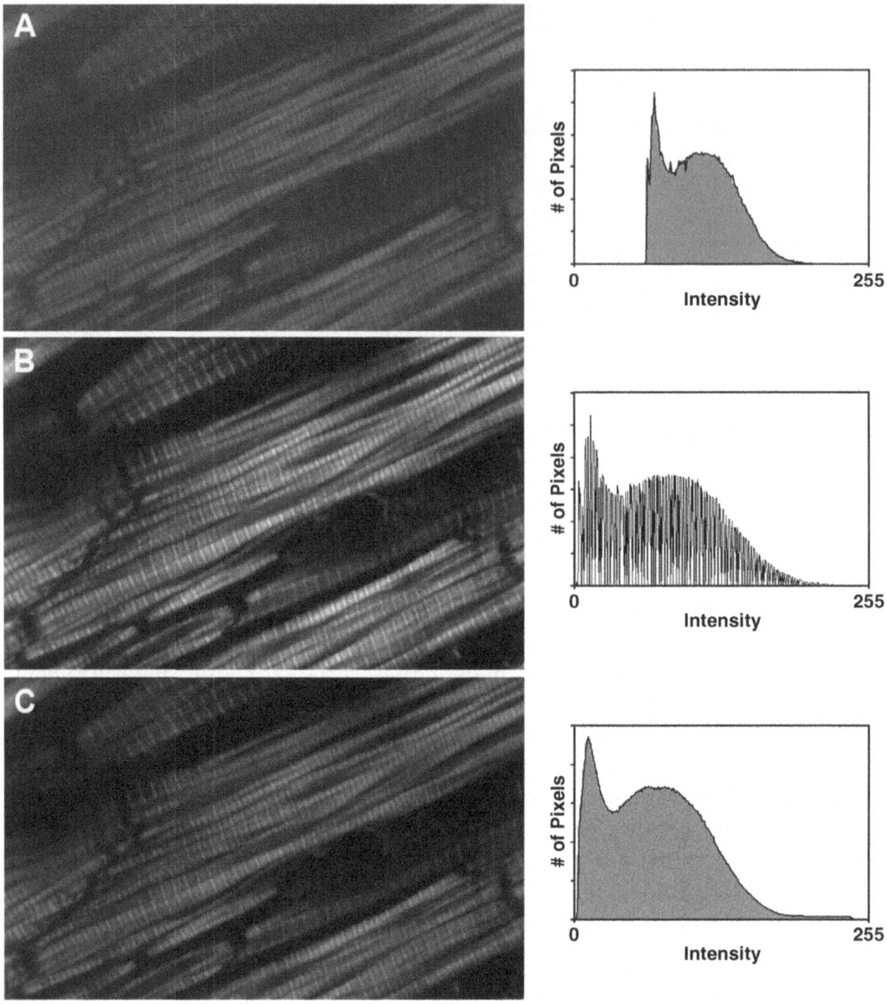

Fig. 6.7 Histogram stretch

shows the additional detail that is present when the image is correctly captured. Chapter 9 discusses how to set confocal imaging parameters to insure that the full possible dynamic range is captured in the image.

6.5.1.1 Digital Image Gamma

Gamma is another useful, and relatively benign, parameter that can be used to increase the visibility of specific parts of an image. Gamma is the relationship between the detected gray level (input value) and the gray level that is rendered in the final image (output value) as displayed on the computer screen or in a digital print (Fig. 6.8). Gamma is a power function. In an 8-bit image, gamma determines the output (rendered value) based on the equation:

$$y = 255\left(\frac{x}{255}\right)^{\gamma} \tag{6.1}$$

where y = output value
x = input value
γ = gamma

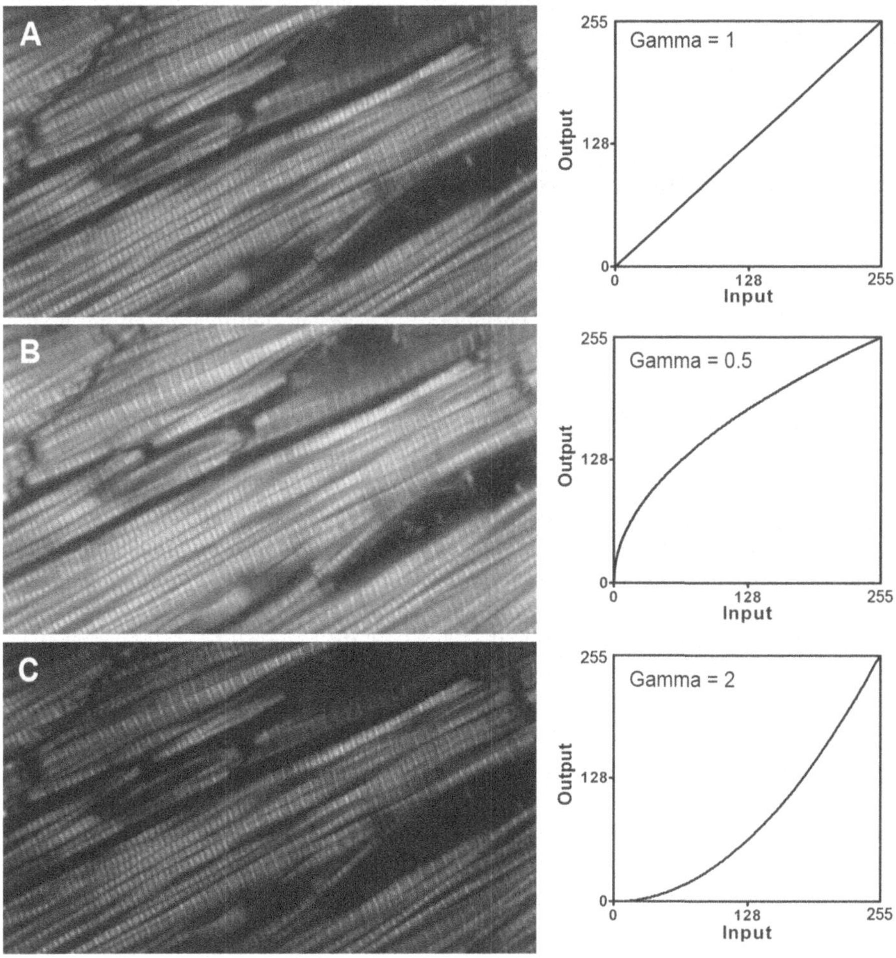

Fig. 6.8 Alteration of image gamma

With a gamma of 1, there is a one-to-one relationship of detected level and rendered level (Fig. 6.8a). An input gray value of 50 would be displayed as 50, and a gray value of 215 would be displayed as 215. In contrast, with a gamma of 0.5 (Fig. 6.8b), an input value of 50 would be displayed as 113, and 215 would be displayed as 234. A gamma less than one helps spread out the information in the darker regions to make it more visible but reduces the difference in intensity between values in the lighter regions. Conversely, a gamma greater than one (Fig. 6.8c) spreads out values in the lighter regions and compresses the darker values. As with histogram stretching, a change in gamma does not change the relative order relationship of the original data. Moreover, with gamma, the original image can be regained by doing the inverse operation[3]. The inverse is called a gamma correction. Gamma correction routines in some software packages are also called Gamma, so don't be perplexed by this confusing usage. Importantly, changes in gamma do not affect the lowest and highest values. A value of 0 remains 0, and a value of 255 remains 255. Thus, changing the gamma does not alter the dynamic range of the image. When necessary, histogram stretch should be done before changing the gamma.

6.5.2 *Avoid Digital Contrast and Brightness Functions*

Histogram stretch and gamma allow fine control of what detail is displayed in the output image without loss of dynamic range. This is not true of most functions that allow the adjustment of contrast and brightness. These can significantly compress the dynamic range, and we suggest that use of these contrast and brightness functions be avoided. Figure 6.9 demonstrates the effect on dynamic range of a standard method of digital brightness (Fig. 6.9b) and contrast (Fig. 6.9c) control found in a number of popular imaging software packages. To increase contrast, the software shifts the entire register of pixel values toward brighter pixels. However, since you cannot get brighter than white (digital value 255), all the top shifted values become 255. Thus, there is no contrast difference, and what was a difference in the very light gray values in Fig. 6.9a have become white and are lost. To decrease brightness, all the values are reduced. In this case all of the very dark gray pixels become black, and information is lost. Similarly, to alter contrast, the slope of the gamma line is increased or decreased (Fig. 6.9c). Both increasing or decreasing the slope to increase or decrease contrast in the output image results in a loss of dynamic range. To increase contrast, the dynamic range is compressed. To decrease contrast in the output image, values at the low and high ends of the range are converted to black or white, respectively. Once these contrast and brightness functions are run on the image, they cannot be reversed and the lost information cannot be regained. The loss of dynamic range and irreversibility of the functions is why we encourage use of histogram stretch and gamma correction functions to enhance specific details in an image.

---

[3]Actually, due to rounding errors, there may be a small difference between the original image and the gamma-correction recovered image, but these differences are generally very small.

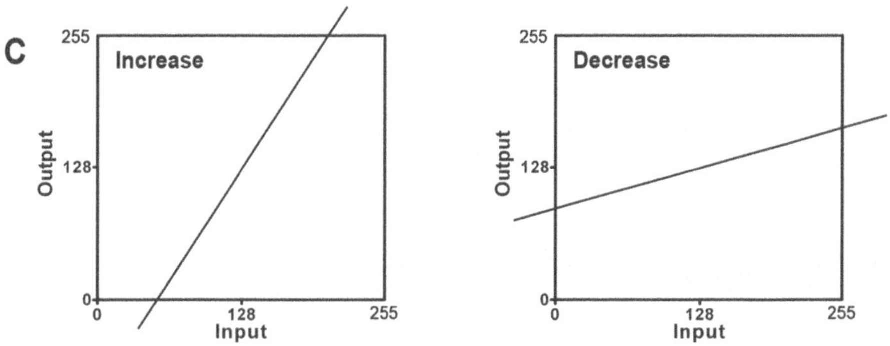

Fig. 6.9 Image brightness and contrast control. (**a**) Dynamic range of unchanged image. (**b**) Effect on dynamic range of increasing or decreasing brightness. (**c**) Effect on dynamic range of increasing or decreasing contrast

6.6 Use of Histogram Stretch and Gamma Changes in Confocal Imaging

Histogram stretch and alterations in gamma are useful tools for making subtle details in an image more visible to a human observer. Gamma changes are particularly useful. The human eye does not have a linear response to light. Our eyes are more sensitive to variations in dark gray levels than equivalent differences in the brighter registers. The eye responds with a gamma of about 2.2. Scientific digital cameras are generally linear capture devices. This is critical for quantitative work but not necessarily for presenting images to an audience. The makers of computer monitors, printers, consumer-grade cameras, and other digital hardware recognize this and often add a gamma function which converts the linear captured information into a presentation that more closely matches what is seen through the microscope. This is a useful tool, but one needs to be aware of the degree to which a correction is employed when using image information to make quantitative or semiquantitative assessments. The gamma correction employed by a piece of software or hardware is information that is generally not made readily available to the end user. For recreational imaging, it is probably not very important. However, for scientific evaluations, it is always worth digging into the product literature or contacting the manufacturer to determine what corrections a software program, monitor, or printer are imparting to the image data and how to turn those features off, if necessary. Alternatively a test sample with known characteristics can be imaged, and the resultant image examined to determine how the collection software has "corrected" the image. This is a tedious process, but it only has to be done once for each camera system.

In practice, histogram stretch and gamma functions should be used to enhance the scientific usefulness of an image by bringing out additional detail. However, any quantitative analysis or comparison of pixel brightness must be done prior to applying any histogram stretch or gamma functions. Moreover, these and all other image processing steps should be done on non-compressed copies of the original image. The original image should always be stored, unaltered, on archival quality media such as a "write once" CD or DVD. The image should be stored in a lossless image format such as the Tagged Image File Format (TIFF). Image formats are discussed later in Sect. 6.9.

6.7 Image Voxels

A major strength of confocal microscopy is the ability to image defined planes within the sample. Although the planes are thinner than those of a widefield imaging system, they do have a finite height. Thus, confocal images produce 3-D data. The depth of the image (not to be confused with pixel gray-scale depth), combined with the X and Y dimensions, defines a digital volume element (voxel). A voxel is the 3-D analog of the 2-D pixel (Fig. 6.10). In the same way that a pixel represents the average luminance within an area, a voxel's density will be the average of the luminance values within a defined volume. As such, a voxel represents the smallest unit of volume information that can be depicted in a 3-D digital reconstruction of an analog object. Also, like pixels, the dimensions of voxels are defined in whole multiples of a unit value (quantum). However, again like pixels, the X, Y, and Z quanta defining the voxel do not have to be the same. In most cases, we want to use the

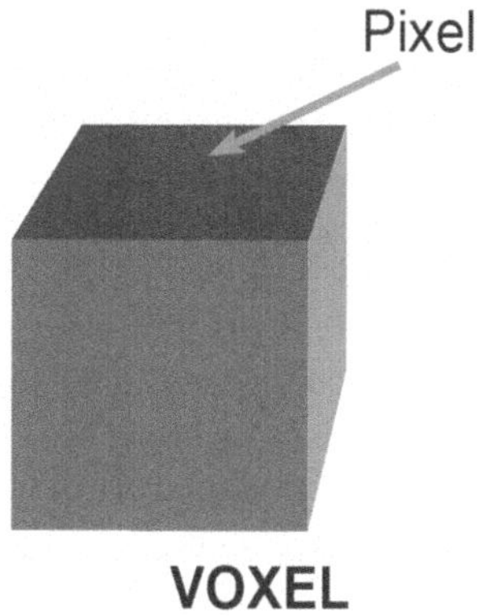

Fig. 6.10 Voxel and pixel relationship

same value for X and Y. However, as discussed in Chap. 7, in microscopy, the Z quanta is usually restricted to a larger value by the physics of microscopy imaging.

Since voxels have three dimensions, voxel information can be used to reconstruct a single plane of the sample, and it also provides information for 3-D reconstructions of larger volumes. If sequential planes are collected, they can be stacked on top of each other to produce a 3-D map of the entire specimen volume. Moreover, the digitized value for a voxel can be transformed in the same way pixel values can be transformed, and the voxel histogram and gamma can be manipulated using the same techniques used for pixels. The methods by which the confocal microscope collects voxel information are discussed in Chap. 9. The constraints on voxel size imposed by the physics of microscope imaging are described in Chap. 7, and the methods of working with voxels to produce digital 3D reconstructions are covered in Chap. 10.

6.8 Color Images

So far we only have dealt with gray-scale images, those that are composed of pixels or voxels with various shades of gray. Digital color microscope images are produced by combining color information encoded as multiple gray-scale images. The two most important methods (color spaces) for microscopists are the Red, Green, and Blue system (RGB color space) and Cyan, Magenta, Yellow, and Key (black) system (CMYK color space). RGB is the mechanism used for computer screens and the one closest to human and microscopic visualization. It is the method implemented on most commercial confocal microscopes. In contrast, CMYK is the system used by most printers.

The RGB system is an additive color system. Black is the absence of any color information, while adding equal contributions of red, green, and blue produces white. Having only red information, and no green or blue, produces red. Likewise, having only green or only blue produces green or blue, respectively. However, by adding certain proportions of red, green, and blue, we can produce a variety of colors. This is equivalent to combining three beams of different colored light. In practice, in confocal microscopy, three gray-scale images representing red, green, and blue, respectively, are combined. The hue and brightness of each resulting color pixel is the result of the addition of the information contained in each of the three overlain pixels. Thus, for a gray scale of 0–255, an RGB image will have 3 bytes of information for each pixel: one byte for red, one for green, and one for blue. As discussed in Chaps. 7 and 9, in confocal microscopy, we assign the gray-scale density value based on the number of photons collected by the detector, and the hue is assigned arbitrarily but usually based on the specific color filters used to collect the signal.

The CMYK system is a subtractive process. It is equivalent to mixing paint. Conceptually, the process begins with a blank white sheet on which the image will be produced. In the case of a digital image, this means the starting pixel is white. To this white pixel, a certain amount of cyan, magenta, and yellow color is added. Equal contributions of each produce black. Since it mimics the printing process of applying ink to paper, the CMYK system is used by printers. In most printing processes, black ink is substituted for equal amounts of cyan, magenta, and yellow. This saves money on ink and produces deeper dark tones.

Microscopy uses additive light and so is best done in the RGB color space. However, some journals require CMYK images for publication because this system matches the printing process. Although there are similarities between the RGB and CMYK systems, they are not identical. When converting between the two systems, such as when preparing micrographs for publication, the results of the conversion should be carefully checked to make sure the converted image retains accurate color information. Printers also convert the RGB information in a stored image to CMYK when printing. This sometimes requires some readjustment of the images to make sure the printed image is faithful to what was viewed in the microscope or on the computer screen.

A 24-bit RGB color image (8 bits of red information, 8 bits of green information, and 8 bits of blue information) is capable of coding for approximately 16.7 million different colors. Unfortunately, limitations of the capture and rendering devices (cameras, scanners, printers, monitors, etc.) generally are not capable of accurately capturing or depicting 16.7 million different values. Since the deficiencies are not consistent across capture and rendering platforms, this further complicates the accurate reproduction of color information. For most scientific imaging, however, it is sufficient to realize that accurate color transfer is not achievable without a great deal of effort but that this level of accuracy is generally not required for confocal studies, especially since confocal imaging usually collects each set of color information separately. This is explained further in Chaps. 7, 8, and 9.

In order to reasonably accurately depict the colors of the object under investigation, the microscopist needs to understand a few additional concepts about color rendition and color management. These concepts include Hue, Saturation, Brightness, and Luminance.

Hue is what defines the color. All blues have the same hue.

Saturation determines how intense the color appears. A fully saturated color is deep and brilliant. Less-saturated colors appear faded compared to the fully saturated color.

Brightness is a measure of how light or dark the color appears. A brightness value of 0 indicates black.

Luminance describes the perceived brightness. The difference between luminance and brightness is that luminance takes into consideration the color sensitivity of the human eye. For instance, the human eye is much more sensitive to light in the yellow-green range than it is to blue.

Of course, any color in a confocal image is usually artificially added through the use of look-up tables (LUTs). The high-resolution detectors used in modern confocal microscopy detect only the presence of photons. They do not measure the wavelength of those photons and so cannot assign color to the photon. The color component is added to each pixel or voxel by the use of LUTs available in the software based on parameters the microscopist has defined. If a red filter is in place, it is useful, but not obligatory, to display the value for the photons collected as the red information in an RGB image.

6.9 File Formats

The file format is the set of instructions for storing and retrieving the image information. Formats can be broadly categorized into generic and proprietary formats. Proprietary formats are those that are system or software dependent. The Adobe PSD format is an example of a proprietary format. Many of the confocal microscope manufacturer's software also include proprietary formats. The Zeiss LSM and newer CZI format, the Leica LIF, and Nikon ND2 formats are examples. These machine-specific formats allow the software to acquire and store additional information about the image as part of the stored image. In the case of confocal microscopes, this often includes a wealth of information about the microscope settings used to collect the image. Having information about the pinhole diameter, magnification, laser settings, filters, etc. used during image collection is extremely useful. However, the downside of proprietary formats is that the proprietary software is required to decode the stored image. This makes it difficult to share the images with others or to import the images into other software programs. Most confocal microscope manufacturers now also make available a free browser that will recognize and display their proprietary formatted images. Although these free browsers have limited capabilities, they do at least allow basic processing of data and sharing of images with others.

The alternative to proprietary image formats is generic formats. These are formats that are independent of machine, operating system, file system, compiler, or processor. Moreover, images stored in generic formats can be outputted to any hardware (printers, monitors, etc.) that recognize the format. Most modern image formats also maintain backward compatibility as the format evolves. Luckily, most confocal software also allows saving a copy of an image in one of these generic formats.

There are numerous generic formats, but the two most important for scientific imaging are the TIFF and JPEG (Joint Photographic Experts Group) formats. These formats are recognized by almost all imaging programs, and most proprietary software allows the option of saving in these formats. For scientific imaging, the key difference between the two is that, in the TIFF format, the image information is not changed when the stored image is created. In contrast, the JPEG image format compresses the image so that less computer memory is required to store the image. When computer memory was expensive, compression was very useful. JPEG allows compression of the image at ratios >1:200. This compression, though, comes at the expense of image information. The goal of JPEG storage is not to store every bit of information in the digital image. Its goal is merely to store and reproduce the format in a "photorealistic" form. This means that the JPEG image, when rendered from storage, will "look" almost identical to the original. However, since your image is your data, just having an image look like the original is usually not sufficient. Figure 6.11a shows a TIFF image, and Fig. 6.11b shows that same image after conversion to JPEG format. Figure 6.11c shows the subtraction of the two images from each other. All the nonblack pixels in Fig. 6.11c indicate pixels that were altered by the JPEG compression. It should be clear from Fig. 6.11c that scientific imaging should primarily employ lossless storage techniques like TIFF. We suggest TIFF because it is the most serviceable and widely accepted format for scientific image storage and use.

Because it degrades image information, the JPEG format should be avoided for most scientific imaging applications. Although the JPEG format allows one to choose the degree of compression (the higher the compression ratio, the more information that is discarded when storing the image), there is no option in the JPEG format for a lossless saving of the image data. It is also important to note that each time a JPEG image is saved, the program again runs the compression algorithm. Thus repeated saving of an image in JPEG format leads to additive degradation of the image information. The image information loss is the reason why the Ninth Commandment of the Confocal Ten Commandments described in Chap. 1 is "The JPEG image format is Evil." However, sometimes you need to dance with the devil. This is admissible as long as you recognize the consequences and confirm that the changes occurring do not affect the conclusions drawn from the image. For instance, emailing a 10-megabyte image to a colleague for review may not be accommodated by many email systems. In this situation, JPEG compression to a 50-kilobyte image would be extremely useful. However, quantitative analysis of image information should never be done on a JPEG image. Moreover, as discussed above, it is a good rule of thumb never to resave a JPEG image. When you need to revisit the image, always go back and make a copy of the original TIFF image, and do any post-collection analysis on the TIFF copy.

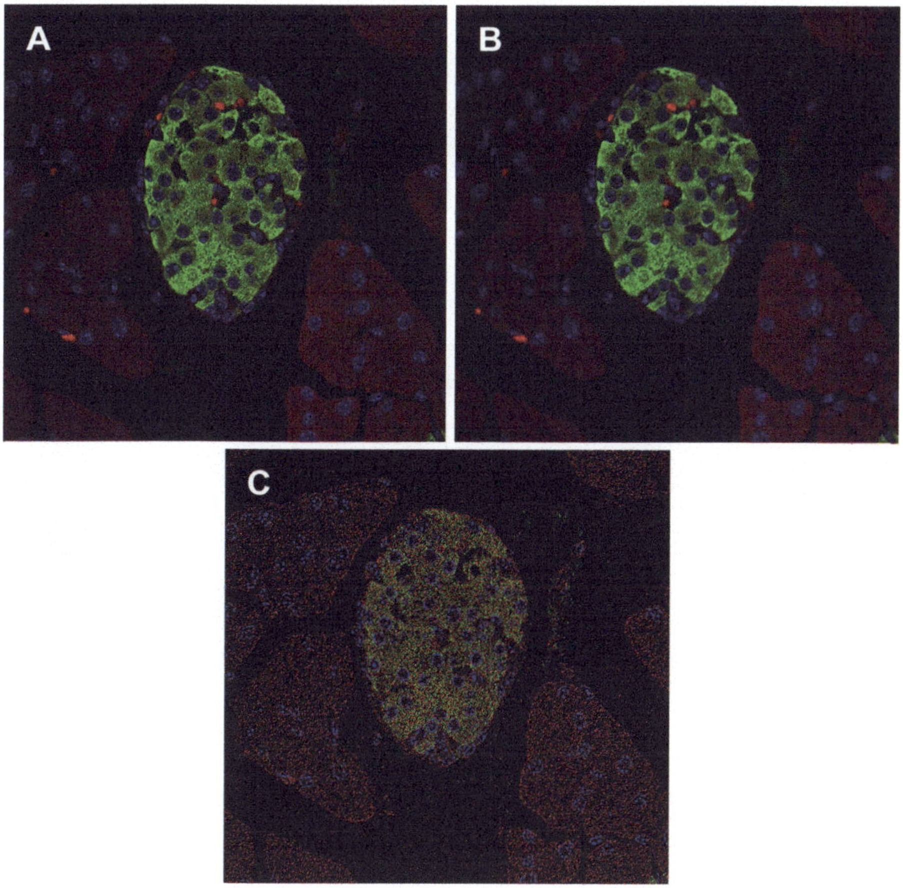

Fig. 6.11 Effect of JPEG compression on image information. (**a**) Tiff image of fluorescently labeled pancreas. (**b**) The same image as in Fig. **a** but saved in JPEG format. (**c**) Subtraction of image **b** from image **a** showing pixels altered by JPEG compression algorithm

It is worth reiterating here that "your image is your data!" Most modifications done to your image, including JPEG conversion, are irreversible. For this reason, it is critical that you maintain the originally collected image on archival media and only work from copies of the original for any image analysis or conversion routines that alter, in any way, the image information.

Literature Cited

Shannon C (1998) Communication in the presence of noise. Proc IEEE 86:447–457

Chapter 7
Confocal Digital Image Capture

W. Gray (Jay) Jerome

7.1 Introduction

Good microscopy includes optimizing the information obtainable from your confocal image with the minimum amount of effort and time. A key to achieving this goal is to match, as closely as possible, the parameters of all hardware and software in the image collection train. For reasons outlined later in this chapter, it is particularly important to match the resolution of your digital image to the output resolution from the microscope. In Chap. 6 we discussed the resolution parameters of the digital image. In this chapter we cover the optical resolution obtainable from fluorescence confocal light microscopes, we review the important parameters of the image acquisition hardware that influence image information, and we illustrate how to match these two parameters to optimize confocal image acquisition.

For images, two related types of resolution are important: spatial resolution and contrast resolution. Spatial resolution defines the size (or fineness of detail) of objects that can be reliably detected. In addition to spatial resolution, to "see" an individual feature, there must be a sufficient contrast difference between the feature and its surroundings. Blue writing on blue paper will not be visible, even if the letters are quite large, because there is not a suitable difference in contrast. The contrast difference can be a difference in color or in brightness. Thus, bright yellow letters on a bright blue background will be easily distinguishable as will light gray letters on a dark gray background. Since an RGB color image is really three grayscale images combined (see Chap. 6 for details), we will limit our contrast discussion to grayscale values (densities), but it should be remembered that in color images, differences in color are just as valid as differences in brightness. Chapter 6

W. G. Jerome (✉)
Department of Pathology, Microbiology and Immunology, Vanderbilt University School of Medicine, Nashville, TN, USA
e-mail: Jay.Jerome@Vanderbilt.edu

W. G. Jerome, R. L. Price (eds.), *Basic Confocal Microscopy*,
https://doi.org/10.1007/978-3-319-97454-5_7

discussed the dynamic range of a digital image. A larger dynamic range allows for smaller shading gradations and thus higher digital contrast resolution. Of course, our eye cannot distinguish the small differences between two adjacent gray values in the 4096 value dynamic range of a 12-bit image. However, the computer can, and so post-image processing can be used to selectively highlight specific values as long as the values were accurately captured during initial image acquisition.

The contrast resolution of a digital image collected with a fluorescent or confocal microscope is dependent upon multiple factors, (1) the inherent contrast differences in areas of the specimen, (2) the ability of the microscope to maintain those differences through the light path, (3) the camera's ability to capture those inherent differences, and (4) the image storage and display system's ability to accurately render those differences. In other words, even if a specimen generates a bright fluorescent signal that is easily distinguished from the background noise, that information will not be resolved in the final image if those contrast differences are degraded as the light passes through the specimen and microscope components, or if the camera system cannot detect the signal differences or represent those differences as discretely different values during the analog to digital conversion process. This is true even if the structures are well resolved spatially. A strength of the confocal microscope is that by reducing or eliminating fluorescence signal from out of focus planes, the contrast difference between in-focus fluorescent objects and the background is increased.

7.2 Basics of Optical Microscopy Resolution

A highly magnified image is not very useful if it is not sharp. A fuzzy image lacks fine (high-frequency) detail. Once lost, this detail cannot be regained no matter how high the image is magnified ("Resolution is a One Way Street!"). Increased magnification will only make the fuzziness more apparent. This excess magnification is referred to as "empty magnification" because it achieves no gain in image information. Since you cannot gain back information, it is imperative during image capture to obtain as much information as possible, or at least as much as is necessary for your experimental purpose. We can define the maximum image information obtainable under specific conditions as the highest resolution. If we could obtain infinitely fine resolution, we could produce an infinitely sharp image of infinitely small objects. Unfortunately, the physical properties of light constrain the size of objects that can be sharply imaged using photons with wavelengths within the visible spectrum; objects smaller than this limit will be fuzzy. Cameras and the analog to digital conversion process introduce additional constraints that potentially can reduce the resolution of the final image. For this reason, the good microscopist must understand how the microscope hardware, the digital capture device, and the image storage method can contribute to the loss of image information. The goal is to limit loss as much as possible.

7.2.1 Lateral Spatial Resolution

The limitation in sharpness of a two-dimensional microscopic image is termed the lateral (or angular) spatial resolution. The lateral resolution is usually defined in terms of how close two objects can be and still be recognized as two distinct objects. As the objects get closer together, it becomes harder to see the space between the spots. At some point, the space is no longer visible, and the two spots appear to merge into a single object. At this distance, the space is no longer resolved. The corollary is that structures with dimensions smaller than this distance will also not be resolved because these fine details will blur together.

This definition works very well for a bright-field image where multiple points are illuminated at the same time and light from adjacent objects can interfere. However, in fluorescence confocal microscopy, we generally project a focused point (or series of discrete points in the case of multipoint scanning instruments) into a sample and then image only that small illuminated area. Thus, each point in the final image is independently collected. For this type of imaging scheme, the important questions are: (1) what is the smallest excitation spot size (probe diameter) that can be produced by focusing a beam of photons, (2) how does the imaging system affect the information arising from that illuminated volume as it is projected to the detector, and (3) how faithfully does the analog to digital conversion system (camera) collect the information projected by the microscope? This section deals with the first two of these questions.

Confocal microscopes illuminate a three-dimensional volume in the specimen, but, since the constraints on the size of that volume are slightly different in the lateral (X,Y direction) and the axial (Z) direction, we will deal with the lateral and axial dimensions separately.

The key reason for the lateral image blurring in a microscope is diffraction of light waves. If the photons from a small point are focused through a lens (including the lens of our eyes), the two-dimensional image of the original point will not appear as a point in the lateral plane but rather will be seen as an Airy pattern (Fig. 7.1a). This pattern is named for the astronomer George Airy, who first described the phenomenon. The Airy pattern has a central bright circle (Airy disk) surrounded by a series of halos (maxima). The halos are separated from the central disk by dark rings (minima). As shown in Fig. 7.1b, which plots the intensity of light along the line x traversing the Airy pattern, most of the light from a point is focused into the Airy disk (about 97% of the light). The first maxima (first halo) contains most of the remaining light (about 1.7%). Each successive maxima after the first has progressively less light. In most situations we can ignore the first maxima, but you should always remember it is there. In some low light situations, it can contribute to the image.

It is obvious then that the Airy disk places a finite limit on how small an area we can illuminate in our specimen. This phenomenon also limits the size of the image of the illuminated area sent to a detector since this path also involves passage of light waves through a lens. The brightness at any particular point away from the

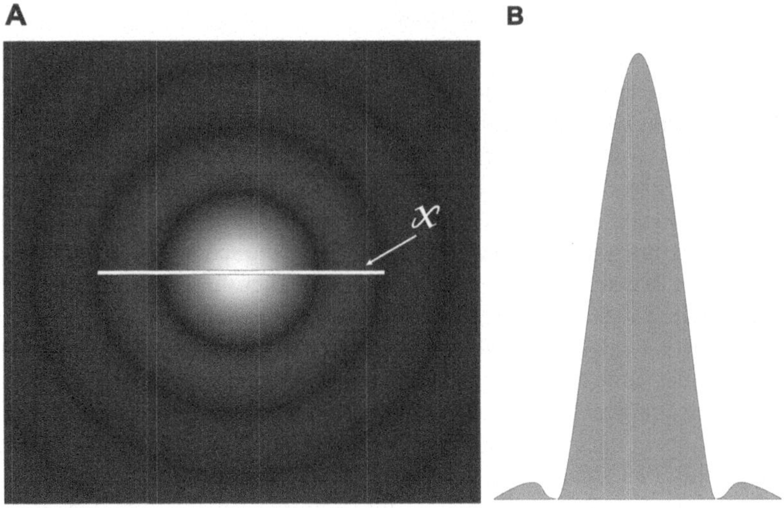

Fig. 7.1 (**a**) Image of an Airy pattern. The central spot is the Airy disk. (**b**) Plot of intensity along line X in 7.1A

center of the Airy pattern is dependent upon the strength of the light waves as they merge and the mixture of phases of all the merged wave fronts. The graph in Fig. 7.1b illustrates that there is not a sharp demarcation between very bright and very dark regions of the Airy pattern but rather a continuous change in intensity from the very brightest point of the Airy disk to the very darkest region of the first minimum.

Although it is not critical to understand all the details of why the Airy pattern is formed, it is worth a brief discussion of the general theory behind the effect. An Airy pattern results from the diffraction of light. When electromagnetic waves, such as light, meet a solid edge, the waves bend around the edge to form a new wave front. According to Huygen's principle, each point on the wave front emits vibrations in multiple directions (Huygen's wavelets). This concept is illustrated in Fig. 7.2a. At the point of focus of a lens, depicted by A in Fig. 7.2b, these wavelets arrive in phase, and so their amplitudes are additive (Fig. 7.2c). However, wavelets also arrive at other points in the same image plane (S_i) away from the focus point A. In regions where wavelets arrive exactly out of phase, such as point B in Fig. 7.2b, their amplitudes will cancel each other out as illustrated in Fig. 7.2c. This subtraction will form the first dark ring (minima). Waves converging even further from the focal point will again arrive in phase and form the first halo (maxima). Most of the light will fall within the central (Airy) disk, and this will be the brightest point. By comparison, the first maxima will be considerably dimmer than the Airy disk.

The size of the Airy disk and spacing of the halos are dependent upon the wavelength of the light and the proportion of the total light coming from the specimen that is collected by the lens (measured as the collecting angle). The radius of the disk (r) can be defined by Eq. 7.1.

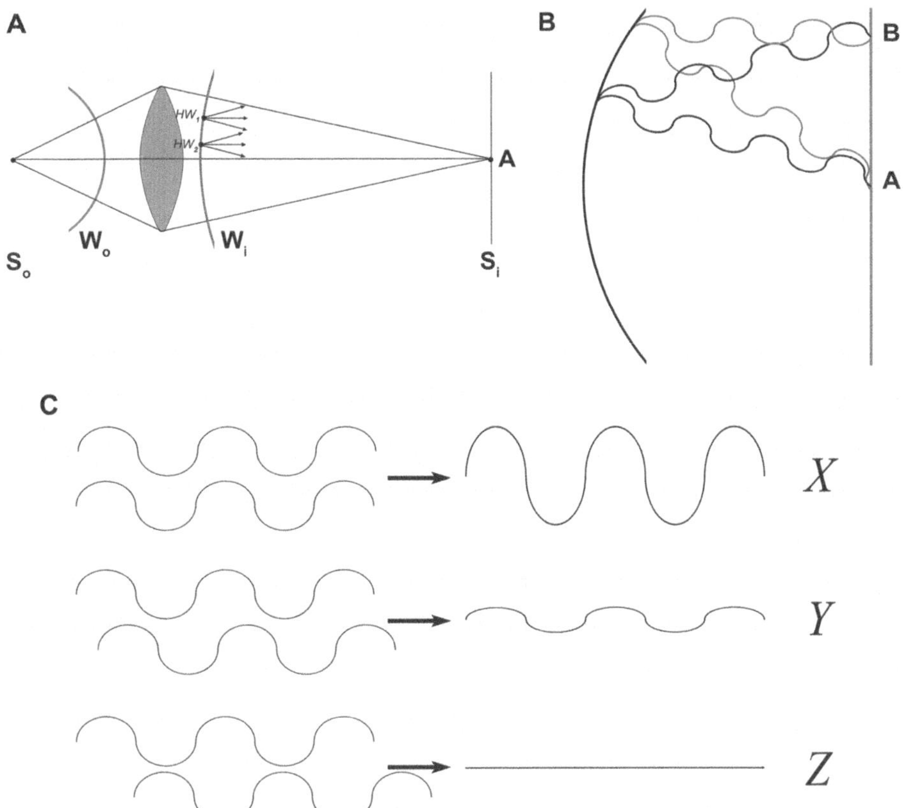

Fig. 7.2 Theory of Airy pattern formation. (**a**) Huygen's wavelets (HW1 and HW2) are formed all along the wave front (W1). (**b**) At the focal point of the lens (point **a**), the Huygen's wavelets arrive in phase and so are added together. At other points in the same plane (such as point B), the wavelets arrive out of phase and so are subtractive. (**c**) Waves that are exactly in phase (example X) will add together, and the resulting amplitude will be the sum of the individual amplitudes; waves that are exactly out of phase (example Z) will cancel each other out, and the resulting amplitude will be zero; waves that are not totally in or out of phase (example Y) will produce a signal with an amplitude in between that of X and Z

$$r = 0.61\lambda / \sin\theta \tag{7.1}$$

where λ = wavelength

θ = half the collecting angle of the lens

The equation tells us that the radius of the Airy disk is equal to a constant times the wavelength of the illumination divided by the sine of half the collecting angle. Half the collecting angle of the lens is the angle between the optical axis and the edge of the lens (Fig. 7.3).

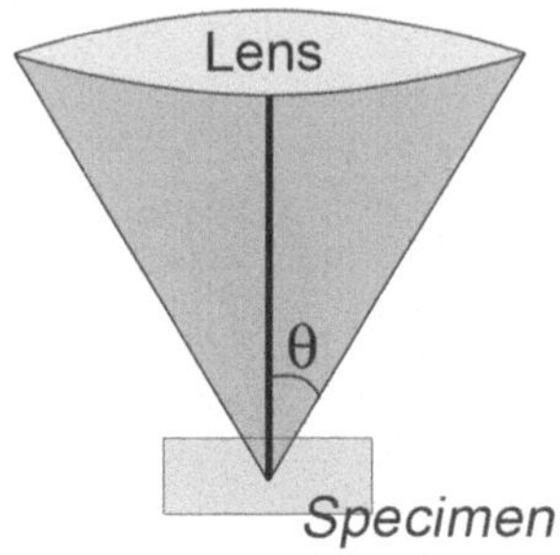

Fig. 7.3 Diagram of the cone of light collected by a lens. θ represents one half of the collecting angle of the lens

This equation defines an absolute value for the area of the specimen illuminated by visible photons. However, since we illuminate discrete volumes separately, in confocal microscopy we have the potential to localize the presence of individual small molecules that are smaller than the diameter of an Airy disk. Nevertheless, because we have illuminated an area larger than the molecule, we cannot pinpoint the precise location. We can only determine that the emitter was contained within the area illuminated. Likewise, when the object we are viewing is smaller than an Airy disk, we have no information about its size and shape. In order to get this information, the object must be larger than an Airy disk. So the Airy disk provides practical limits to the fineness of the detail we can obtain from the optical images of an object.

In recent years some newer, specialized imaging techniques have been developed that use multiple lenses or multiple illuminations to independently collect data below the diffraction limited level (Demmerle et al. 2017; Heddleston and Chew 2016; Lippincott-Schwartz and Manley 2009; Sahl et al. 2017). These increased resolution fluorescent microscopy techniques (discussed in Chap. 8) are essentially methods for gaining further information in order to pinpoint the location of an emitter smaller than an Airy diameter. As such, they present very exciting extensions of our imaging capabilities for very specialized situations. However, each of these methods comes with tradeoffs that restrict their uses to specific situations. This is why, for the majority of confocal microscopy studies, resolution is limited by the size of the Airy disk, and we will use this limit for discussion throughout the remainder of this chapter.

7.2.2 Lateral Contrast Resolution

As discussed earlier, our ability to "see" two points as separate and distinct is dependent upon both the lateral and contrast resolution. So, the observed resolution greatly depends on the detection method. In the early days of microscopy, the eye was the detection method, and so early definitions of resolution were based on the ability of our eyes to discriminate contrast differences. In the nineteenth century, John William Strutt (Lord Rayleigh) examined the resolution problem and came up with a method of approximating the resolution limit in standard fluorescence microscopy. He proposed that there would be sufficient contrast between the centers of the Airy disks of two objects if the center of one Airy disk lies at or further away than the first minimum of the other object's Airy disk (Fig. 7.4a). In other words, we can determine the resolution limit with Airy disk in Eq. 7.1. This is true in air or in a vacuum. However, in a material with a higher refractive index, the velocity of light slows down. Thus, when using immersion oil or water, we need to account for this change in velocity because wavelength is related to velocity. Equation 7.2 shows this relationship.

$$\lambda = {}^{v}\!/_{f} \tag{7.2}$$

where: λ = wavelength
v = velocity
f = frequency

The frequency of light waves does not change. So, as velocity decreases so does wavelength. We can account for this in Eq. 7.1 by adding a refractive index term to the equation. The refractive index of a medium is the ratio of the velocity of light in a vacuum compared to the velocity of light in the medium of interest.

$$n = {}^{V_v}\!/_{V_m} \tag{7.3}$$

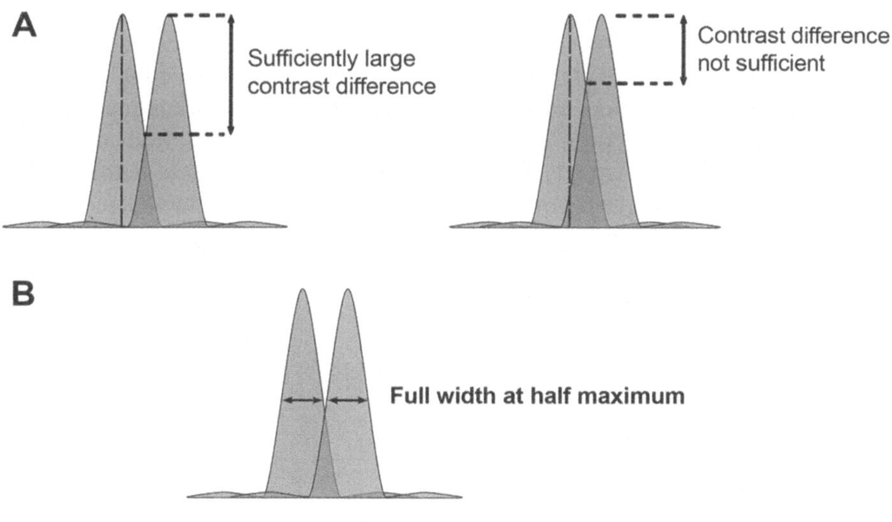

Fig. 7.4 (**a**) The Rayleigh criterion for contrast resolution. Two objects can be resolved only if the contrast difference is sufficiently large enough. (**b**) Illustration of full width at half max value for the Airy disks of two adjacent objects

where: n = refractive index
v_v = velocity of light in a vacuum
v_m = velocity of light in a medium

We can define a further term, the numerical aperture of a lens, as the sine of ½ the collecting angle times the refractive index of the medium used with the lens. This value is provided for all good quality lenses and is usually designated as the numerical aperture. Taking into consideration numerical aperture, we can formalize the Rayleigh criterion as the following:

$$r = {}^{0.61\lambda}\!/_{(\sin\theta)(n)} = {}^{0.61\lambda}\!/_{\mathrm{NA}} \tag{7.4}$$

where: r = minimum resolved distance
λ = wavelength
θ = half of the collecting angle of the lens
n = refractive index
NA = numerical aperture

Some typical values for microscopy imaging would be light at a wavelength of 450nms, $\sin\theta = 0.92$ (for $\theta = 67°$), $n = 1.515$. The NA would be 1.515×0.92 which equals 1.4. Given this, Eq. 7.4 allows us to determine that r (minimum resolved distance) would be 196 nm.

The Rayleigh resolution criterion is a relative one and based on the ability of our eye to discriminate contrast. If we have a detector that exceeds our eyes' capacity, the value for r can be decreased further. A related criterion partially takes detector efficiency into consideration. In this criterion, usually referred to as "full width at half max," it is assumed that objects can be discriminated if they are at least as far apart as the full width of their image at the half-maximal brightness value (Fig. 7.4 b). The double arrows in Fig. 7.4b indicate the full width of each curve at half the maximal intensity value (FWHM). For the eye, the FWHM distance will be only a little less than that determined by the Raleigh criterion. It is important to remember that neither the Rayleigh criterion nor FWHM represent absolutes. Rather, they are general rules of thumb. Methods that can increase contrast further or detectors that can discriminate smaller levels of contrast can resolve even smaller distances. One strength of the confocal microscope is the increase in image contrast. Based, at least partly, on this, the confocal provides a gain in lateral resolution. This improvement in lateral resolution is defined in Eq. 7.5.

$$r(\mathrm{confocal}) = {}^{0.4\lambda}\!/_{\mathrm{NA}} \tag{7.5}$$

where: r = minimum resolved distance
NA = numerical aperture
λ = wavelength

No matter what resolution criteria we choose, the equation for the Airy disk describes the minimum area that can be illuminated by an electromagnetic wave such as light. Since the image of that illuminated area is also the product of electromagnetic waves traversing a lens, the equation for the Airy disk also describes how the microscope affects the image of that illuminated volume. As the prominent early microscopist Ernst Abbe pointed out, we can think of a large object as a series of points that either emit or diffract light. Thus, the Rayleigh criterion defines the level of fine detail we can discern within a larger object. Resolution is, thus, the key determinant of how sharp detail will appear in our image. Just as we describe the pixel as the smallest resolvable unit of a digital image, we can define a resolved element (resel) as the smallest resolvable unit of a microscopic image. Of course, there is also a role for magnification, the other major component of a microscope. For instance, our eye cannot discriminate distances much smaller than about 0.2 mm. Thus, we need to magnify the resel in order to see it. We will have more to say on this later in this chapter.

It is worth reiterating that all criteria for resolution are intimately dependent upon the Airy disk size. We can circumvent some of its limitations, but if we use a lens for imaging, we cannot completely get rid of the consequences of the diffraction of light.

7.2.3 Lateral Contrast Transfer Function

Just as a point of light is altered as it traverses the optical path, the contrast inherent in an object is also altered by the microscope and other components of the imaging system. The effect of the imaging system on contrast can be described by the contrast transfer function (CTF). Since resolution is intimately dependent on contrast, the CTF is just as critical as the size of the Airy disk for understanding how an optical system affects resolution. Figure 7.5 depicts a theoretical CTF for a microscope. A full discussion of the CTF is more appropriate for an advanced optics book.

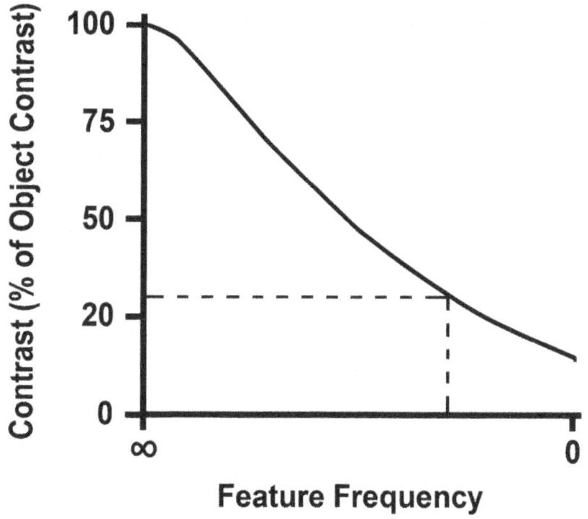

Fig. 7.5 Typical contrast transfer function (CTF) for a lens. The *dotted line* represents the Rayleigh criterion cut off point for contrast resolution

However, a few key points can be easily seen from the graph. First, there is always a loss of contrast during the imaging process. Thus, the microscope does not faithfully reproduce an object down to the Abbe-defined diffraction limit and then the image rapidly degrades; there is degradation in contrast at all levels. Second, the CTF is inversely related to feature size, e.g., small features will be reduced more in contrast than larger features. Throughout this chapter, we have reiterated the critical interdependence of spatial and contrast resolution in determining our ability to "see" an object. In fact, the Rayleigh criterion sets the resolution limit at that point where the image contrast is reduced to 25% (dotted line on graph in Fig. 7.5).

Figure 7.5 only depicts the CTF of a theoretical optical system. However, all aspects of the imaging train can also affect the CTF. Thus, we also must concern ourselves with the ability of our eye or camera to detect contrast differences. In other words, we need to know something about the CTF of each of these components. Since confocal detectors are generally grayscale instruments, the contrast resolution of a detector is related to how many separate gray tones between black and white can be discriminated. For a digital image, this is the bit depth discussed in Chap. 6. A minimum requirement for most microscopy is 8 bits (256 gray levels), but as discussed later in this chapter, higher bit depth (12-bit) detectors provide some significant advantages, particularly for quantitation.

7.3 Image Production and Resolution in a Scanning Confocal Microscope

The confocal microscope sequentially scans a focused point of light through the specimen and measures the fluorescence generated at each point. Details of this process are described in Chaps. 8 and 9. An image of the specimen's fluorescent intensity at each spot is then sequentially determined and stored as pixel information. The brightness of individual pixels in the resulting image is a representation of the fluorescence intensity at the corresponding spot in the specimen. The laser scanning microscope usually does this with a single point of light, while the spinning disk uses multiple, but still independent, points at once.

To convert the signal from an excited area in the specimen into a pixel gray value in the image requires the collection and measurement of the photons generated from the excited area by a detector. The resulting image pixel, representing a defined spot in the specimen, will have three critical values: the X and Y position of the pixel, the X and Y dimensions of the pixel (pixel size), and the gray value assigned to the pixel. Since the digital image of the scanned area is formed by summing all of the photons coming from the area illuminated into a single value that becomes the brightness value (grayscale value) for that pixel, it is clear that the resolution of an image cannot be any smaller than the size of the illumination spot projected into the specimen or onto the detector. Since, except for specialized situations, the radius of the illumination spot cannot be any smaller than the radius of the Airy disk, the Airy disk defines the smallest area we can illuminate and thus the smallest object we can visualize correctly or, for objects larger than an Airy disk, the finest detail we can visualize. This is depicted in Fig. 7.6. Thus, except for very specialized situations, this places a theoretical limit on the lateral (X-Y) resolution that can be detected.

However, we must have a detector that is capable of detecting these closely spaced signals. This means that the detector's spatial resolution must match or exceed the spatial resolution of the microscope image (Fig. 7.6). In addition, the detector must be sensitive enough to discriminate the signal differences (e.g., have sufficient contrast resolution). Detectors for light signals run the gamut from film to diode arrays. However, most commercial confocal microscopes utilize one of two types of detector schemes. For point scanning instruments, the detectors usually quantitatively analyze the photons coming from the specimen without discriminating from where the photons were generated. The location in X, Y, and Z dimensions is inferred from the location of the excitation bean. Spinning disk confocal microscopes, on the other hand, generally employ an array of detectors that discriminate the X and Y location of the signal (the Z location is inferred from the position along the Z axis of the focused excitation light). As explained in the next sections, each of these approaches brings with it certain assumptions that must be recognized and considered in collecting and interpreting the resultant confocal image.

Throughout this chapter, the term spinning disk will refer to disks that contain an array of pinholes. However, for completeness, it should be pointed out that another variety of spinning disk microscope (often referred to as a slit-scan confocal) uses a disk with a series of narrow slits rather than pinholes. As the disk spins, the slits

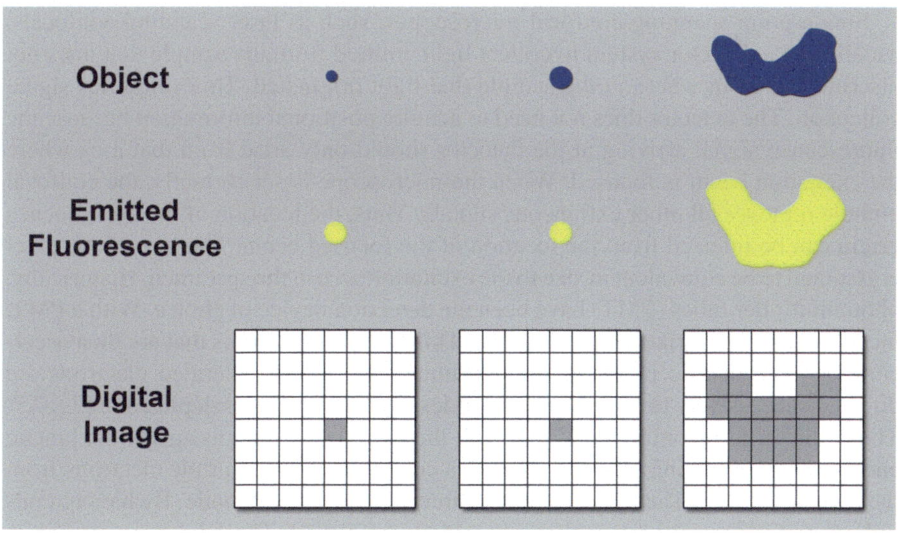

Fig. 7.6 Comparison of detection (yellow) and digital display (gray) of objects (blue) smaller than the Airy excitation area (diffraction limited fluorescence), just as large as the Airy excitation area, or much larger than the Airy excitation area illuminated by focused beam of photons

scan the excitation light over the surface of the specimen and also restrict the emitted light returning to the detector from out of focus planes. These instruments increase the light transmittance onto the specimen but at the cost of a decrease in the axial spatial resolution. Spinning disk confocals are discussed more thoroughly in Chap. 8.

7.3.1 Image Production in a Single Point Scanning Confocal Microscope

Sensitivity is a key to determining the appropriate detection device for a microscope setup. Fluorescence emission occurs over all angles, not just in the direction of the collecting (objective) lens. Thus, only a small percentage of the emitted photons enter the optical path of the microscope. Higher NA lenses will collect from a wider angle and so will gather more of the photons, but even high NA lenses collect only about 30% of the emitted light. Loss of photons within the optical train due to absorption, reflection, and deflection lowers the number of photons that reach the detector even further. In a well-maintained system with high-quality optics, it is still doubtful that more than about 20–25% of emitted photons will reach the detector. Thus, highly efficient detectors are necessary to avoid attenuating the signal any further. Besides sensitivity, a detector needs to have low noise to maintain a high SNR (good contrast) even when the signal strength is low. A good detector also should be equally sensitive to all wavelengths and should have a linear response and broad dynamic range to accommodate both faint and bright signals.

Single-point scanning confocal microscopes, such as laser scanning confocals, usually use a detector system to collect light emitted from the sample that does not discriminate from where in the sample that light originated. This simplifies signal collection. The detector does not need to acquire positional information because the fluorescence signal arriving at the detector should only arise from that area where the excitation beam is focused. When the microscope is set correctly, the confocal pinhole removes all other extraneous signals. Thus, the location of the fluorescence origin can be inferred from the location of the focused beam. The size of the pixel is assumed to be equivalent in size to the excitation area in the specimen. Historically, photomultiplier tubes (PMT) have been the detection device of choice. With a PMT, incoming photons strike a photocathode. This liberates electrons that are then accelerated by an electric potential in a vacuum tube. The accelerated electrons are directed successively to a series of electrodes (dynodes). This is depicted in Fig. 7.7. The collision of incoming electrons with the dynode surface dissipates the kinetic energy of the incoming electrons, and this energy liberates multiple electrons from the dynode surface. These electrons are directed to another dynode. By having multiple dynodes, the initial signal can be greatly amplified. The final generated electrons interact with an anode, and the consequent current is measured. The strength of the output signal is proportional to the input signal. The software then translates

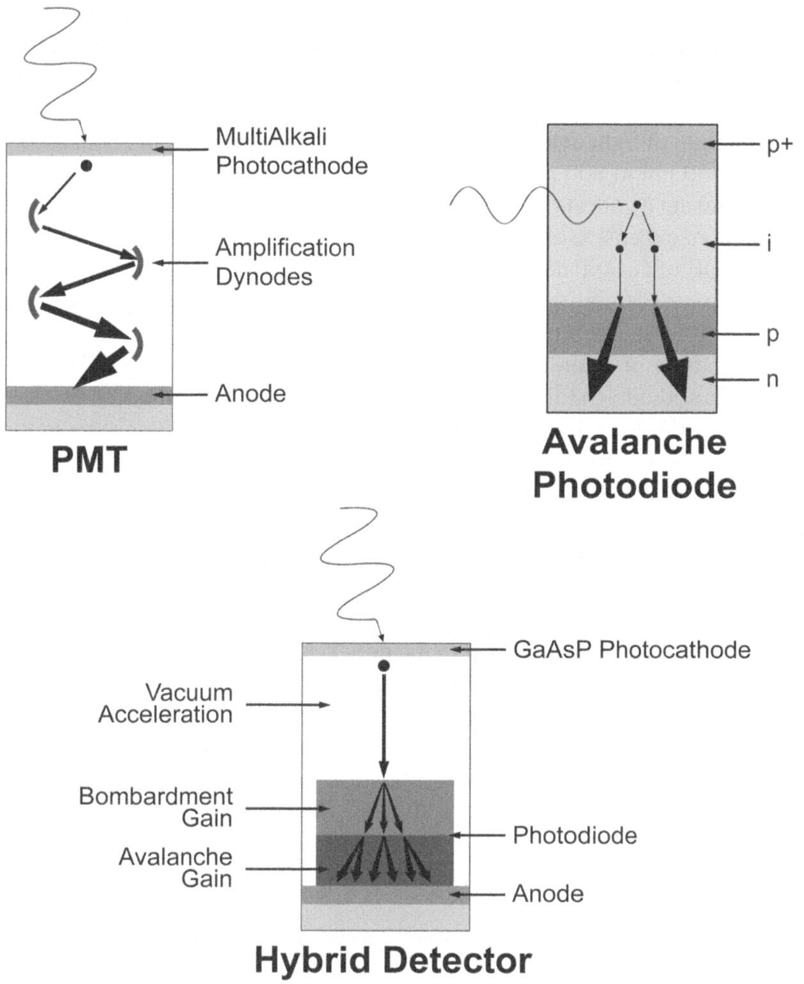

Fig. 7.7 Photon detectors for point scanning confocal microscopes. Wavy arrow indicates photon interacting with detector. Increasing size and number of arrows indicates multiplication of electrons generated from initial photon detection event

the detected signal into a grayscale value for each individual pixel of the digital image. The grayscale value of the pixel is assigned based on the signal strength, which is a function of the number of photons absorbed by the photocathode. Thus, the conversion of the analog information into a digital signal is relatively direct and straightforward with PMTs.

Standard PMTs with a multi-alkali cathode are a good robust solution for general point scanning confocal microscopy. They are relatively inexpensive compared to other types of photon detectors, and they are robust and have a high dynamic range, excellent signal-to-noise ratio, and a fast response time. However, multi-alkali

PMTs have a low quantum efficiency (QE). Quantum efficiency (QE) is a measure of the effectiveness of the conversion of photons to electrons. If every photon hitting the cathode was converted to an electron, the QE would be 1 (100%). Also, the QE of a detector usually is not uniform across the spectrum of wavelengths so will vary with wavelength of light detected. Multi-alkali PMTs have a QE of about 25% in the range from 350 nm to 800 nm. The QE of PMTs can be increased to between 40% and 50% by using a gallium arsenide phosphide (GaAsP) cathode. However, GaAsP PMTs are more expensive and more susceptible to damage compared to multi-alkali PMTs. The QE of GaAsP detectors also falls off dramatically at wavelengths below 400 nm or above 650 nm. For this reason, confocals with GaAsP detectors usually also provide at least one multi-alkali PMT as well.

A second type of photon detector is the avalanche photodiode (Fig. 7.7). An avalanche photodiode (APD) is essentially a PIN diode with an additional layer to allow signal amplification. As in a standard PIN diode, there is a positively doped P layer (p+) and a negatively doped N layer (n+) with a silicon Intrinsic absorption layer (i) between them. However, in avalanche photodiodes a thin p layer is added adjacent to the n + layer. The pn + region creates a signal amplification. During operation, a high reverse bias voltage is applied. When an incident photon creates an electron-hole pair in the i region, the hole moves toward the p + layer and the free electron migrates toward the n + layer. Successive impact ionizations of the initial electron and those generated from it in the pn + layer create additional electron-hole pairs, thus amplifying the initial free electron into multiple free electrons. This is known as the avalanche effect. APDs have high sensitivity and high speed and can have a high quantum yield (>50%) and relatively low noise, although noise level increases as reverse bias (and thus amplification) increases. APDs are very good in situations where the specimen signal is dim and quantitative single photon detection is needed. However, APDs have very low dynamic range, so they are not suited for general types of confocal microscopy or any operations where signal strength from the sample will be bright.

Recently, a new detector design has evolved that takes advantage of the high dynamic range of PMTs and the high sensitivity of APDs. These hybrid detectors (Fig. 7.7) are vacuum tubes, like PMTs, and use a GaAsP photocathode for high QE of initial photon detection. The generated electrons are then accelerated in a high voltage within the vacuum tube. However, the hybrid detectors use an APD to amplify the signal. Hybrid detectors have an overall gain that is less than a standard PMT and thus not as good a dynamic range, but they have significantly better dynamic range than APDs alone.

Hybrid detectors have many advantages for use in specific situations. Hybrid detectors have an active region that is smaller than a PMT but significantly greater than a standard APD. This compromise decreases thermal noise, but the larger active region still allows precise focusing of the sample photons onto the target region. Besides better SNR, hybrid detectors have fast response time, better pulse resolution than PMTs, and very limited after pulse. After pulse is a false signal generated within the detector by stray radiation. Hybrid detectors are excellent choices in situations where linearity of detection, wide dynamic range, and fast response time are critical such as for two photon microscopy, fluorescence lifetime imaging, or fluorescence correlation spectroscopy.

Since the first edition of this book, the use of spectral detection has become common in confocal microscopy. Spectral detection is not really a new type of detector but rather a different approach to photon detection that allows collection of more information about the photons being directed toward the detector. In a spectral detection system, light coming from the sample passes through a dispersive element that spreads the light out spatially along its different energies (wavelengths). In this way, the various components of the signal are spectrally resolved before detection. Detection is usually by PMTs or hybrid detectors. In this way the system can analyze the number of photons generated at a precise region of the sample and determine the wavelength of the photons (usually to within 5–10 nm resolution) of those photons. Different types of spectral systems employ different methods for spreading and detecting the energies, but the result remains the same; the information is combined into a high-resolution map of the component energies emitted from the specimen. Figure 7.8 depicts an array of detectors, but other solutions to capturing spectral information, including the acousto-optical beam splitter (AOBS) Leica spectral system, have been developed. The Leica detector will be discussed further in Chap. 9. Each type of spectral system has their advantages and drawbacks. If considering purchasing a spectral detection system, the best approach is to test the different types on relevant samples to make sure the performance is consistent with your needs.

One benefit of spectral detection is the ability to estimate the contribution of two or more fluorochromes with partially overlapping emission spectra to the multifluorochrome-produced final signal. This is depicted in Fig. 7.9. Figure 7.9a shows a potential complicated emission spectrum from two fluorochromes with overlapping emission spectra. By knowing the emission spectrum for each fluorochrome (Fig. 7.9b) derived from ideal samples, one can mathematically estimate how much each fluorochrome contributed to the intensity is measured at each wavelength (Fig. 7.9c). In this way, some bleed through can be removed to get a more accurate analysis of the intensity of each fluorochrome in the sample. Different confocal manufacturers use different algorithms to unmix the collected spectrum, but all do a reasonable job of separating signals when the overlap of emission spectra is not too great.

7.3.2 Image Production in a Multipoint Scanning Confocal Microscope

In contrast to single-point scanners, detectors for multipoint scanners such as spinning disk confocals must be able to collect spatial information. Although these microscopes do excite individual points, multiple points are excited at once, and it is critical to discriminate photons coming from one point in the image from those

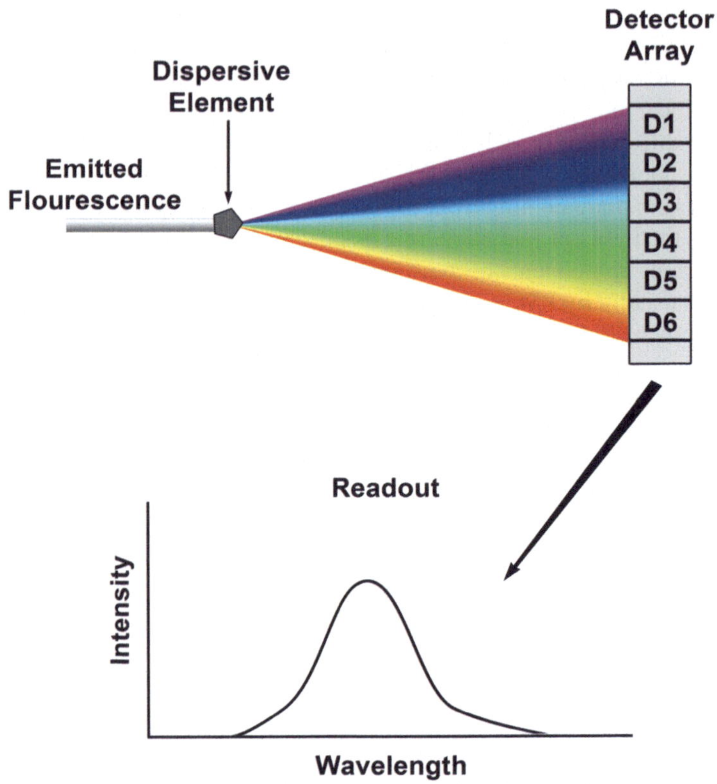

Fig. 7.8 Typical spectral detection scheme. Emitted fluorescence is divided into component wavelengths and the components sampled individually to produce a graph of intensity values at different wavelengths

photons generated in a different location. The bases for the modern simultaneous detectors required for multipoint confocals are solid-state electronic devices composed of arrays of discrete photodetectors. Each detector captures photons coming from their respective area of the specimen and converts this information to an electrical signal that is proportional to the amount of light sensed by the detector. The electrical signal is converted to digital information, and that digital information can be stored or displayed. Thus, unlike detectors for single-point scanning instruments, the capture device not only measures the signal from the specimen but also its location along the X-Y axis. This means that the size of the individual discrete photodetectors can limit the size of an object that can be visualized. This key point will be discussed later in this chapter.

The most widely used microscope photodetector systems have some form of charge-coupled device (CCD) or complementary metal oxide semiconductor (CMOS) for sensors (Fig. 7.10). Although there are a number of similarities, there are also dramatic differences between these two approaches (Magnan 2003). The

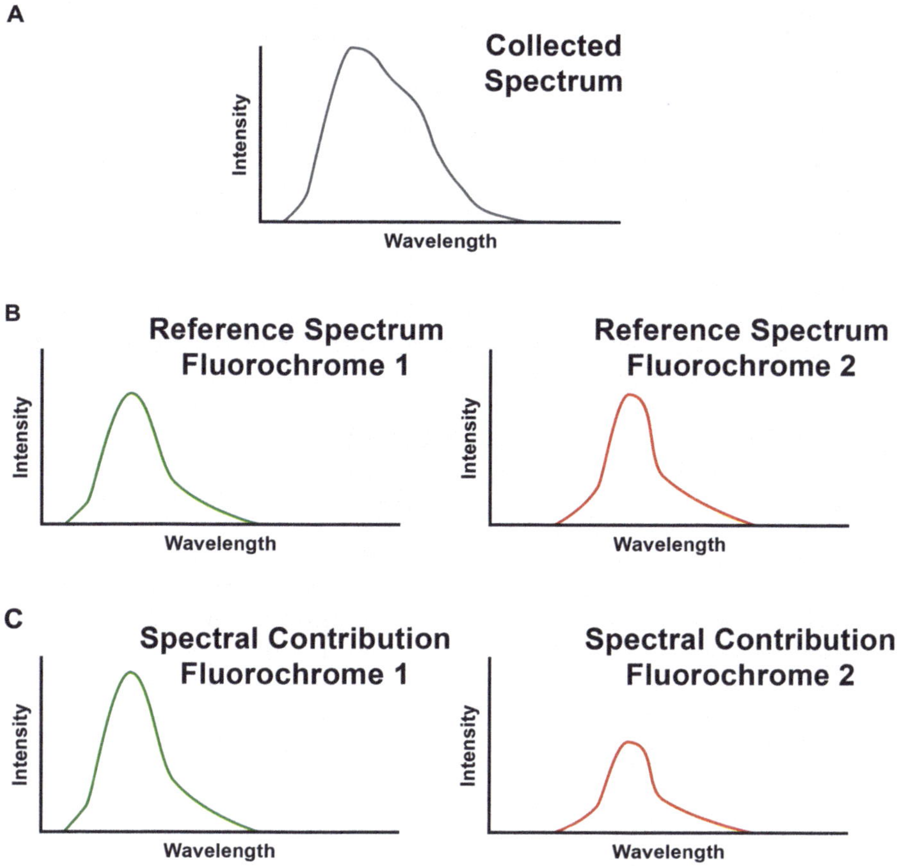

Fig. 7.9 Spectral detection allows the discrimination of the contribution to the total fluorescence of two fluorochromes with somewhat overlapping emission spectra

cooled CCD[1] (Fig. 7.10a) contains a two-dimensional array of photodiodes capable of sensing photons. The charge-sensing region of the photodiode is coupled to a charge storage region. In Fig. 7.10a, each gray square represents an individual photodiode, and each photodiode defines a single pixel. As photons from the specimen hit the photodiode, they are converted to charge. Thus, the charge strength is related to the number of photons hitting the detector during a short exposure time period. When the image capture period is ended, the charge information is read out sequentially from each photodiode. There are various schemes for reading out, but essentially the charge cloud of each photodiode is transferred one at time to an amplifier which measures the value of the charge and converts it into a voltage. Associated electronics then reduce noise, digitize the values, and output the digital value for

[1] All modern scientific-grade CCD and CMOS cameras are cooled to reduce noise and increase signal to noise levels.

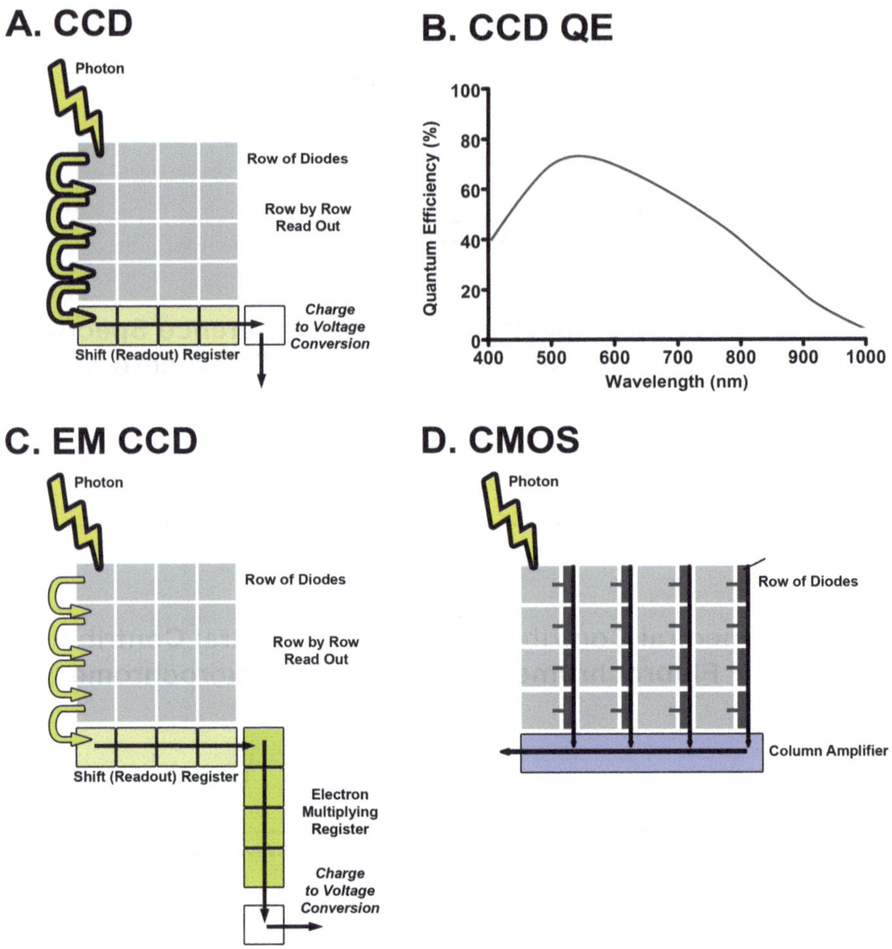

Fig. 7.10 CCD and CMOS design. (**a**) Design of a CCD camera. (**b**) Typical QE curve for scientific-grade cooled CCD. (**c**) Design of an EM-CCD camera. (**d**) Design of a CMOS camera. (*Based on Jerome. 2017. Microscopy Today doi:*https://doi.org/10.1017/S155192951170061X. *Reprinted with permission*)

storage, usually in computer memory. CCD detectors have high QE across the visible spectrum (Fig. 7.10 b) and large dynamic range. Noise levels, however, can be high in very low light situations. For these situations, electron multiplying CCDs (EM-CCD) have been developed. They increase signal output without substantially increasing noise by passing the output charge through an electron multiplier consisting of multiple electrodes before analog to digital conversion (Fig. 7.10 c). The electrodes generate additional electrons by impact ionization, increasing the signal from 100- to 1,000 fold. However, the individual photodiode size in EM-CCD cameras is usually larger than that for CCDs, and so lateral (X-Y) resolution is less for most EM-CCDs. CCDs are very serviceable and have been the backbone of microscope cameras for years. The main drawback of the CCD design, however, is that the sequential transfer takes time and thus limits how fast individual images can be acquired because all photodiode charge transfer processes must be completed before a new round of image collection can begin.

CMOS cameras also have a two-dimensional array of photodetectors that converts photons to electrical charge in a manner proportional to the photon energy absorbed. However, on a CMOS chip, the amplification and digitization steps happen at each individual sensor in parallel before the information is passed off the chip for storage (Fig. 7.10 d). This parallel processing approach dramatically reduces the time it takes to relay the image information and prepare the sensor array to capture another image. CMOS cameras have traditionally been the camera of choice for live cell imaging because their fast speed limits exposure of the specimen to the potentially damaging excitation beam. One of the tradeoffs for this increased speed, however, is that there is increased pixel-to-pixel variability since each pixel has its own separate amplifier circuit.

Even just 5 years ago, choosing between a CCD and CMOS camera was fairly simple. One chose a CCD for high resolution or a CMOS system if you needed fast acquisition to capture rapid events or limit exposure. With current developments, particularly advances in CMOS systems, the choices are not as clear cut. One needs to delve deeper into the specifics of each camera system in order to match a particular camera to your needs. Among the many parameters of camera design that can be investigated, we feel a few of these provide the most information regarding whether a system will meet your needs.

One of the key components to consider is the size of the individual photodiode sensors (pixels). This determines the spatial resolution limit of the sensor. For a given magnification, smaller photodiode size allows higher frequency information to be captured. The effect of pixel (photodiode) size is discussed in detail in the next section. Photodiode size, however, is not the only determinant of spatial resolution, so specific tests should be run to determine the highest spatial resolution the sensor is capable of obtaining. As individual photodiode size gets smaller, its photon capacity (how many photons before it becomes saturated) decreases. Thus, smaller pixels have lower potential SNRs and lower potential dynamic range (Chen et al. 2000). Today, most high resolution microscope cameras (both CCD and CMOS) with a photodiode size of ~6.5 um will provide full microscope resolution with good dynamic range and signal to noise characteristics. Table 7.1 compares some critical parameters of modern cooled CCD and scientific CMOS (sCMOS) cameras with pixel size of 6.5 um^2. CMOS systems have a little more noise than equivalent size CCD systems, but with the latest round of scientific CMOS (sCMOS) cameras, I have not found the difference to be very noticeable, except at very low light levels.

Resolution is often stated as the number of megapixels or the pixels per inch of a system. I find these parameters much less useful than the individual pixel dimensions. Megapixels and pixels per inch do not provide critical information regarding the smallest object in a sample that can be effectively identified and captured with the camera system. The number of megapixels is not totally useless as it does

Table 7.1 cCCD and sCMOS cameras: comparison of key parameters

Parameter	cCCD	sCMOS
Pixel size	6.5 um × 6.5 um	6.5 um × 6.5 um
Number of pixels	1392 × 1040	2048 × 2048
Peak QE	70% at 600 nm	72% at 560 nm
Readout speed	Up to 17 fps	Up to 100 fps
Readout noise	2e	2.5e
Dynamic Range	3,000:1	30,000:1

indicate how large a field of view can be captured, but it provides no information about the spatial resolution within that field of view. Pixels per inch is derived from the pixel size, so sensor size from pixels per inch can be calculated, but why not just look at the sensor size directly for comparing maximum achievable spatial resolution between different camera systems?

The spectral response of the camera refers to how well specific wavelengths are detected by the sensor. Both cooled CCD and high-end (scientific) CMOS sensors are sensitive to wavelengths between 400 nm and 1000 nm. However, one should always check the spectral response curve of the specific system being considered. There is usually some sensitivity in the near IR range, and CMOS sensors tend to be better at detecting IR compared to CCD sensors.

The frame rate of a camera refers to the number of full frames that can be captured per second. Digital camera systems allow adjustment of the frame rate up to some maximum. This is where the parallel processing design of CMOS cameras clearly has the edge. Maximum frame rates for CMOS systems are faster than for CCD systems, although some clever tricks have improved CCD frame capture rates. However, for both systems, since faster frame rates mean reduced exposure time, the SNR decreases as frame rate increases.

Quantum Efficiency (QE) is a measure of the efficiency of conversion of photons to charge. As with PMTs and hybrid detectors, the QE of photodiode sensors is not uniform across the spectrum. Both CCD and CMOS detectors tend to have less QE at both ends of their usable wavelengths. Several years ago, the QE of CMOS cameras were significantly worse than for CCD systems, but the most recent sCMOS designs rival the QE of CCD sensors. At their maximum, QE values range around 60%.

The dynamic range of a camera indicates how many separate and distinct gray levels the camera is capable of effectively detecting and storing. In other words, what is the lowest light level the sensor can detect and how many photons can a sensor collect before it is saturated. If there is a large difference between the lowest light detectable and saturation, there is the possibility to discriminate more subtle differences in the light coming from different areas of the specimen. For instance, if the lowest detection possible above noise is 5 and the sensor saturates at 50, then 5 defines what the sensor determines as black and 50 as white. That leaves the values of 6–49 to define shades of gray between black and white. So this sensor can discriminate 44 different charge levels and assign them to 44 different gray values. However, if the range is 2–200, more subtle differences in the number of photons falling on one photodiode versus an adjacent one can be discriminated. Therefore, the greater the dynamic range the better the camera is at detecting subtle differences in shading within a specimen. The dynamic range is the full well capacity of the sensor divided by the read noise. The dynamic range is usually reported as how many bits of information can be discriminated. The dynamic range is different at different frame capture rates, so camera systems should be compared at the frame rate needed in requisite experiments. Dynamic range is another area where CCD systems used to excel, but sCMOS technology is rapidly catching up.

Read noise is the inherent electronic noise in the system. The camera noise level is important because it helps determine the dynamic range of the sensor and how efficiently the camera can capture very dim light. Read noise increases as frame-rate increases. Thus, even though CMOS cameras are capable of very high frame rates (100 frames per second), these rates are really only useful when a very strong signal is present so that the increased noise does not interfere with detection.

Dark noise is noise that is generated by temperature excitation of electrons within the photodetectors. Cooling a camera reduces the dark noise. For most situations, the dark noise of cooled CCDs is negligible, not so for cooled sCMOS cameras. However, dark noise in sCMOS cameras has been steadily going down as sCMOS manufacturers improve performance. It may come about that soon cooled sCMOS cameras will rival cooled CCDs in terms of dark noise.

Camera manufacturers provide a long list of specifications for their systems, especially for those parameters for which their products excel. However, we have found that the few parameters described above are sufficient to narrow down the potential list of cameras that are useful for a specific purpose. Then, despite what the specifications sheet indicates, one should always compare systems on their own microscopes with their specimens to make sure that the camera system is appropriate for their needs. This practical test is necessary because there is chip-to-chip variability and also because each manufacturer includes hardware and software in their camera that alters the initial photon signal collected by the photodiode. This signal processing is an integral part of the camera, but this means that two cameras from different manufacturers with the same CCD or CMOS chip can have a different level of performance.

7.3.3 Matching Microscope Image Lateral Resolution to Photodiode Size for cCCD and sCMOS

CCDs and CMOS systems have high QE, and their dynamic range and linearity are excellent. However, their sensitivity is not as good as PMTs, and the time to collect an image and transfer it to computer memory is slower than that of PMTs. However, because multiple photodiodes are arrayed together in a single camera, the overall frame capture rate can be faster than PMTs. The most important feature of photodiode array cameras that must be considered in microscopy is that, unlike PMTs, array photodiodes do have a physical size. This must be taken into account in order to optimize image resolution.

Table 7.2 Magnification of Airy disk radius by lenses with different NA and magnifications

Lens	NA	Airy disk diameter (micrometers)	Magnified Diameter (micrometers)
40× dry	0.93	0.3	12.0
63× water	1.3	0.2	12.6
100× water	1.3	0.2	20

In addition to considering the size of the photodiode in a CCD or CMOS system, matching the microscope image resolution to the photodiode size also involves knowledge of the resolving power of the microscope lens. The photodiode sensor size should be available from the camera manufacturer. A sensor size of 6.45 square micrometers is a popular size for high resolution microscope cameras. This turns out to be a very convenient size for microscopy.

From Eq. 7.5 we can determine that the resel obtained from an emitted beam of 550 nm focused with a lens having a NA of 1.3 (a pretty standard NA for the type of high resolution water immersion lens often used in confocal microscopy) will be a little less than 0.2 micrometers in diameter. We will round this number to 0.2 micrometers for the purposes of this discussion. If the lens had a magnifying power of 63 times, the smallest resolved objects (0.2 micrometers) would be projected onto the chip at a size of 12.6 micrometers (Table 7.2), and this is roughly twice the size of the sensor. A 40× dry lens with NA = 0.93 will resolve structures of approximately 0.3 micrometers which would then be magnified onto the chip at a size of 12.0 micrometers; Again, roughly twice the size of a single sensor.

The projection to twice the size of the sensor is important because of a key concept in sampling statistics called the Nyquist-Shannon theorem (Nyquist theorem, Nyquist criterion). This theorem is a fundamental principal in sampling theory. Obviously, the fidelity of the analog to digital conversion of signals, such as the irregular analog signal depicted in Fig. 7.11a, is dependent upon how small we set our sampling interval. In the case of analog to digital conversion, the sampling interval sets the quanta size. Figure 7.11c has a smaller sampling interval (digital quanta) than Fig. 7.11b and, thus, produces a better reproduction of the original analog signal. The Nyquist-Shannon theorem shows that an analog signal can be reasonably well reproduced if the smallest element (frequency) of the signal is sampled at least twice. This means we can only have confidence in structures that are at least twice as large as our sampling interval. In the case of a digital image reconstruction, Nyquist-Shannon tells us that our sampling probe (the photodiode elements of the camera) must have a diameter of half the size of the object in both the X and Y directions. Thus, a sensor of 6.45 micrometers is very close to the Nyquist criterion of 6.3 micrometers to sample a resel of 12.6 micrometers. In practice, the sampling interval is often set to sample the smallest detail 2.3× to be sure of capturing adequate information.

Table 7.2 also describes the magnified resel for a 100×, NA 1.3 water lens. You can see that a 6.45 micrometer sensor far exceeds the Nyquist criterion. We can take advantage of this fact and combine, or bin, the information from four adjacent sensors to make a sensor with a size 12.9 × 12.9 micrometers. This is again close to the Nyquist criterion (10 micrometers), although it slightly under samples the resel.

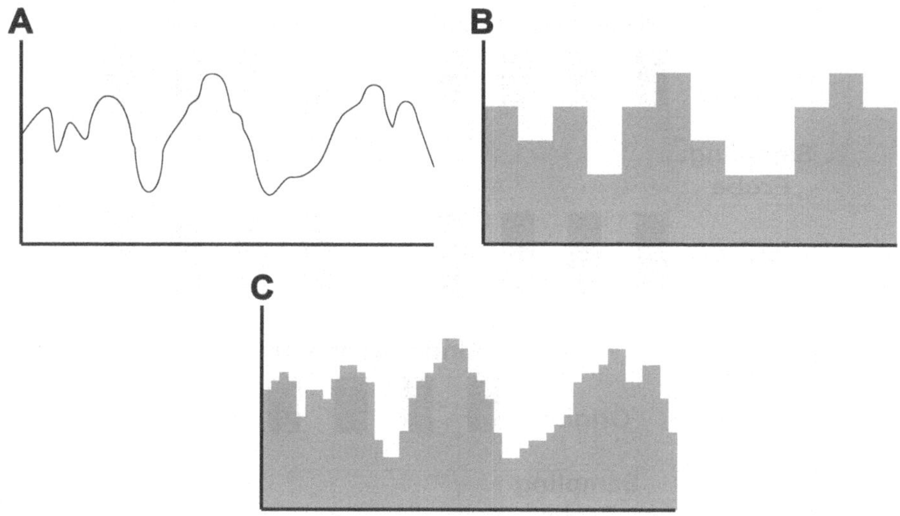

Fig. 7.11 Effect of digital sampling interval on fidelity of signal reproduction

Binning of CCD cameras has the advantage that the image acquisition time can be enhanced and the SNR decreased since we are collecting 4 times the signal with relative little decrease in image resolution (there would be no decrease in image resolution if we exactly matched the Nyquist criterion). Since CMOS cameras do processing at each photodiode, binning pixels of a CMOS camera does not increase readout time, but it does reduce signal to noise. We will have more on binning in Sect. 7.4.

Of course, all of these calculations do not take into consideration that in confocal microscopy, we are generally imaging a thick specimen. Interactions of both excitation and emitted photons with the specimen can reduce the resolution from that which is theoretically calculated. Moreover, there are other factors at play when using high NA lenses and laser beams are polarized, which can distort the point spread function (PSF) in one direction (see Sect. 7.5 for more information on the PSF). Despite these practical issues, it is always better to match as close as possible those parameters you can determine than to ignore physics and hope for the best.

7.3.4 Effect of Missampling

Incorrect sampling frequency can have significant effects on image quality. These are best illustrated by imaging a periodic test specimen, such as a grating with uniform spacing. Comparing the original grating to the image produced as depicted in Fig. 7.12, it is clear that sampling at the Nyquist frequency produces an accurate representation of the original. Oversampling adds no additional spatial information to the final image but does increase the exposure of the sample to a potentially

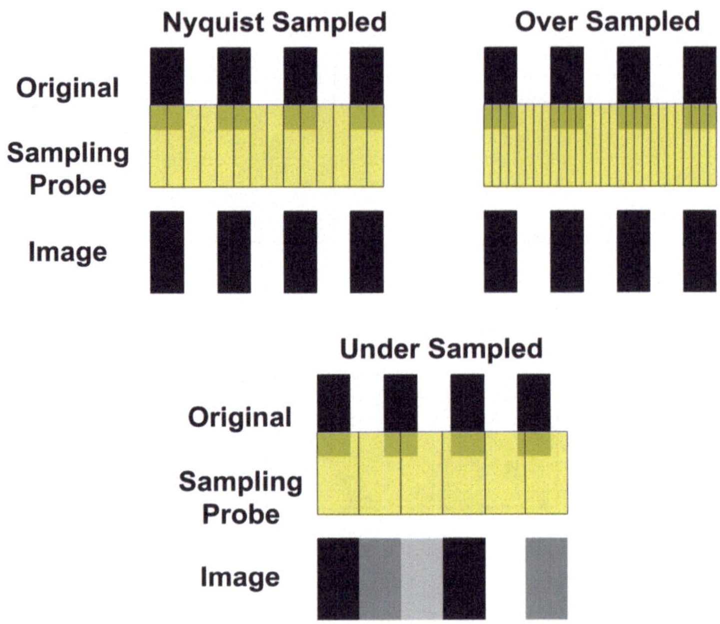

Fig. 7.12 Effect on image fidelity of under- and oversampling a specific frequency

harmful beam. This can exacerbate photobleaching and damage to the sample. In contrast, undersampling significantly alters the representation such that the original pattern is obscured in the image. If this were a real biological sample, this obscuring would blur the detail of structures smaller than twice the sampling frequency. A potentially more problematic effect of incorrect sampling frequency, however, is spatial aliasing. Spatial aliasing is the occurrence in an image of structural information that was not in the original object. This occurs when periodic structures are undersampled at a frequency that leads to misinterpretation of the structural periodicity. Moiré patterns (Fig. 7.13) are an example of spatial aliasing. Figure 7.13b shows a Moiré pattern formed when an image of a brick wall is collected at too low a sampling frequency. It is easy to spot such artefacts when viewing a known structure such as a brick wall. However, when these patterns arise in a biological sample of unknown structure, they can be misinterpreted as specimen fine detail. Spuriously jagged outlines are another example of spatial aliasing.

7.4 CDD Dynamic Range and Array Size

It should be clear by now that the diode size and correct sampling are very important in determining the resolution and quality of the final digital image produced by a CCD or CMOS camera connected to a confocal microscope. However, image

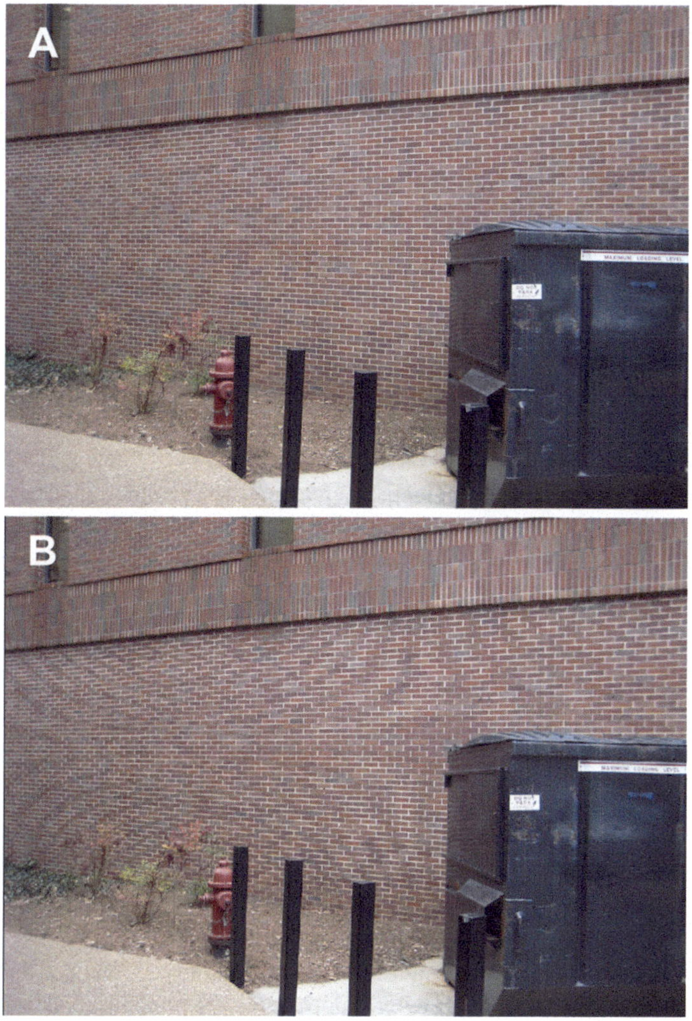

Fig. 7.13 Aliasing artifact caused by undersampling

contrast is just as important to the ability to recognize and record information from closely spaced objects. Thus, the dynamic range of the camera (number of gray values between black and white that the camera can accurately detect) is also critical. The dynamic range is affected primarily by the inherent noise level and the full well capacity of the sensor.

When photons hit the photodiode sensors of the camera, electrons are generated. The well capacity, or well depth, is the number of electrons that can be generated before the capacity of the sensor is exceeded. Obviously, large capacities are desired. However, the smaller the sensor size the smaller will be the capacity. As introduced above, the other major influence on dynamic range is readout noise. Scientific-grade array cameras for microscopy are now cooled, and these, coupled with electronic advances and reproducibility of signal readout, bring the noise in these cameras to very low levels of <10 electrons per sensor at reasonable readout rates.

The dynamic range of a photodiode array camera can be determined as the full well capacity divided by the noise (SNR). A photodiode with a full well capacity of 41,000 electrons and a noise level per well of 10 electrons would be able to detect 4,100 different gray values. In other words, it could produce a 12-bit image ($2^{12} = 4096$ gray levels).

$$\text{Dynamic range} = \frac{41{,}000}{10} = 4{,}100\,\text{gray values}$$

Thus, even if the A to D conversion produced a 16-bit gray value (16 binary digits long), 4 of those digits would be spurious information because the sensor was only capable of an accuracy of 12 bits.

There is, however, a way of actually increasing the dynamic range of a photodiode array system. This is by binning several sensors together into a single virtual sensor. Since read out noise is not additive, binning of sensors produces huge increases in dynamic range. For instance, binning 4 pixels (2 sensors along the X direction and 2 along the Y direction) for the sensors in the example above makes the camera capable of detection of 16,400 gray levels.

$$\text{Dynamic range} = \frac{41{,}000 \times 2 \times 2}{10} = 16{,}400\,\text{gray values}$$

The array size (number of pixels in the X and Y direction) determines the size of the image that can be collected and thus the area of sample (field size) from which information can be collected for each image. Currently available CCD and CMOS cameras for scientific use have array sizes in the range of about 1,000 × 1,000 pixels to upwards of 5,000 × 5,000 pixels and larger. A 1,000 × 1,000 array of sensors with an individual sensor size of 6.45 micrometer would produce a camera that can capture an area 6,450 micrometers × 6,450 micrometers. With a 63x lens in place, this means the confocal image would contain information from 102 micrometers × 102 micrometers (10,404 square micrometers) of sample area. Larger cameras collect more sample information, but for CCD cameras, this comes at the expense of slower readout rates.

7.5 Image Capture in 3-D

The most important aspect of a confocal microscope for most biological studies is the ability to acquire high-resolution information about the three-dimensional architecture of the specimen. As discussed throughout this book, confocal microscopy isolates a defined planer thickness of the sample. In the case of a single slice, the third dimension is the thickness of the optical slice. In cases where a larger volume is built up by adding together multiple slices, the thickness of an individual slice remains a critical component of the 3-D data set. With this in mind, we need to consider the characteristics of all three dimensions of a point source projected into a sample. Just as the lateral dimensions of the image of a point source are blurred to form an Airy pattern, the axial dimension of the image is also blurred. The degree of blurring is referred to as the axial PSF. The PSF describes how the optical system affects the image of a point source along the X, Y, and Z axis. The Airy disk is the lateral component of the PSF. In this section, we will consider the remaining axis: the optical axis (Z axis).

Confocal microscopy of fluorescent objects is an incoherent imaging system. This means that each point of information is collected independently, and if the system is designed and aligned well, there should be no interference between the light waves coming from one point and those from a different point. In other words, the light waves are independent. In the single-point scanning confocal, incoherence is assured because each point is temporally separated. In a multipoint scanning system, each point collected at a single time point is sufficiently spaced as to be optically spatially separate.

The image of a point source along the axial plane (the axial PSF) is an hourglass shape with the smallest point being at the focal point of the lens (Fig. 7.14a). By removing out of focus light, the confocal microscope converts the hourglass to an oblate spheroid (ellipsoid), as shown in Fig. 7.14b. Note that this means the PSF is not a sphere; the axial dimension is larger than the lateral dimension. Thus, the limit of resolution in the axial (Z) direction will not be as good as that in the lateral (X-Y) plane.

Like the Airy disk, the radius of the PSF along the optical axis is dependent on the illumination wavelength and the NA of the lens. Several equations have been generated to define confocal PSF resolution. Equation 7.6 describes the most often used equation and highlights the relationship between wavelength and NA when the pinhole aperture is set to exactly capture the full diameter of the Airy disk focused onto the aperture.

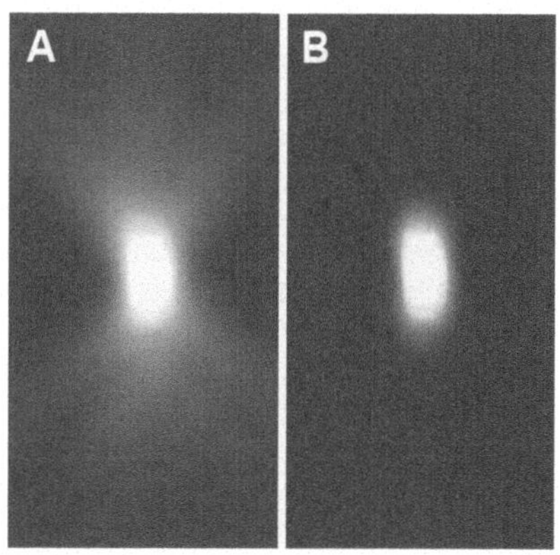

Fig. 7.14 Z axis of point spread function of a 0.15 um bead imaged by a microscope without a pinhole (**a**) and one with a pinhole set to 1 Airy unit (**b**)

$$d_{\text{axial}} = \frac{1.4\left(\lambda^{*} n\right)}{\text{NA}^{2}} \quad (7.6)$$

where: λ = wavelength
n = refractive index
NA = numerical aperture

Just as the Airy disk determines the lateral resolution, the PSF defines the volume resolution (3-D resel) and, therefore, determines the minimum digital voxel size that can be accurately represented in a 3-D digital image collected by a confocal microscope. Table 7.3 presents some practical dimensions for the lateral and axial radii of the PSF for a lens with NA 1.4 in a system using immersion oil with refractive index 1.515. It is worth producing similar tables for the lenses and wavelengths most often used for a confocal microscope and having the data close by the microscope for quick reference. Of course, the resel dimensions listed are the minimum dimensions obtainable. Unlike the lateral resolution, reducing the pinhole below one Airy unit produces little gain in axial resolution but does significantly decrease the number of photons passing through the aperture. In contrast, if one opens the pinhole of the microscope greater than one Airy unit, then more photons pass to the detector. This is because the photons passing are from the out-of-focus regions. This will blur the edges of objects and so will degrade both the lateral and axial resolution. However, opening the pinhole is useful when the signal is too low for efficient detection.

Since the PSF determines the volume being illuminated and imaged, an understanding of the PSF is important for determining how to correctly sample an object along the Z axis. The Z axis sampling rate is set by instructing the microscope how far to move the focal point along the Z axis between collection of each image plane (slice thickness). Taking into consideration the Nyquist criterion, in the optical system defined in Table 7.3 using light of wavelength 510 nm, the slice thickness should be set to 0.276 nm in order to correctly collect the details of the smallest resolvable structures in the sample volume. Luckily, the software that is provided with most modern confocal microscopes can automatically determine the volume being excited and detected based on the pinhole setting.

Often constraints such as acquisition time, fluorophore bleaching, and specimen beam sensitivity preclude one from sampling an object at theoretical resolution limit. For instance, it would take about 72 scans to collect the full volume of a 20-micrometer-thick specimen at a rate of 0.276 micrometers per slice (the frequency required for Nyquist sampling of the smallest resolvable structures). This could well be impractical, particularly if the fluorophore was prone to photobleach-

Table 7.3 Calculated X, Y, and Z dimensions for PSF at different wavelengths of an optical system with refractive index 1.515 and NA 1.4

Color	Wavelength	X and Y	Z
Blue	442 nm	0.129 micrometers	0.478 micrometers
Green	541 nm	0.158 micrometers	0.585 micrometers
Red	690 nm	0.202 micrometers	0.747 micrometers

ing. It is perfectly permissible to use a different sampling frequency, but the pinhole diameter should be adjusted accordingly. The ease of changing the pinhole diameter to match the situation is an advantage of laser scanning microscopes not shared by multipoint scanners.

There are multiple reasons why it is important to match the optical slice thickness to that which is actually being collected by the instrument during image formation. The first, and most obvious, is that if you image two very thin planes that are widely spaced apart, the resultant 3-D image will either be flatter than the specimen (thus distorting 3-D relationships) or the computer will have to fill in the empty space with interpolated information. A second reason to match the optical slice thickness to the collection frequency is that it avoids the possibility of aliasing artifacts. Since detail smaller than the optical slice thickness will not be resolved, very little axial resolution is gained by maintaining a smaller pinhole (although an improvement in lateral resolution is gained by increased contrast). Finally, by opening up the pinhole, a larger signal is collected which can dramatically improve the SNR of the image. All of the compromises made when selecting the pinhole size are addressed further in Chap. 9.

A critical point to recognize from Table 7.3 is that the slice thickness collected using blue excitation light (when the pinhole is set to one Airy unit for blue light) is not the same as when red light (with pinhole set to one Airy unit for red light) is used to excite the specimen. Thus, the axial information in the 3-D resel will not be equivalent. This is illustrated in Fig. 7.15. It compares the axial slice of a cell illuminated with blue light (A) or red light (B). The bright region at the tip of the excitation beam arrow indicates the expected axial PSF. If the resulting 3-D images of sequential blue scans and red scans through the cell are simply combined without consideration of the different depth of each, the software will distort one of the images to match the thickness of the other. This can introduce artifacts into the image. If you want to combine two images collected with different excitation wavelengths, the easiest solution is to set the pinhole to 1 Airy unit for the longer wavelength and then open the pinhole for the shorter wavelength so that the slice thickness matches that of the longer wavelength scan. Most modern point scanning confocal microscopes allow you to do this easily. Chapter 9 provides more details on this process.

The discussion above outlined the factors determining the theoretic minimum PSF. However, alignment errors, lens aberrations, refractive index mismatches or discrepancies, and contamination can increase the PSF or introduce asymmetries in the PSF. The true PSF will also include the interaction of the light waves with the

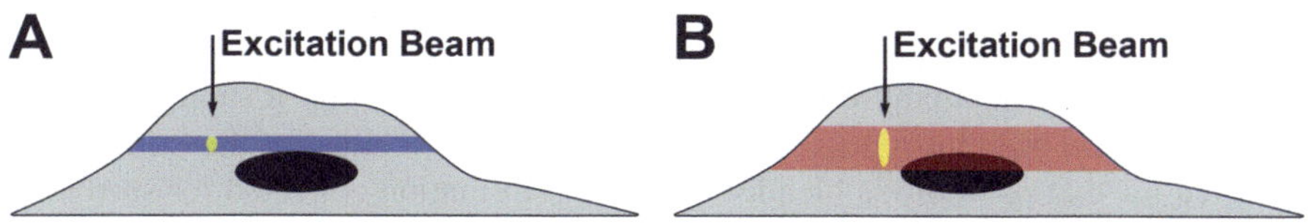

Fig. 7.15 Depiction of difference in axial (Z axis) resolution when a sample is scanned with a beam of blue light versus scanning with red light

sample. However, the effect of interaction with the biological sample is very difficult to predict. This is because the specimen content is rarely homogenous and the composition is usually changing with time. The specimen interaction generally causes further blurring of the image, and this effect becomes more pronounced as thickness of the specimen increases. Luckily, for quick calculations of optimal sampling frequency, we can normally ignore the effect of the sample and use the theoretical calculations based mostly on what we know about the objective lens. Our lack of knowledge about the true PSF does, however, indicate we must be aware that our image may contain artefactual information.

7.6 A Word About Color Image Capture: Don't!

The cameras for confocal microscopy are almost universally black and white cameras. These convert signal strength to a grayscale value within the dynamic range of the instrument. The image does not contain color information. Color information is provided separately by the use of specific lookup tables (LUTs) selected by the user. The color assignment is usually based on knowledge of the filters used to restrict the wavelengths reaching the detector. However, other colors could also be assigned. The green fluorescent protein signal could just as easily be displayed as yellow or red. Specifics of how this is done are provided in Chap. 9.

PMTs are inherently gray scaled and so color must always be assigned. On the other hand, color diode array cameras are available. At present, though, it is generally best to avoid these for high resolution work. The reason for this is that current color cameras acquire color information either at the expense of image lateral resolution, image dynamic range (because of degradation of the signal to noise ratio) or by greatly lengthening exposure time. As discussed earlier, any increase in the exposure of the sample to the beam increases the potential for specimen degradation.

To understand why current color diode array cameras are problematic for high resolution work, we need to discuss the various ways that CCD cameras can capture color information. The most common type of color CCD chip places an array of color filters in front of the photodiodes of the CCD. In this way, only narrow wavelengths of light get through to be detected by the photodiode. The most common type of color filter array is the Bayer filter. This places a red, green, or blue color filter in a specific pattern (Fig. 7.16) in front of each individual sensor. Each unit cell of the filter encompasses four sensors: two collecting green light and one each for red and blue. The color assigned to each pixel in the image is a mixture of red, green, and blue based on the levels of red, green, and blue detected by the sensors. Depending upon the algorithm used for interpolation and anti-aliasing, this binning of pixels reduces the lateral resolution by a factor of two to four. It should be noted that the hue and brightness value assigned each pixel is a blend of the information collected for that photodiode blended with the information collected for surrounding pixels. Another approach is to divide up the output signal into red, green, and blue components and detect them separately with three different detectors and to

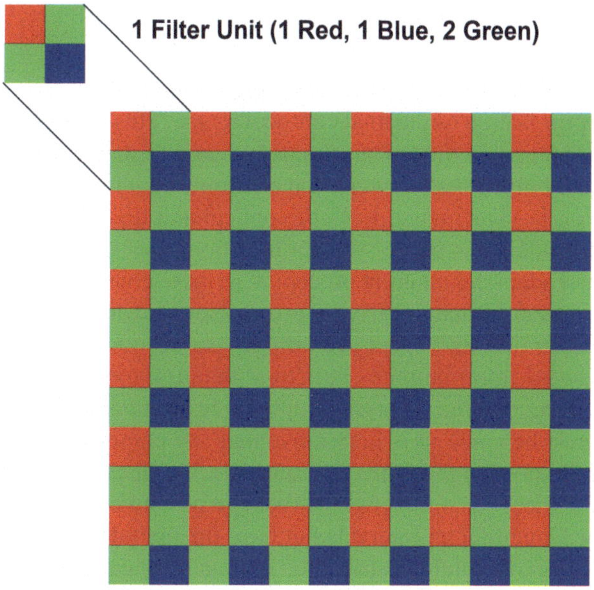

Fig. 7.16 Diagram of typical Bayer filter pattern present on a color CCD array

then recombine the image. This introduces additional elements into the light path that can degrade the signal.

A new development in color sensing arrays has been designed by Foveon (Foveon Inc., Santa Clara, CA). Their design uses three arrays stacked sequentially on top of each other with filters in between each array. Since red light penetrates deeper into silicon than blue light, the first layer captures the blue information. A filter prevents the blue light from traversing through to lower layers. In the same manner, the second array captures the green information, and the bottom layer captures the red information. This scheme allows the full resolution of the chip and avoids problems of post-collection alignment of the red, green, and blue images. So far this approach has only been implemented on photographic cameras and is not commercially available for confocal microscope cameras. However, the technology shows promise if it can be adapted for high resolution (small diode size) confocal microscopy work. For the reasons above, the best approach for confocal microscopy using diode array cameras remains sequentially placing filters in front of the CCD array. This allows the sequential collection of the red, then green, and then blue information separately. The three images can then be combined. This approach uses the full resolution of the camera. It is not without drawbacks, however. One must remember that full color information requires at least three separate exposures of the specimen and, thus, increases the possibility of photodamage. The exposure time for image collection could be reduced to reduce damage, but this increases the noise in the image. Sequential imaging also introduces issues of correct alignment of the three images that can reduce image resolution. Despite these drawbacks, the sequential approach remains the most useful that retains full resolution if full color imaging is required. Thankfully, most of the time, color information can be obtained or implied in other ways.

7.7 Conclusions

It should be clear, now, that there are numerous user selectable options that allow optimization of the final confocal image. This is extremely useful, but it does require diligence on the part of the user to assure image quality and fidelity. Optimization of the information in an image requires maximizing the spatial and contrast resolution in your image. Both the optical and digital restrictions on resolution must be considered as part of the optimization process. Failure to recognize the constraints and adapt the imaging procedure to these constraints at the least can result in loss of image information. Even more problematic, however, is that failure to recognize the constraints and adapt accordingly can lead to the inadvertent presentation of artefactual information as structural detail. Experimental and sample limitations may dictate that the theoretical optimal resolution is not feasible and that compromises must be made. It is imperative that any compromises are made rationally and with full knowledge of the consequences. For instance, it makes no sense to collect optical slices (movement of the focal plane between collecting individual image planes) at 1 micrometer intervals but have the pinhole set to resolve 0.5 micrometer slices. Rather, the pinhole should be set to a resolution along the Z axis which assures that the sampling frequency along the Z axis meets the Nyquist criterion. Without proper consideration of the various resolution constraints imposed by the sample, microscope, and detector, the data contained in an image could be incorrectly interpreted.

Literature Cited

Chen T, Catrysse P, Gamal A, Wandell B (2000) How small should a pixel size be? In: Blouke M, Sampat N, Williams G (eds) Proceedings SPIE, vol. 3965, Sensors and camera systems for scientific, industrial and digital photography applications. SPIE, pp 451–459

Demmerle J, Innocent C, North AJ, Ball G, Muller M, Miron E, Matsuda A, Dobbie IM, Markaki Y, Schermelleh L (2017) Strategy and practical guidelines for successful structured illumination microscopy. Nat Protoc 12:988–110

Heddleston JM, Chew T-L (2016) Light Sheet microscopes: novel imaging toolbox for visualizing life's processes. Int J Biochem Cell Biol 80:119–123

Lippincott-Schwartz J, Manley S (2009) Putting super-resolution fluorescence microscopy to work. Nat Methods 6:21–23

Magnan P (2003) Detection of visible photons on CCD and CMOS: a comparative view. Nucl Inst Methods Phys Res A 504:199–212

Sahl SJ, Hell SW, Jakobs S (2017) Flourescence nanoscopy in cell biology. Nat Rev Mol Cell Biol 18:685–701

Chapter 8
Types of Confocal Instruments: Basic Principles and Advantages and Disadvantages

John Fuseler, W. Gray (Jay) Jerome, and Robert L. Price

8.1 Introduction

As noted in the Historical Perspectives section of Chap. 1, several types of confocal laser scanning microscopes (CLSM) have been introduced over the years. At this point, these can be broken down into four basic categories: single-photon point-scanning confocal systems, multiphoton (nonlinear) point-scanning confocal systems, spinning disk confocal microscopes, and super-resolution systems. New developments are continually being added to the hardware and software of these microscopes to improve their performance, but the majority of confocal systems will fall into one of these groups. Table 8.1 compares many of the features of single-photon, multiphoton, and spinning disk systems.

J. Fuseler
Department of Pathology, Microbiology and Immunology, University of South Carolina School of Medicine, Columbia, SC, USA

W. G. Jerome (✉)
Department of Pathology, Microbiology and Immunology, Vanderbilt University School of Medicine, Nashville, TN, USA
e-mail: Jay.Jerome@Vanderbilt.edu

R. L. Price
Department of Cell Biology and Anatomy, University of South Carolina School of Medicine, Columbia, SC, USA
e-mail: Bob.Price@uscmed.sc.edu

W. G. Jerome, R. L. Price (eds.), *Basic Confocal Microscopy*,
https://doi.org/10.1007/978-3-319-97454-5_8

Table 8.1 Comparison of several characteristics of single-photon CSLM, multiphoton CSLM, and spinning disk confocal systems

Characteristic	Single-photon CSLM	Multiphoton CSLM	Spinning disk
Photon source	High-intensity short-wavelength lasers	Adjustable lasers capable of long wavelengths (IR)	Lasers, mercury, xenon, or metal-halide lamps
Acquisition speed	At full-resolution and frame size scan speeds can be several seconds per frame. Fast scans of regions of interest at reduced resolution available on some systems	Similar limitations as single photon	Rapid, high frame rates of 20^{+} frames per second
X-Y resolution	Up to 4096 × 4096 on some systems	Up to 4096 × 4096 on some systems	Depends on CCD elements and binning
Z- resolution	~0.6 μm optimum based on 63× NA1.4 oil objective with pinhole at 1 airy unit		Typically fixed (micro-lenses)
Detectors	Photomultiplier tubes (PMTs) or related systems (see Chap. 7)	Photomultiplier tubes (PMT) or related systems (see Chap. 7)	Charge-coupled device (CCD) or complementary metal-oxide semiconductors (CMOS)
Multi-channel imaging	Simultaneous or sequential	Simultaneous or sequential	Sequential
Photobleaching and phototoxicity	May be problematic	Typically not a problem	Reduced based on fast imaging parameters
Region-specific bleaching	Possible on instruments that allow region of interest control of excitation areas	Possible on instruments that allow region of interest control of excitation areas	Requires additional hardware available for some systems
Depth of imaging	Limited based on optics and wavelength of laser	Greatly improved over single photon based on long-wavelength excitation	Limited

8.2 Single-Photon Point-Scanning Confocal Microscopes

These instruments, by far, represent the majority of confocal microscopes in today's laboratories and include basic instruments with two or three lasers all the way up to very advanced spectral imaging microscopes such as the Zeiss LSM 880, Leica SP8, Nikon A1^{+}, and Olympus Fluoview FV3000 instruments. Resonant scanning systems, designed to increase the rate of scanning for live cell imaging, also fall within this group of instruments.

Figure 8.1 illustrates a simplified diagram of the path that photons take prior to and after interaction with the specimen in a single-photon point-scanning confocal

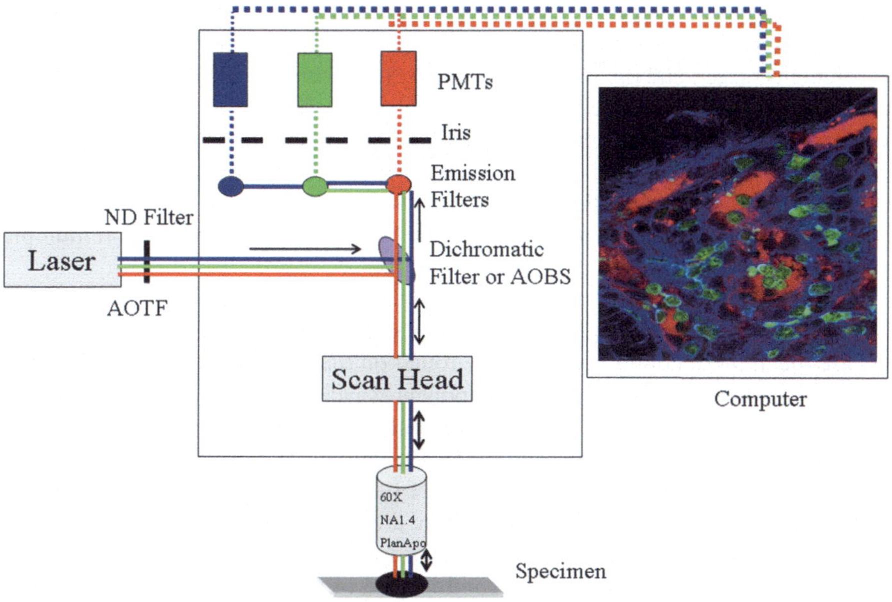

Fig. 8.1 Generalized diagram of the mechanism for sequentially exciting and collecting signal from a sample labeled with three fluorochromes

system. After exiting the laser, photons pass through a neutral density filter or an acousto-optical tunable filter (AOTF) that allows control of the laser intensity that contacts the specimen. The laser beam then passes through a dichroic filter present in most confocal systems or an acousto-optical beam splitter (AOBS) in Leica microscopes. This filtering mechanism separates the higher-energy short-wavelength excitation photons from the lower-energy long-wavelength photons emitted from the specimen. The laser beam then passes through the scan head and objective lens before interacting with the specimen. As the laser is scanned across the specimen, fluorophores are excited resulting in emission of photons from the specimen.

Photons emitted from the specimen take the reverse path (epifluorescence microscopes) of the excitation photons and first pass through the objective and then through the dichroic filter or AOBS. In the example shown in Fig. 8.1 blue, green and red laser lines are used to excite the specimen. Emitted photons are then sorted by placing a dichroic mirror and long pass filter in front of the first detector, so only red light passes through to the detector. Shorter wavelengths are redirected. In most laser scanning confocal systems, the detector is usually a photomultiplier tube (PMT), although many systems are now incorporating higher sensitivity GaAsP detectors (Sect. 7.3.1). Many laser scanning systems, such as the one depicted in Fig. 8.1, have multiple detectors, and in newer systems, combinations of PMTs and GaAsP detectors are becoming common. Confocal detectors do not recognize color; they collect and record photons of all wavelengths. It is the combination of filters between the specimen and detector that allow us to assign specific colors (wavelengths) to the detected photons. In the case of a three-PMT detector system, when the long (red) wavelengths are passed on to a PMT, the remaining shorter wavelengths (green and blue) are redirected by the dichroic or AOBS to other detectors. A second dichroic directs the green wavelengths through a band-pass filter to a second detector, while shorter (blue) wavelengths are redirected to a third detector, and long (any remaining red) wavelengths are blocked and do not progress further. The third PMT receives only the short wavelength (blue) light left in the emitted light. Combinations of filters and detectors can be used in this manner to sort multiple wavelengths of light.

When the first edition of this book was published, spectral detectors were a relatively new development in laser scanning confocal microscopy. Today spectral detectors are much more common and can be found on many confocal systems. Spectral detectors separate the light into its component wavelengths and have the ability to detect very narrow ranges (10 nm or less) of light, allowing the microscopist to more precisely analyze the spectrum of wavelengths emitted from a specimen. However, in practice, selection of too narrow a band of light results in a poor signal-to-noise ratio (SNR), so often bands of 20 nm or more are used for imaging.

When planning an experiment, it is very important to know the filters available on the instrument before selecting a combination of fluorochromes to use for multiple labeling experiments. Improper selection of fluorochromes and filters may result in problems such as bleed through and poor signal levels as discussed in Chap. 3.

Prior to reaching the detector, a pinhole (aperture, iris) is placed in the light path at the light beam's focal point. This is a very critical component of a confocal microscope because this is the point where out-of-focus light is separated from light coming from the focal plane. In some systems, a pinhole is placed in front of each detector, while in others, a single pinhole is used to remove out-of-focus light prior to the photons being sorted and sent to the various detectors. Proper use of the pinhole is important in achieving the essential feature of confocal microscopy so that only light from the focal point of the objective is used in image formation. Thus, *the focused point of the specimen and detection of the signal is in conjugate focus (confocal)* resulting in the improved resolution, contrast, and SNR characteristic of confocal images. Once the emission photons are sorted, the information is converted to a digital image as discussed in Chaps. 6 and 7.

The ability to remove out-of-focus light from the image and the use of high-resolution and very sensitive detection devices, along with microscope stages that can be tightly controlled in movement in the Z direction, provide the capability to collect a series (Z-series) of in-focus images throughout the depth of a specimen. This feature allows high-resolution scanning in the X, Y, and Z axes for the collection and reconstruction of true three-dimensional data sets, which is a major advantage of CSLM systems when compared to widefield fluorescence systems (Fig. 1.4).

Choices made by the operator at each point along the optical path affect both the amount of specimen damage that may occur and image quality. Factors that increase the intensity of the laser contacting the specimen or the length of time the laser remains in contact with a point in the specimen will increase the number of photons (signal) available for formation of the image but will result in increased specimen damage in the form of quenching of fluorescent reagents and/or cell death if imaging live cells. Other factors, such as increasing the diameter of the pinhole, will alter image parameters such as SNR and resolution but will not directly affect specimen damage since interaction of the laser with the specimen has been completed. Understanding the multiple compromises made in collection of an image is essential since nearly every change made in operating parameters will affect specimen damage, image quality, or both. Chapter 9 provides details on how the microscope should be set up to maximize specimen preservation and image quality and discusses many of the compromises in imaging that must be made.

8.2.1 Limitations of Single-Photon Confocal Systems: Cost

An obvious limitation in the use of confocal microscopes is the cost of acquisition and maintenance. Depending on configuration, commercially available basic instruments often approach $300–400 K, and advanced systems very often exceed $1 M in cost. Instruments of this magnitude typically require service contracts that are in the neighborhood of $20–25 K per year for basic instruments and often more than $50 K per year for high-end instrumentation. For this reason most systems are placed in core facilities or within groups of well-funded investigators.

8.2.2 Limitations of Single-Photon Confocal Systems: Difficult to Operate

Confocal microscopes are also relatively difficult to operate requiring, at a minimum, a part-time operator for routine maintenance and training of new users. The importance of a skilled operator for training cannot be underestimated as the operating parameters of a confocal system can easily be set up incorrectly resulting in the collection of data that does not accurately represent the sample. This can lead to artifacts with regard to the size of structures, three-dimensional reconstructions, and interpretation of fluorochrome co-localization as discussed in Chaps. 10 and 11. Helping the microscopists identify these critical parameters and understanding how to properly set them are key goals of this book.

8.2.3 Limitations of Single-Photon Confocal Systems: Speed of Acquisition

Another distinct disadvantage of the point-scanning systems is that image acquisition is a relatively slow process. In these systems the lasers are scanned across the specimen in an X-Y direction by galvanometer mirrors in the scan head. The speed of the mirrors is limited which in turn limits the rate the laser can be rastered across the surface of the sample. As a result, it often takes 2–3 s to collect a single frame. If the goal of the experiment is to image rapidly occurring events such as a calcium flux, imaging must be completed in milliseconds rather than seconds.

Spinning disk (discussed in detail in Sect. 8.4), slit scanning, and resonant scanning systems are modifications of confocal technology that have been developed to address the problem of slow image acquisition. Slit scanning systems use a wedge of excitation light rather than a point of light and slit apertures rather than pinhole apertures to decrease the time required to scan a sample. Nikon has also produced a Swept Field System that is a combination of point and slit scanning technology so that the option of high-resolution or high-speed imaging can be selected based on the goal of the experiment.

In high-speed resonant scanning systems, rather than rastering the beam, the galvanometer mirrors vibrate at a fixed frequency to rapidly move the beam across the specimen. This significantly increases the speed of scanning and results in scans that take approximately 125 μs. At this speed resonant scanners are capable of collecting 30 frames or more per second which is fast enough to record most events in live cell imaging. The speed of acquisition also reduces many of the phototoxicity problems associated with using high-intensity lasers since exposure is for a very short period of time.

8.2.4 Limitations of Single-Photon Confocal Systems: Photobleaching and Phototoxicity

Associated with the problem of slow speed of image acquisition are the problems of photobleaching of fixed samples and phototoxicity if imaging live cells. As noted above, the high-intensity lasers of single-photon systems may result in considerable photodamage resulting in photobleaching of fluorescent dyes and the death of cells if imaging living samples. Both multiphoton and spinning disk systems discussed below minimize these types of damage.

Much of the photobleaching problem with single-photon CSLM systems is that, as is the case with epifluorescence microscopes, the entire volume of the specimen is exposed to the high-intensity laser. This means that as the focal plane of the specimen is being scanned, the optical planes above and below the focal plane are also exposed to the high-intensity photons of the laser (Fig. 8.2). This may result in specimen damage throughout the specimen. In addition, emitted photons from the out-of-focus planes can potentially contribute to the final image resulting in decreased resolution and image quality. As described in Sect. 8.3, multiphoton excitation can help minimize the beam interaction with fluorophores in the out-of-focus planes.

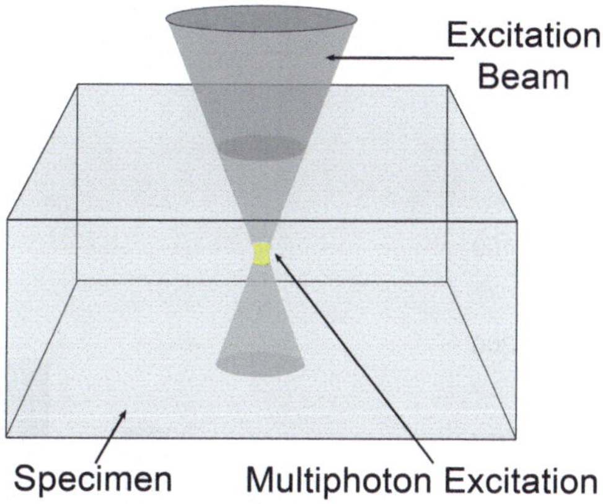

Fig. 8.2 The photon density is only high enough at the focal point of the excitation beam for two-photon absorption to occur. Fluorophores above and below the plane of focus remain unexcited

8.2.5 Limitations of Single-Photon Confocal Systems: Resolution

Resolution was discussed extensively in Sect. 7.2 and the equation for calculation of resolution given in Eq. 7.1. Traditionally the resolution obtained with a confocal microscope is typically compared to that of a widefield epifluorescence system and is therefore considered to be excellent. However, even with confocal systems, the imaging source is comprised of light which has a long wavelength (λ) compared to electrons, and with the introduction of several super-resolution instruments, realization of resolution limitations in traditional confocal microscopy has become more evident.

The diffraction limitations on resolution when using photons as the imaging source were important factors in development of several super-resolution light microscopy techniques that received the 2014 Nobel Prize in Chemistry for Drs. Betzig, Hell, and Moerner (https://www.nobelprize.org/nobel_prizes/chemistry/laureates/2014/press.html). A full discussion of super-resolution techniques goes beyond the topic of basic confocal microscopy addressed here, but a brief introduction to the technique and reference to several reviews (Leung and Chou 2011; Yamanaka et al. 2014; Sydor et al. 2015; Laine et al. 2016) that discuss the topic are warranted. The importance of super-resolution techniques is easily recognized by examining the number of recent publications using the technology. A March 2018 search of NIH PubMed using "super-resolution microscopy" as the keyword search term showed a rapid growth in use of the technology as reported in publications (Fig. 8.3).

Most super-resolution techniques reduce the point spread function (PSF) generated from a fluorochrome and deconvolution to supersede diffraction-limited resolution. Reduction of the PSF may be by use of lasers to deactivate part of the PSF as in STED (stimulated emission depletion microscopy) and ground-state depletion (GSD) microscopy, through the use of sparsely labeled photoactivatable fluorochromes such as photoactivated localization (PALM) and stochastic optical recon-

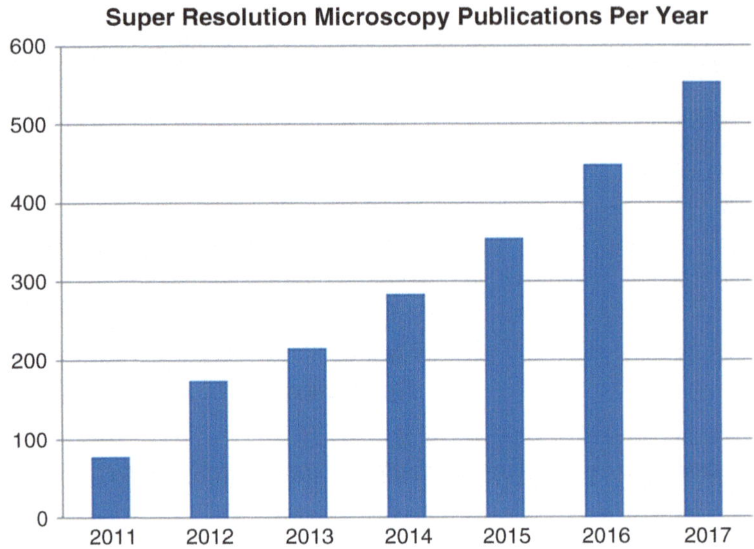

Fig. 8.3 Chart showing the rapid increase in super-resolution papers published since 2011

struction (STORM) microscopy, or through computational reconstruction of several images as in structured illumination microscopy (SIM). There are also many variations of these and other techniques that are being developed to circumvent the PSF. These include combinations of techniques such as saturated SIM (SSIM) that uses a combination of computation reconstruction in conjunction with the principle of STED. All of these techniques are reviewed in detail in the reviews cited above.

The above techniques all achieve resolution in the range of tens of nanometers and have various limitations such as the use of specific lasers and or fluorochromes. Several manufacturers have developed systems that improve on the practical resolution limits (200 nm or more) of confocal microscopy and achieve resolution in the range of 140 nm. A major advantage of these systems, such as the Zeiss Airyscan, the Leica HyVolution, and the Olympus Fluoview 3000-OSR is that they can often be used with specimens prepared with standard specimen preparation fluorochromes and protocols previously discussed. The primary advances in technology used in these enhanced resolution systems are high sensitivity GaAsP and Leica HyD detectors that require very little signal to saturate. The high sensitivity allows the pinhole to be decreased to approximately 0.7 Airy units (AU) improving resolution from the standard optimum setting of 1 AU. In combination with deconvolution, the reduced pinhole size attains the improved resolution.

The significance of super-resolution techniques, whether it is true super resolution in the range of 50 nanometers or less or in the range of 140 nanometers, is our understanding of biomolecular mechanisms. Many co-localization studies that were previously performed with standard confocal imaging systems that experienced diffraction-limited resolution in the range of 200 nm or more must now be questioned. Figure 1.4 showed a comparison of traditional confocal and Airyscan images from the brain hippocampus. Figures 8.4 and 8.5 show super-resolution images that further illustrate the improved resolution associated with PALM and STORM instruments. Structures that previously could not be resolved due to the PSF are now seen as distinct structures.

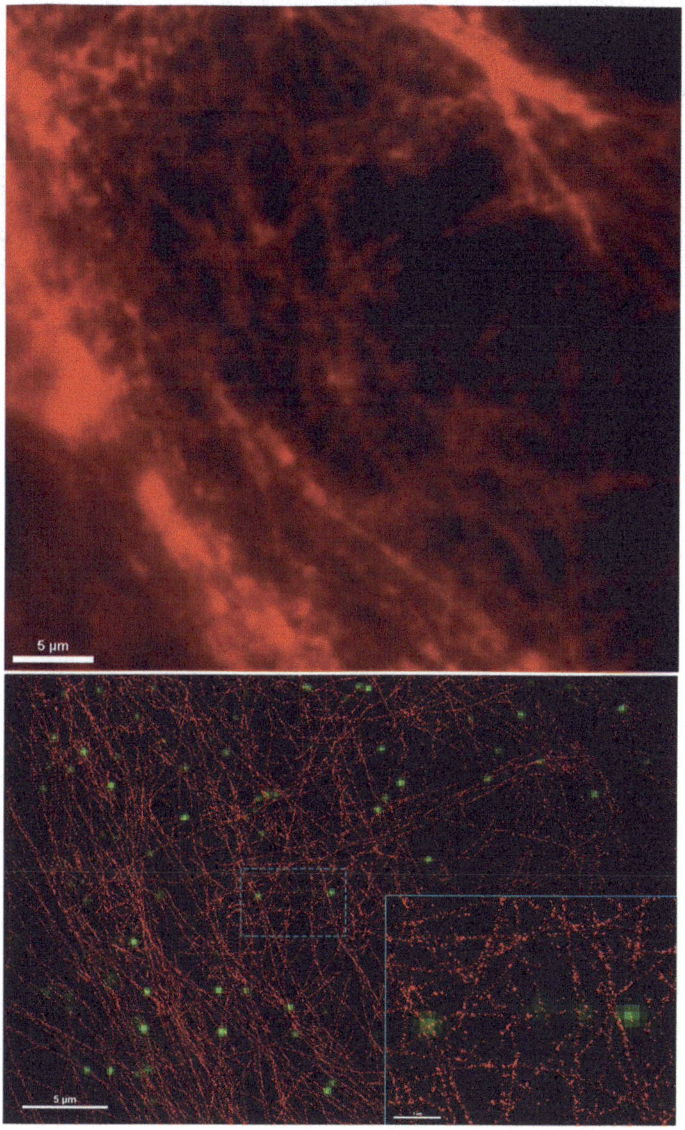

Fig. 8.4 Comparison of widefield (top) and direct stochastic optical reconstruction microscopy (dSTORM) images showing increased resolution of alpha-tubulin labeling in CO-1 cells. Images provided courtesy of John Allen from Nikon and Bryan Mills and Meagan Postema from Vanderbilt University

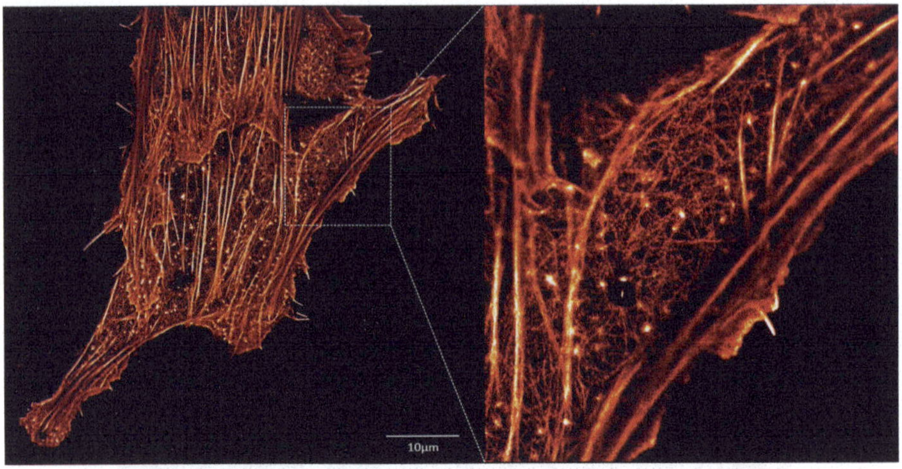

Fig. 8.5 Super-resolution interferometric Photoactivation and Localization Microscopy (iPALM) image of actin in NIH-3 T3 cells. Images provided courtesy of Teng-Leong Chew: https://www.aicjanelia.org/ipalm-techspecs

Table 8.2 Comparison of the wavelengths of visible light available for confocal imaging and electrons when an electron microscope is operated at 60 kV

Color/source	Wavelength in nm
Electrons	0.005
Violet	380–450
Blue	450–490
Green	490–560
Yellow	560–590
Orange	590–630
Red	630–760

Even with the development of super-resolution imaging techniques, the use of photons continues to limit the ultimate resolution that can be obtained with light microscopy based on the physics of photon wavelengths. In comparison, the resolution available with an electron microscope is orders of magnitude higher. As shown in Table 8.2, the wavelengths of visible light are significantly longer than those of electrons with the shortest wavelength of light typically used in confocal microscopy being ~ 400 nm. When operated at an accelerating voltage of 60 kV, which is relatively low for transmission electron microscopy (TEM), the wavelength of electrons is 0.005 nm. When these wavelength values are inserted into Eq. 7.1 for resolution, the theoretical resolution of a TEM operated at 60 kV is 0.005 nm, although in practice due to astigmatism and spherical and chromatic aberrations the actual resolution is closer to 0.1 nm. This is still an order of magnitude better than the reported 10 nm resolution obtainable with some super-resolution microscopes. Figure 8.6 illustrates the difference in available resolution when imaging cardiac myocytes labeled with fluorochromes by CSLM and with colloidal gold particle

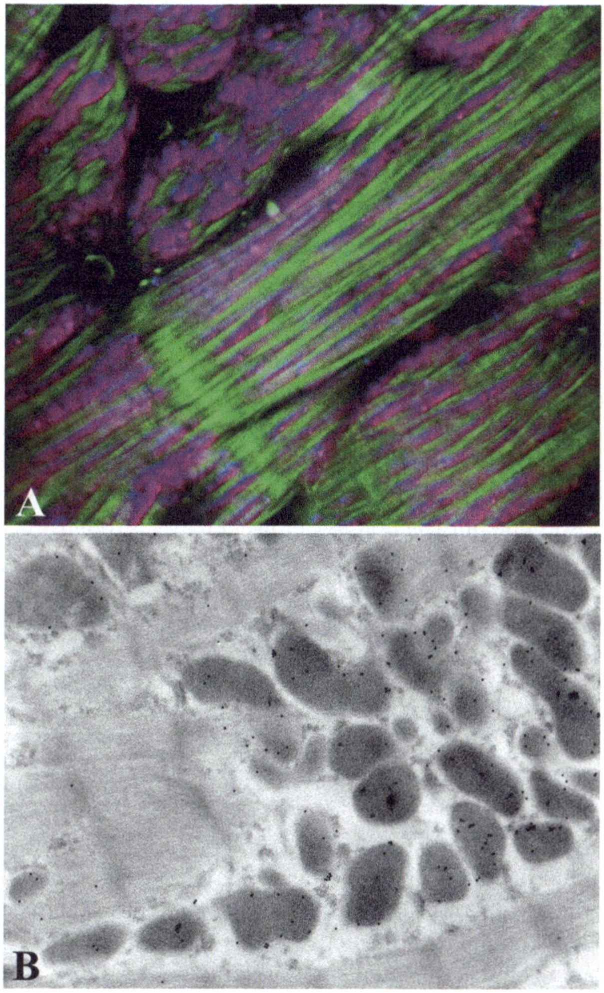

Fig. 8.6 Comparison of resolution levels with confocal microscopy (**a**) and transmission electron microscopy (**b**). In (**a**), cardiac myocytes are labeled with phalloidin for f-actin (green), MitoTracker (red), and cytochrome C (blue). Cytochrome C is primarily localized to the mitochondria creating the purple color. In (**b**), even with minimal fixation for subsequent immunocytochemistry, actin fibers can be seen in the cardiac myocytes cytochrome C labeled with 10 nm colloidal gold particles show clear localization in the mitochondria

labeling for imaging by TEM. While specimen preparation for immunocytochemistry at the TEM level of resolution is much more difficult than for confocal microscopy, TEM does provide an essential correlative technology to confocal and super-resolution imaging for high-resolution labeling. Additional details and examples of probes that can be used for both CSLM and electron microscopy for correlative studies were presented in Chap. 5.

8.3 Multiphoton Point-Scanning Confocal Systems

As described in Chap. 2, in addition to excitation by a single photon, fluorochromes can be excited if more than one photon is absorbed simultaneously. Although the potential for multiple absorptions of photons with different energies exists, in practice, it is usually absorption of two photons of similar energies that is most often exploited. Thus, for the remainder of this section, we will focus on two-photon excitation microscopy. The quantum possibility of two-photon absorption was first proposed by Maria Göppert-Mayer in 1931. However, technical limitations delayed a practical test of the theory. As will become clear later in this section, the invention of appropriate lasers provided the needed excitation source for two-photon excitation. With the development of these laser microcopy systems has come the development of two-photon microscopy based on the pioneering work of Winfried Denk and others in Watt Webb's laboratory at Cornell University (Denk et al. 1990).

The general theory of two-photon excitation is relatively simple. The excitation of a fluorochrome requires the absorption of energy from an impinging photon. Once energy is absorbed, the fluorochrome is refractory to additional energy absorption for a finite amount of time. However, if two photons arrive within the absorption cross section of the fluorochrome simultaneously, the energies of both can be absorbed as long as their combined energies match the energy needed to excite the fluorochrome. In this respect, simultaneously means within about 10^{-18} s. Thus, the absorption of two photons of lower energy is roughly equivalent to the absorption of a single photon of higher energy. Since the energy of a photon is inversely proportional to its wavelength, the two lower-energy photons should be about twice the wavelength of the corresponding single higher-energy photon. For example, a fluorophore that normally absorbs at 400 nm can also be excited by two photons of wavelength about 800 nm. Of course, neither of the 800 nm photons alone will have sufficient energy to excite the fluorochrome, so unless both photons arrive essentially simultaneously, the fluorochrome will not be excited.

Because the two-photon excitation requires simultaneous absorption, the generation of fluorescence emission depends on the square of the excitation intensity. In practice, the probability of two photons falling within the excitation cross section of a fluorophore is very small except at the focal point of the excitation beam. At this point, the photons are sufficiently crowded together so that a significant number of two-photon excitations occur. This results in localized excitation within a narrow region around the focal point of the excitation beam as depicted in Fig. 2.4. The fact that a highly focused beam of light greatly increases the probability of two-photon excitation means that the scanning confocal microscope would be an ideal tool for two-photon imaging.

The probability that two photons will hit a fluorochrome simultaneously and be absorbed is dependent upon the localized photon density. The density required for two-photon absorption is estimated to be about a million times greater than the density required for single-photon absorption. Thus, two-photon excitation requires very high photon densities. Unlike arc lamps, high-power lasers can provide sufficient photons for two-photon excitation. However, as discussed throughout this book, overly high excitation fluxes can damage both specimen and fluorophore. To get around this, the laser excitation for two-photon excitation is pulsed. In this way, very high photon densities are delivered during the pulse, but the average laser energy over time is low (Denk et al. 1995). Pulse durations are ultrashort ranging from around 100 femtoseconds to 1 picosecond with duty cycles (pulse duration divided by time between pulses) of about 10^{-5}. This is accomplished using mode-locked lasers. Although short, the pulses are still longer than the time scale of absorption events (~10^{-18} s), and so two-photon excitation is facilitated.

Several distinct advantages accrue from two-photon excitation. Since a signal is only generated at the focal point, there is no fluorescence from the out-of-focus planes. This means that the pinhole can be dispensed with in cases where you need to maximize signal. However, better resolution is still obtained by using a confocal pinhole (Gauderon et al. 1999), and so, in practice, it is useful to use an optimized pinhole unless emission intensity is limiting.

An additional advantage of two-photon excitation related to the localized excitation is that by limiting excitation to the focal plane, two-photon excitation can decrease excitation-induced specimen damage. With single-photon excitation, the entire beam path (Fig. 8.2) within the specimen is subject to potential specimen damage and photobleaching (discussed in detail in Chap. 9). In contrast, with two-photon excitation, only the small focal volume is excited. Damage and photobleaching are limited to this small excitation region. This is of limited utility if you are only imaging a single plane, but it is a tremendous advantage when collecting a stack of focal planes for 3-D reconstructions. With single-photon excitation, fluorochromes above and below the focal plane are also excited. Thus, with multiple passes, each fluorochrome is excited multiple times, and the probability of specimen damage or photobleaching is increased. However, with multiphoton excitation, each fluorochrome is excited only once. Moreover, since there is no out-of-focus absorption, more of the excitation beam reaches the plane of focus and is available to excite fluorophores. This can lead to a 2–three-fold increase in the depth of tissue that can be successfully imaged (Centonze and White 1998). Finally, the two-photon excitation wavelengths used are, by necessity, longer wavelengths, usually in the red to far-red range. These wavelengths of light have the ability to penetrate deeper into wet tissue and produce less tissue damage than shorter wavelengths. Added together, these benefits mean that two-photon excitation provides a very distinct advantage for imaging thick living tissue (Piston 1999).

Of course, all of the advantages of multiphoton excitation are negated if the specimen is prepared incorrectly or the microscope is not set up to maximally exploit the multiphoton capabilities. Chapters 4, 5 and 9 discuss proper specimen preparation and optimization of microscope parameters. Here we will only point out two important considerations which are sometimes overlooked when doing two-photon imaging of live material. The first is matching refractive indices. Using the PSF of a fluorescent bead to measure effective beam penetration, Gerritsen and deGrauw (1999) showed that two-photon excitation allowed imaging at much greater depths compared to single-photon laser scanning confocal, but this advantage was greatly reduced if oil immersion optics were employed. The degradation of the image was due to chromatic aberration introduced because of refractive index mismatches between oil and the essentially water environment of the specimen. In contrast, the advantages of two-photon excitation were maximally exploited when water-immersion optics were used.

The second mistake that one should be diligent to avoid is using the wrong optics. Since two-photon excitation involves long wavelength photons, the objective lens (when using an epifluorescence setup) must be suited to using far-red and infrared light. Ideally, the lens should have high transmittance in these wavelengths, provide minimal pulse broadening, be corrected for infrared wavelengths, and have a suitably long-working distance for imaging thick samples. Luckily, many of the confocal microscope manufacturers are now making very good lenses suitable for two-photon microscopy, so finding suitable lenses is not as difficult as it was in the past.

8.4 Spinning Disk Systems

Unlike laser scanning systems, where the image is formed by moving a diffraction-limited spot of laser light across the object plane of the specimen in a raster pattern, in spinning disk systems, the image is generated by the simultaneous illumination of multiple spots in the object plane. Each spot is generated by a series of micro-lens or pinhole apertures in a rapidly spinning disk (Chap. 1, Fig. 1.2).

8.4.1 The Nipkow Disk

In 1884 two seminal events in the progress of the development of light microscopy occurred. At about the same time as Abbe (1884) published his milestone work which provided the basic foundation for modern light microscopy, a young student in Germany, Paul Nipkow (1884), discovered how to encode a two-dimensional optical image into an electrical signal that could be transmitted as a one-dimensional or serial, time-dependent signal, similar to that of Morse code. This signal could then be transmitted over a single cable. Prior attempts at accomplishing this feat had resulted in the development of a highly complicated apparatus requiring multiple detectors and cables which gave very limited results. Nipkow's genius in solving this problem was to dissect the image by scanning it in a raster pattern using a spinning disk. The disk was opaque to light except where perforated with a series of rectangular holes spaced under specific geometrical conditions. The holes in the disk were placed at a constant angle relative to the center of the disk and at a constant, progressively decreasing radius to generate the raster pattern. Such a geometrical arrangement of apertures in the disk constitutes an Archimedes spiral (Fig. 8.7a). When such a disk is rotated at a constant velocity, the brightness of each image element passing through the apertures in the disk produces an image in a raster pattern. The brightness or intensity of each image element is essentially constant along the radii from the center to the margin of the disk. In addition, the arrangement of the pinholes describes the image as parallel concentric raster arcs. The arc of the pinholes also progressively traces out the image as adjacent parallel straight lines along the radii at any given X-Y position over the specimen. The image output resulting from the raster pattern is captured by an appropriate device such as a charge-coupled device (CCD) camera.

The arrangement of the pinholes defined by the Archimedes spiral results in a pattern which expresses equal peripheral pitch along the spiral pattern of the pinholes. In addition, the pattern produces equal radial pitch which results in equal illumination and scanning independent of the radius and rotation speed of the disk. Thus, there is minimal to no distortion of the image when it is obtained by the Archimedes spiral arrangement of pinholes in the Nipkow disk.

Alternate patterns of pinholes in the spinning disk are unsatisfactory for confocal image acquisition. When the pinholes are arranged in an equal fixed angular spiral or circular pattern, the light intensity at the outer margin of the disk is less than that in the inner regions of the disk. This unequal distribution of light intensity across the disk is the result of the pitch between the pinholes becoming wider toward the outer margin of the disk. The increased pitch results in an increasing distance between the pinholes at the periphery of the disk. Thus, in a given unit of time, fewer pinholes collect light information at the outer margin of the disk resulting in progressive degradation of the image in this region. If the pinholes are arranged in an equally spaced tetragonal pattern, there is no loss of light intensity across the disk. Since the pinholes are equally spaced, there is no imbalance of light intensity between the inner and outer peripheries of the disk. However, when the disk is rotated, the scanning pitch of the pinholes is not equal which results in the generation of light and dark stripes across the image (Tanaami et al. 2002).

The modern Nipkow disk (Fig. 8.7b) consists essentially of two major components. The core of the disk is an optically flat glass disk which is rigid enough to withstand rotation speeds of up to 2000 rpm. On this glass disk, the Nipkow arrangement of pinholes is produced by photolithographic methods. This surface consists of a reflecting black chrome layer (< 1.0 μm thick) with a reflectivity of only a few percent laid down on a glass disk. The most important criteria in the design of the Nipkow disk for confocal microscopy are the size and spacing of the pinholes.

8.4.2 Nipkow Disk Pinhole Size

In spinning disk systems, collimated light impinges on the disk and the effective aperture of the light rays entering the microscope objective lens are determined by diffraction at the disk pinhole. The magnitude of diffraction by the pinhole is a function of the light wavelength and pinhole size (R-radius). Calculations have demonstrated that resolution in the Z direction (d_z) is directly proportional to the square of the radius (R) of the pinhole (Kino 1995). If the pinholes in the disk are too small, loss of illumination intensity by light rays being diffracted outside of the objective lens aperture occurs, and image information from emitted photons is lost as these photons are not able to enter the pinhole (Fig. 8.8a).

When the pinholes are too large (Fig. 8.8b), resolution of the image in the X-Y and Z directions is significantly reduced. In this case, the collimated light rays a

b

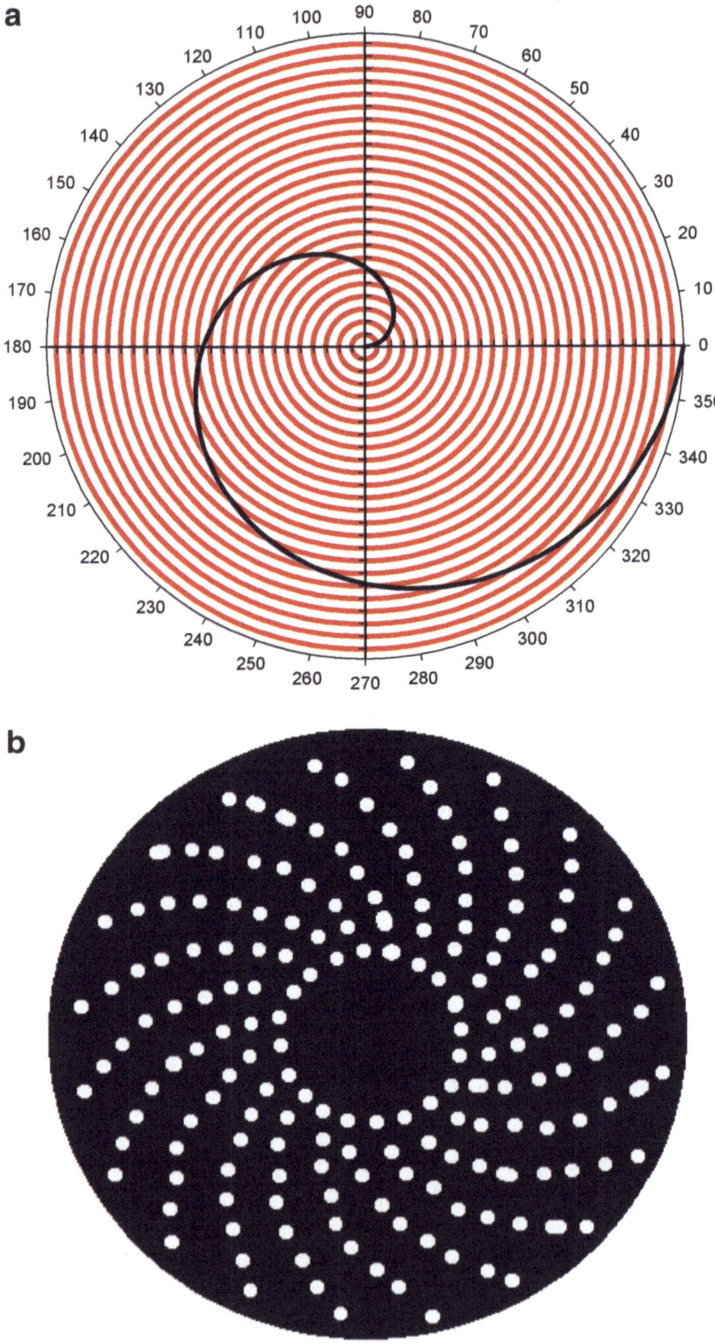

Fig. 8.7 The Archimedes spiral. (**a**) shows the Archimedes spiral plotted in polar coordinates (black line). When pinholes are place on the Archimedes spiral at the intersections of the black and red lines and the Archimedes spiral rotated the pinholes sweep out concentric parallel raster lines (red lines). When multiple Archimedes spirals are present the entire image field is continuously scanned without the presence of intervening scan lines. (**b**) shows a modern Nipkow disk which is a blackened glass disk with pinholes arranged in Archimedes spirals

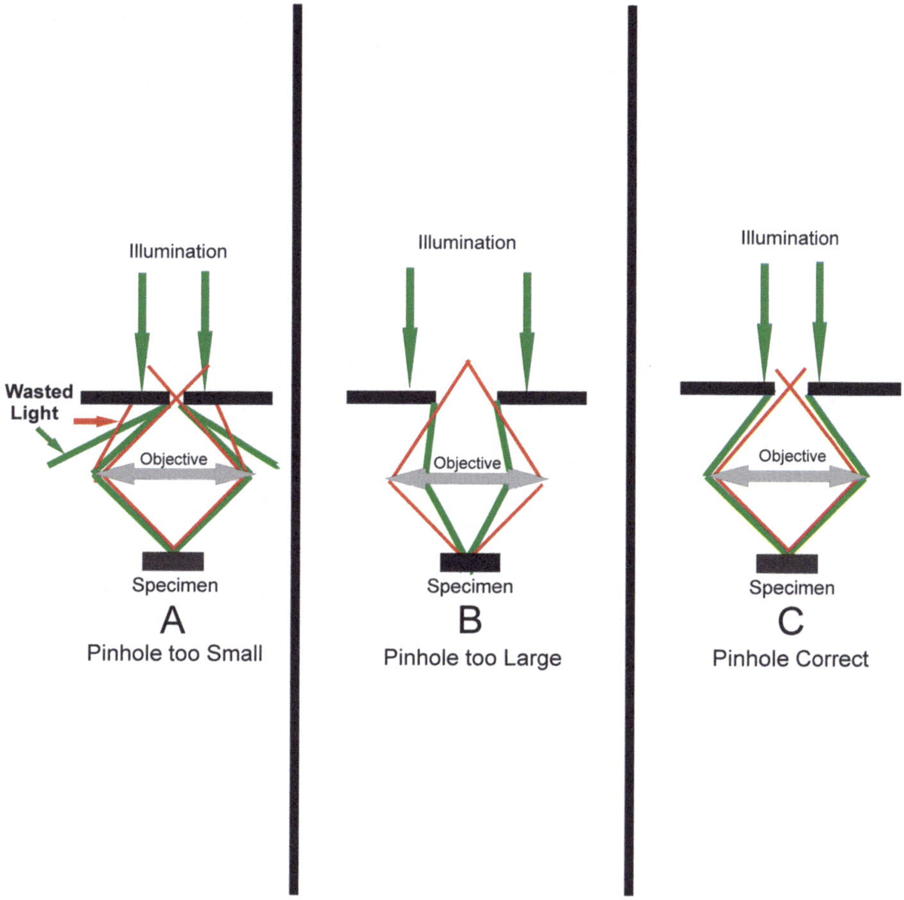

Fig. 8.8 Effect of Nipkow disk pinhole size (diameter) in Nipkow disk on image quality. In (**a**) the pinhole in the Nipkow disk is too small. Image quality is degraded by loss of higher-order diffracted light rays not being collected by the pinhole. In (**b**) the pinhole in the Nipkow disk is too large. Image quality (resolution) is lost because the light rays are insufficiently diffracted and do not completely fill the aperture of the objective. In 8.78 the pinhole in the Nipkow is of the proper radius. Under these conditions diffracted light from the pinhole fills the aperture of the objective, and the refracted or light emitted by the specimen is maximally collected by the pinhole

which enter the pinhole aperture are diffracted into a relatively narrow beam which does not completely fill the aperture of the objective lens. This causes the objective lens to function at a lower NA than designed for and results in loss of image resolution. In addition, emitted light from the specimen may not enter the appropriate pinhole. This results in further loss of resolution in the X-Y and Z directions (Fig. 8.8b). In a pinhole of correct size (radius), the diffracted light from the pinhole fills the objective aperture, and the refracted light from the specimen is maximally collected by the pinhole (Fig. 8.8c).

An equation which estimates the optimal pinhole radius was derived from the Fraunhofer diffraction theory for a circular aperture in the focal plane of the objective (see Goodman 1968; Xiao and Kino 1987; Kino and Xiao 1990 for details of the equations and assumptions used). The optimal pinhole radius is determined by the following equation:

$$R_o = \frac{0.25\lambda M}{\mathrm{NA}}$$

where λ is the wavelength of the light used, M the magnification, and NA the numerical aperture of the objective. This equation is applicable to microscopes of fixed tube length and infinity-corrected optics. The R_0 value for a 100×, 1.4NA oil immersion objective with light at a λ of 546 nm is 9.75 um, and for a 10× (0.5 NA) objective under the same conditions, R_o is 2.73 um. This indicates that the ideal situation would require an individual disk with the appropriate sized pinholes to match each objective on the microscope. Since pinholes of these ideal sizes would result in a significant loss of illumination (1% transmitted light or less) and require multiple disks in a system, alternate approaches have been employed.

In general, a single disk is used with the optimal R_0 value chosen to match the objective lens with the highest magnification and NA. The pinhole size used in early experimental designs of spinning disk microscopes was usually 10um. Under this design criterion, the pinhole diameter is larger than optimal for the objectives with lower magnification and NAs resulting in proportional loss of resolution at these magnifications. However, this loss of resolution for the lower magnification objectives may be acceptable if specimen resolution is less critical in this range of magnification. If R_o becomes very large, out-of-focus information in the Z-plane contributes to the image, and the system is no longer confocal and becomes a standard fluorescence microscope. Most modern commercially available Nipkow confocal systems (Yokogawa systems) use a pinhole with a R_o of 20–25 um to insure the collection of sufficient light intensity for observation and image collection. This sacrifices some high-end resolution with high NA objectives, and there is a similar loss of resolution at low magnification. Loss of signal and resolution are often criticisms of spinning disk confocals, but these compromises are balanced by the speed of imaging for live cell applications.

8.4.3 Nipkow Disk Pinhole Spacing

The Archimedes spiral of a spinning disk is governed by the simple polar equation:

$$r = a + b\theta$$

The parameter a refers to the turn of the spiral, and b is the distance between successive turnings. In the Archimedes spiral, the successive turns have a constant separation distance which is equal to $2\pi b$ when θ is measured in radians. Applying this to the Nipkow disk for confocal microscopy, identical pinholes are laid out circumferentially on the rotational arm of an Archimedes spiral at a constant angle (θ) and constantly increasing distances (r) from the center of the disk. Multiple Archimedes spiral arms are present on the Nipkow disk resulting in 20,000–200,000 pinholes depending on the size and manufacturer of the disk.

As the spacing between adjacent pinholes in the disk is decreased, the pinholes become too close together, and the aggregate area of the pinholes approaches the total area illuminated. Under these conditions, the signal intensity emitted from the specimen no long decreases sharply with the defocus distance characteristic of the objective lens. Instead, a portion of the out-of-focus light emitted by the specimen can return to the detector through alternate adjacent pinholes leading to loss of intensity and image resolution. An additional problem also arises when spacing between the pinholes is too small. Here the closeness of the pinholes, equal to or less than 5 pinhole radii apart, can result in interference between images resulting in a speckle effect. This is especially true when a monochromatic laser and narrow band-pass filter are used as the excitation source. This phenomenon can be minimized by the use of a diffuser or phase randomizer placed in front of the illumination source.

Spacing of the pinholes too far apart results in uniform loss of excitation light intensity to the specimen and emitted light from the specimen. Both result in a deterioration of overall image quality.

The ideal spacing of the pinhole along the arch of the Archimedes spiral is a function of the magnification, NA, and Z defocus distance characteristic of the objective lens. Here, as with the size of the pinhole, compromises have been made in pinhole spacing to reduce the complexity of the mechanism and moderate the cost of the instrument. A current typical Nipkow disk has a pinhole spacing of 253 um based on high-magnification (100×), high-resolution (NA = 1.4) immersion objectives. With objectives of lower magnification and NA, the ideal pinhole spacing increases in distance (Fig. 8.9). This adjustment of the pinhole spacing does not produce a significant loss of image resolution provided the emitted light rays from the specimen are focused on the pinhole. However, there is loss of both excitation and emission light intensities with low magnification and NA objectives and short pinhole spacing.

8.4.4 The Petran Microscope

The initial application of the Nipkow disk to optical microscopy was the development of the Tandem Scanning Reflected Light Microscope (TSRLM) by Petran and his co-workers (Petran et al. 1968, 1985). Here the disk is perforated with many holes laid down along the path generated by a multiple set of interweaving Archimedes spirals. The pinholes are separated by a distance large enough so that there is no interaction between the images formed by the individual pinholes. The complete image is formed by moving the pinholes (spinning the disk) so as to fill the

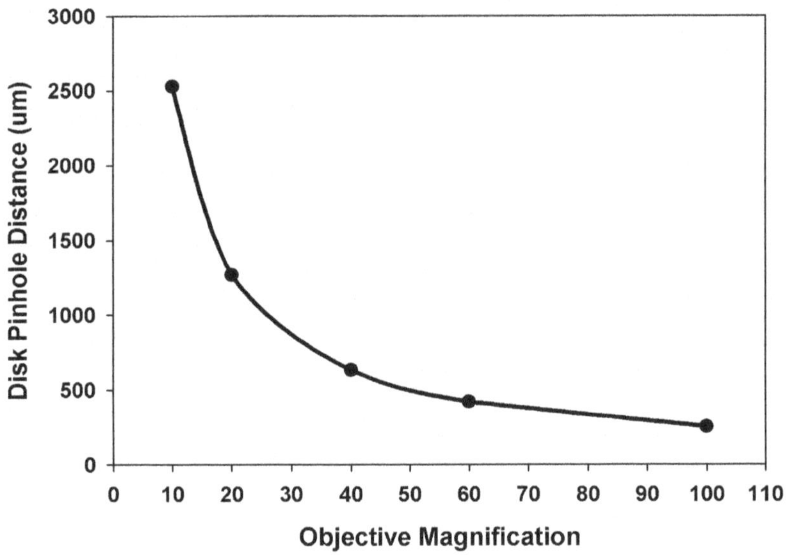

Fig. 8.9 Relationship between ideal pinhole spacing and magnification power of the microscope objectives

space between them. Although TSRLM had the advantages of real-time imaging and good cross-sectioning ability, it possessed numerous disadvantages which limited its usefulness. The most serious problems were a poor light budget with only about 1% of the light being useful, complicated mechanical components which were required to reduce internal reflection from the disk, and considerable difficulty in maintaining the alignment of the large number of optical components required for operation of the system.

8.4.5 The Xiao and Kino Microscope

The next advance in spinning disk-based confocal microscopy was the development of the real-time scanning optical microscope (RSOM) which is able to detect light from the same pinhole from which a given area of the specimen is illuminated (Xiao and Kino 1987; Xiao et al. 1988). This is accomplished by constructing a disk of highly reflective black chrome. The surface of the disk produces a reflected beam which is easily eliminated by an optical stop. The disk is also tilted and the optical stop placed in a position where the light reflected from the disk is focused (Xiao et al. 1990). Tilting of the disk can distort the image due to defocusing of the pinholes near the edge of the field of view. Typically, for a few degrees of tilt, the distortion is minimal. However, when imaging requires high-resolution or extremely accurate metrology, tilting of the disk is inappropriate, and other systems must be used.

Additionally, to further reduce interference by reflected light, the light from the source is polarized and light reaching the eyepiece observed through an analyzer with its plane of polarization rotated 90° to the polarized source. A ¼ wave plate is placed in front of the objective lens so that light reflected from the plane of polarization is rotated 90°, so it can be observed through the eyepiece (Fig. 8.10). This set-up and orientation of optical components enhance the SNR in the image. However, because fluorescent light emitted from a fluorescently labeled specimen is randomly polarized, a small fraction of the desired signal is also eliminated from the image.

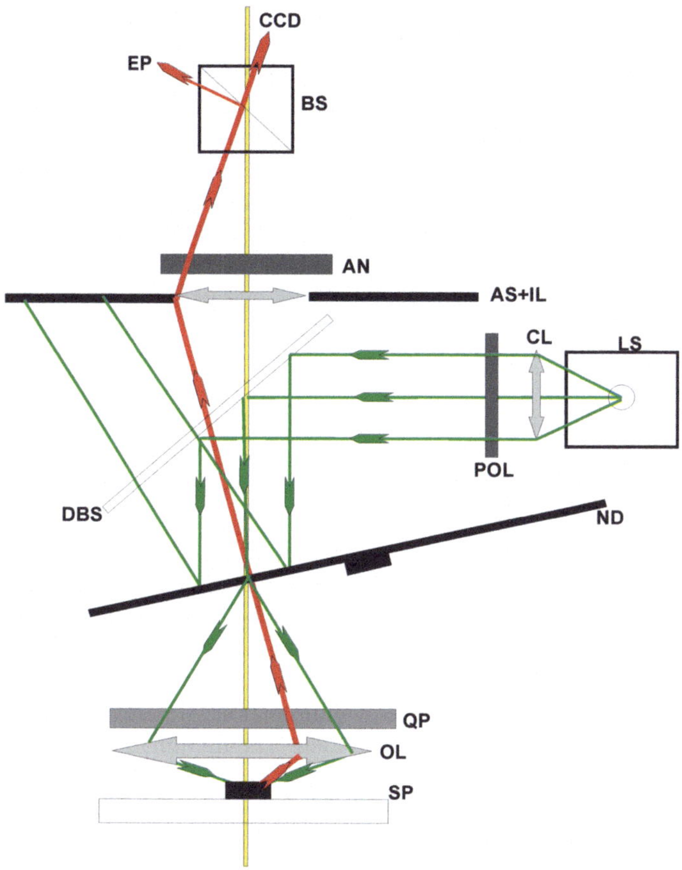

Fig. 8.10 The tilted Nipkow disk confocal microscope. The design of Xiao and Kino. In this design, the Nipkow disk is tilted to eliminate reflected incident light from the disk from interfering with light emitted by the specimen. Diagram traces the ray path for incident and emitted light interacting with one pinhole. Green lines, incident light from light source (LS); red line, emitted light by the specimen (SP); yellow line, optic axis of the microscope. Components: LS light source, CL collecting lens, Pol polarizer, DSB dichroic beam splitter, ND Nipkow disk with pinholes, QP quarter-wave plate, OL objective lens, Sp specimen, AS+ IL aperture stop and intermediary lens, BS bean splitter, EP eyepiece; CCD-CCD camera

This loss of fluorescent signal can be minimized by setting the polarizer and analyzer to a parallel orientation. Additionally, for fluorescence, the reflected light from the disk can be removed by a dichroic beam splitter and a barrier filter.

In real time the illumination of the specimen is accomplished by use of relatively broadband light. This results in low temporal coherence and minimal speckle effects which arise from interference between neighboring layers within the specimen or between reflections from the specimen and reflecting components within the microscope. Because of the low transmittance of light through the disk, light sources of high intensity such as mercury or xenon arc sources, or more recently metal-halide lamps, are used. Metal-halide lamps are particularly useful in this application since they emit at essentially constant illumination power across the objective field.

Application of Kohler illumination to the spinning disk configuration also improves the image quality. Each point on the source is focused to a point in the back focal plane of the objective lens when the disk is not present. When the disk is introduced into the optical path, Kohler illumination insures that the central axis of the diffracted beams passing through the individual pinholes all pass through the center of the back focal plane of the objective. This configuration gives uniformity to the illumination and optimizes resolution over the field of view.

8.4.6 The Corle Microscope

The design of the spinning disk confocal microscope by Corle et al. (1991) further improved image quality by eliminating the tilt of the Nipkow disk and improving the overall optical path of the excitation and emission photons. The initial improvement came by selectively isolating the most stable portion of the mercury or xenon arc light source. These light sources exhibit an inherent instability across the arc, especially at the margins. The most stable portion of the arc is the central region directly between the electrodes. In this microscope design, the spot of maximal intensity of the arc near the cathode is isolated by imaging it on to a pinhole aperture by the condenser lens. This concentrates the maximal intensity of the light source into a bright, essentially point source of illumination. This makes the illumination as uniform as possible across the field of view. The illumination in the plane of the Nipkow disk is collimated by the addition of a collecting lens behind the pinhole aperture. Before reaching the Nipkow disk, the collimated rays of light are further refined by the addition of a field stop to remove aberrations from the outer edge of the illumination and to define the size of the field to be illuminated.

A polarizer is then placed in the collimated beam to eliminate spurious reflected light in the system. The polarizer and analyzer in the Corle system are analogous to that used in the RSOM described above. Light exiting from the polarizer is focused onto a 90° beam splitter and the Nipkow disk as a collimated beam of light by a secondary objective lens. The diffracted light from the pinholes of the Nipkow disk is passed through a field lens situated just below the disk. This lens serves to minimize vignetting which produces darkening of the edges of the image field as a result of light falling outside of the aperture of the lower objective lens. This field lens focuses the central rays of the diffracted beam to the center of a tube lens which collimates the beam to the aperture of the lower objective. This in turn focuses the light onto the specimen (Fig. 8.11). The light emitted by the specimen passes back through this lens system and the pinhole of the Nipkow disk where it is focused by the secondary objective as a collimated beam. This collimated light then passes directly through the dichroic beam splitter, a field stop, and the analyzer and is focused by a tube lens to the eyepiece for direct viewing or to an appropriate camera for image collection.

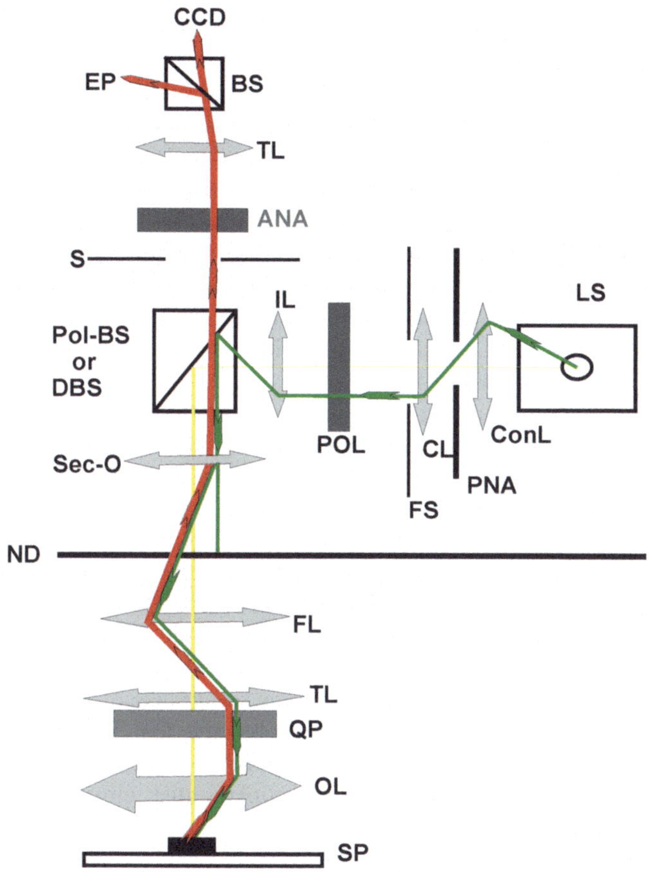

Fig. 8.11 Nipkow spinning disk microscope: Corle design. Diagram traces the ray path for incident and emitted interacting with one pinhole. Green lines, incident light from light source (LS); red line, emitted light by the specimen (SP); yellow line, optic axis of the microscope. Components: LS light source, Cond L condenser lens, PHA pinhole aperture, Col L collimating lens, FS field stop, Pol polarizer, IL intermediary lens, Pol-DSB polarized or dichroic beam splitter, SO secondary objective, ND Nipkow disk with pinholes, FL field lens, TL tube lens 1, QP quarter-wave plate, OL objective lens, Sp specimen, AS aperture stop, Ana analyzer, TL-2 tube lens 2, BS bean splitter, EP eyepiece, CCD-CCD Camera

8.4.7 The Yokogawa Spinning Disk Confocal Microscope

Although spinning disk confocals can obtain images at a very rapid rate, the need for a nonoverlapping spacing (described above) means that a large area of the disk lacks openings. Thus, only a small portion of the available light is used to illuminate the specimen. A significant improvement was made on the spinning disk confocal microscope by the addition of a second spinning disk which contains an array of micro-lenses (Ichihara et al. 1996; Nakano 2002; Genka 1999) which collect and focus more of the available light into the pinholes. This design forms the bases of the compact confocal scan head developed by Yokogawa (Yokogawa Electric Corp). Yokogawa confocal heads are currently used by several confocal manufacturers including PerkinElmer and VisiTech.

In this system, the scanning portion contains a dual spinning disk arrangement in which the first disk has an array of 20,000 micro-lenses arranged in an Archimedes spiral. The micro-lenses of this disk receive the illumination light and are accurately aligned with a corresponding array of pinholes on the second, Nipkow, disk. The micro-lens array disk and the Nipkow disk are physically fixed at a distance equal to the focal length of the micro-lenses and are rotated together so as to scan the field of view at high speed. The typical speed of scanning of this dual disk system is 1800 rpm (30rps) which results in 1200 pinholes scanning the sample at any given time. These spinning rates are 12 times the video rate of 30fps which results in a 12-frame average for each video image.

The constant pitch Archimedes spiral arrangement of the micro-lenses in the first disk and pinholes in the Nipkow disk creates even illumination across the image field without creating scanning artifacts. Light from the laser or arc lamp illumination source is collimated and projected on to the micro-lens array disk. Light focused by the micro-lenses passes through a dichroic beam splitter mounted in the space between the micro-lens and Nipkow disks and enters the corresponding pinhole in the Nipkow disk. First focusing the illumination light by the micro-lens disk prior to it passing through the Nipkow disk improves available illumination at the specimen by about 40%. The incident light is diffracted by the pinhole to fill the aperture of the objective lens where it is focused onto the focal plane of the specimen. Light emitted from the specimen is collected by the objective lens and then focused back on to the pinholes in the Nipkow disk. Transmission of the emitted light from the specimen through the same pinholes in the Nipkow disk eliminates out-of-focus signal and results in a confocal image.

Emitted light from the pinholes in the Nipkow disk is then deflected by a dichroic beam splitter positioned between the Nipkow disk and the upper micro-lens array disk and focused on the appropriate detector (Fig. 8.12). This arrangement of disks has several advantages including the improved light efficiency described above. This has real advantages for real-time confocal imaging of live cells where fluorescence emission is often limiting. The improved light efficiency allows fluorescence imaging to be carried out at low intensity and power per unit area thus reducing or minimizing photobleaching of probes and related phototoxic damage to living cells.

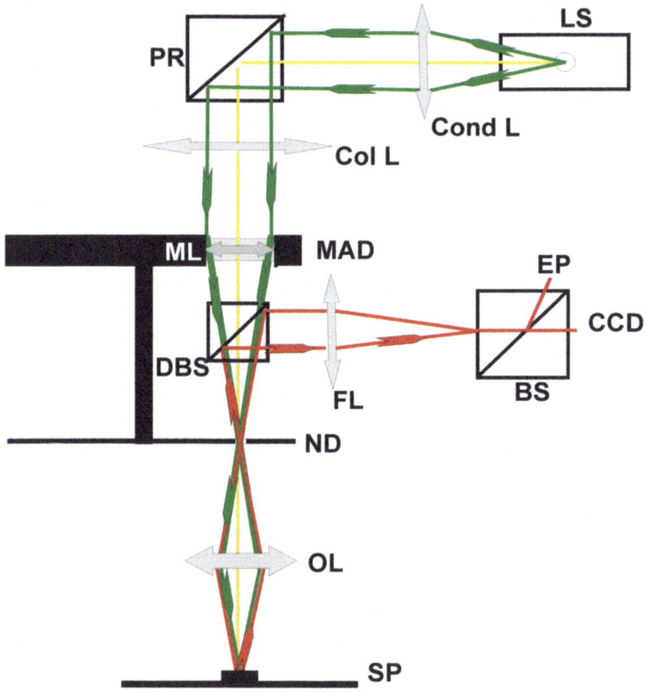

Fig. 8.12 Nipkow spinning disk microscope: Yokogawa design. This design represents a significant improvement over pervious spinning disk microscopes. The principle improvement is the use of a tandem spinning disk system in which the pinholes in the upper disk contain micro-lenses which focus the incident light on to the pinholes in the Nipkow disk. Diagram traces the ray path for incident and emitted interacting with one microlens in the microlens array disk and one tandem pinhole in the Nipkow disk. Green lines, incident light from light source (LS); red line, emitted light by the specimen (SP); yellow line, optic axis of the microscope. Components: LS light source, Cond L condenser lens, PR 90° reflecting prism or front surface mirror, Col L collimating lens, MAD microlens array disk, ML micro-lens, DBS dichroic beam splitter, ND Nipkow disk with pinholes and tandem with microlens array disk, OL objective lens, Sp specimen, FL field lens, BS beam splitter, EP eyepiece, CCD-CCD Camer

By focusing the incident light onto the pinholes in the Nipkow disk, backscatter of the incident light at the surface of the Nipkow disk is also significantly reduced improving the SNR of the confocal image. Improved speed of imaging is also possible with frame scan speeds as high as 1000 frames per second possible.

8.4.8 Slit Scanning Systems

A modification of the pinhole type of spinning disk is the slit scan spinning disk which Olympus has introduced in their Disk Scan Unit (DSU) systems. The DSU systems offer a choice of five exchangeable disks with different slit widths for use with different objectives and specimen thicknesses. This addresses many of the problems discussed above concerning matching of objectives and pinhole aperture for optimum resolution. As discussed above for the swept field slit scanning systems, the point spread function is affected when using slits rather than pinholes, so there is still some sacrifice of resolution, but the overall combined confocal effect and speed of imaging is superior to epifluorescent microscopes.

8.4.9 Image Collection in Spinning Disk Systems

Images generated by spinning disk systems can be viewed directly by looking through a port on the scan head or captured by using the appropriate recording device. Since images are collected at high speed and may also have low SNR levels, CCD or EMCCD cameras remain the most appropriate collection devices in these situations although newer sCMOS cameras are rivaling CCD systems. Although very reliable and inexpensive CCD cameras are available, imaging of live cells using fluorescence microscopy requires very sensitive and thus more expensive CCD cameras. Depending on the spinning disk system, images may be generated at rates between 300 and 1000 frames per second. The actual speed at which the instrument can capture images is dependent on the speed and sensitivity of the attached CCD camera and the intensity of the fluorescent emission signal from the specimen. When high-speed imaging is required, the faster frame rates of CMOS cameras become an advantage. Image collection using sensor arrays such as CDC and CMOS cameras was discussed more fully in Chap. 6. Different experimental applications may require image acquisition at varying speeds or frame rates. The speed of the device is determined by the pixel read rate and the number of pixels which comprise the image.

Literature Cited

Abbe E (1884) Note on the proper definition of the amplifying power of a lens or lens system. J Roy Microsc Soc 4(2):348–351

Centonze V, White J (1998) Multiphoton excitation provides optical sections from deeper within scattering specimens than confocal imaging. Biophys J 75:2015–2024

Corle TR, Mallory CL, Wasserman TD (1991) Improved confocal scanning microscope. U.S. Patent 5,067,805, 26 Nov 1991

Denk W, Strickler J, Webb W (1990) Two-photon laser scanning fluorescence microscopy. Science 248:73–76

Denk W, Piston D, Webb W (1995) Two-photon molecular excitation in laser-scanning microscopy. In: Pawley J (ed) Hanbook of biological confocal microscopy, 2nd edn. Plenum, New York, pp 445–458

Gauderon R, Lukins P, Sheppard C (1999) Effect of a confocal pinhole in two-photon microscopy. Microsc Res Tech 47:210–215

Genka C, Ishida H, Ichimori K, Hirota K, Hirota Y, Tanaami T, Nakazawa H (1999) Visualization of biphasic Ca^2 diffusion from cytosol to nucleus in contracting adult rat cardiac myocytes with an ultra-fast confocal imaging system. Cell Calcium 25:199–208

Gerritsen H, deGrauw C (1999) Imaging of optically thick specimens using two-photon excitation microscopy. Microsc Res Tech 47:206–209

Goodman JW (1968) Introduction to fourier optics. McGraw Hill, New York

Göppert-Mayer M (1931) Uber Elementarakte mit zei Quantenspruengen. Ann Physik (Berlin) 9:273–294

Ichihara A, Tanaami T, Isozaki K, Sugiyama Y, Kosugi K, Mikuriya K, Abe M, Umeda I (1996) High-speed confocal fluorescence microscopy using a Nipkow scanner with microlens for 3-d imaging of a single fluorescent molecule in real time. Bioimages 4:57–62

Kino GS (1995) Intermediate optics in Nipkow disk microscope. In: Pawley JB (ed) Handbook of biological confocal microscopy. Plenum Press, New York, pp 155–165

Kino GS, Xiao GQ (1990) Rea-time scanning optical microscopes. In: Wilson T (ed) Scanning optical microscopes. Pergamon Press, London, pp 361–387

Laine RF, Kaminski Schierle GS, van del Linde S, Kaminski CF (2016) From single-molecule spectroscopy to super-resolution imaging of the neuron: a review. Methods Appl Fluoresc 4:02204

Leung BO, Chou KC (2011) Review of super-resolution fluorescence microscopy for biology. Appl Spectrosc 65(9):967–980

Nakano A (2002) Spinning-disk confocal microscopy—a cutting edge tool for imaging of membrane traffic. Cell Struct Funct 27:349–355

Nipkow P (1884) German Patent no. 30,105. Germany

Petran M, Hadravsky M, Egger MD, Galambos R (1968) Tandem scanning reflected light microscope. J Opt Soc Am 58:661–664

Petran M, Hadravsky M, Boyde A (1985) The tandem scanning reflected light microscope. Scanning 7:97–108

Piston D (1999) Imaging living cells and tissues by two-photon excitation microscopy. Trends Cell Biol 9:66–69

Sydor AM, Czymmek KJ, Puchner EM, Mennella V (2015) Super-resolution microscopy: from single molecules to supramolecular assemblies. Trends Cell Biol 25(12):730–748

Tanaami T, Otsuki S, Tomosada N, Kosugi Y, Shimizu M, Ishida H (2002) High-speed 1 frame/ms scanning confocal microscope with a microlens and Nipkow disk. Appl Opt 41:4704–4708

Xiao and Kino (1987) A real-time confocal scanning optical microscope. In: Wison T, Balk L (eds) Proceedings SPIE, vol. 809, Scanning Imaging Technology, pp 107–113

Xiao GQ, Corle TR, Kino GS (1988) Real time confocal scanning microscope. Appl Phys Lett 53:716–718

Xiao GQ, Kino GS, Masters BR (1990) Observation of the rabbit cornea and lens with a new real time confocal scanning optical microscope. Scanning 12:161–166

Yamanaka M, Smith NI, Fujita K (2014) Introduction to super-resolution microscopy. Microscopy 2014:177–192

Chapter 9
Setting the Confocal Microscope Operating Parameters

Amy E. Rowley, Anna M. Harper, and Robert L. Price

9.1 Introduction

The goal of this chapter is to systematically look at each decision made in obtaining images from a sample stained with either single or multiple fluorochromes on a Leica TCS SP8 Multiphoton Confocal System equipped with a DM6000 CFS upright microscope and 10X, 20X, 25X water immersion, 40X water, 63X oil, and 25X Clarity objectives. Lasers include 488 and 552 nm optically pumped semiconductor lasers (OPSLs), a 638 nm diode laser, and a Coherent tunable IR laser (680 to 1080 nm) for multiphoton and second harmonic generation imaging. While in this chapter most of the specific examples of decisions made in setting instrument parameters are given by using images collected on this instrument, the basic choices made are the same regardless of the manufacturer of the system being used and the available hardware. When operating various confocal instruments, it is often simply a matter of identifying the terminology used, for example, pinhole versus iris, the name of a specific lookup table (LUT), or finding where a particular software function is hidden in the multiple menus necessary to operate the system. Once the basic principles are understood, it is possible to quickly learn to operate any confocal microscope and to obtain publication quality images.

Most manufacturers include extensive image processing capabilities in the system software, this often varies between instruments, and many packages, such as co-localization analysis, deconvolution, and advanced 3D reconstruction, are optional. In this chapter the goal is to cover the basics of image collection and the compromises made to improve image quality while minimizing specimen damage resulting from exposure to the laser. Primary components of the hardware and software, and their effect on image quality and specimen damage, will be discussed.

A. E. Rowley · A. M. Harper · R. L. Price (✉)
Department of Cell Biology and Anatomy, University of South Carolina School of Medicine, Columbia, SC, USA
e-mail: Bob.Price@uscmed.sc.edu

W. G. Jerome, R. L. Price (eds.), *Basic Confocal Microscopy*,
https://doi.org/10.1007/978-3-319-97454-5_9

Several programs such as ImageJ, Photoshop, Imaris, and Amira, in addition to software on various confocal microscopes, including Leica LASX 3D, exist for 2D and 3D processing of confocal images, and the basics of these programs are covered in Chaps. 10 and 11.

Figure 9.1 shows the layout of the system used for this chapter, and the flowchart below illustrates the procedure for starting the instrument and opening an imaging session. It should be noted that for ease of operation, there is considerable redundancy in most operating software. For example, lasers can be started and operated from several different windows, and the objectives can also be changed from several locations in the software. For experienced users multiple access sites are a convenience, but often for novice users, the complexity of the software may be somewhat intimidating. In the sections below, some of these multiple sites are discussed, while in others only the most common method for performing a function is presented. In addition, most programs have several dropdown menus that provide access to multiple choices for setup of instruments. In the following sections, many, but not all,

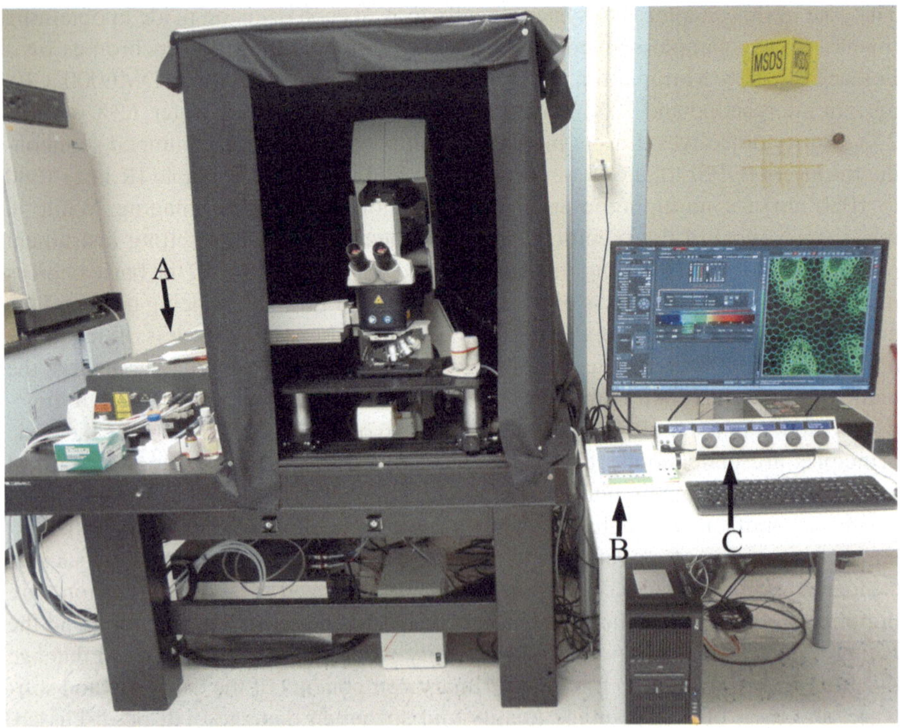

Fig. 9.1 Overview of the Leica SP8 confocal microscope system. This system consists of (from left to right) a Coherent Chameleon tunable laser (**a**), a DM 6000 CFS upright microscope, a with the SmartMove microscope control, the STP8000 controller for setting microscope parameters (**b**), USB control panel for setting imaging parameters (**c**), and the computer/monitor system. The system also includes supply units to turn on the scanner and laser power and a control box to turn on the multiphoton Coherent tunable IR laser

dropdown options are discussed. We hope that sufficient information is provided to assist a beginning operator in efficient use of a confocal microscope.

Flowchart for Starting an Imaging Session on the Leica SP8 Multiphoton System

1. If the IR laser will be used during the imaging session, verify the coolant is ~20 °C; if greater than 25 °C, call for maintenance (Fig. 9.2a). If coolant is operating appropriately, change the IR laser from "Standby" to "On."
2. Most ancillary equipment such as the halogen bulb, microscope, computer, etc. are started through a common power strip(s).
3. Using the toggle switches, turn on the scanner and laser power, and turn the key to "On" to allow laser entry into the system (Fig. 9.2b). Note that this laser control box is mechanically separated from and has no effect on operation of the IR laser.
4. When appropriately logged into the system, start the LASX confocal microscope operating software.
5. From the Configuration dropdown menu, select either "machine_MP_ON" for multiphoton imaging or "machine_MP_OFF" for imaging with only the OPSL or diode lasers (Fig. 9.3a). It is also possible to select loaded or saved settings in this window.
6. If a tiling experiment will be performed, make sure the condenser lens is protected by lowering it to the lowest position and the stage is properly centered.

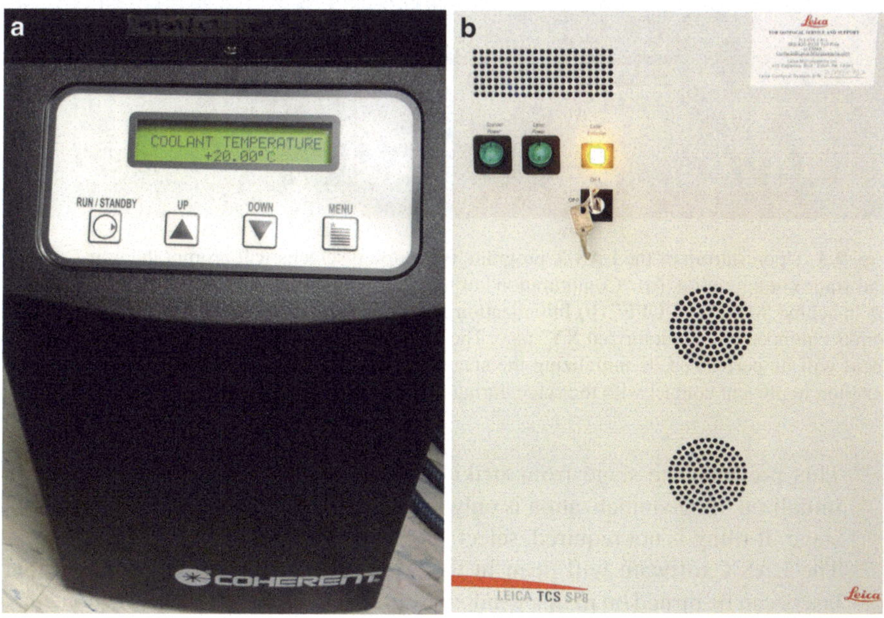

Fig. 9.2 Coolant (**a**) should be verified and read no greater than 20.00° C before operation of the system, and scanner and laser power should be turned on (**b**)

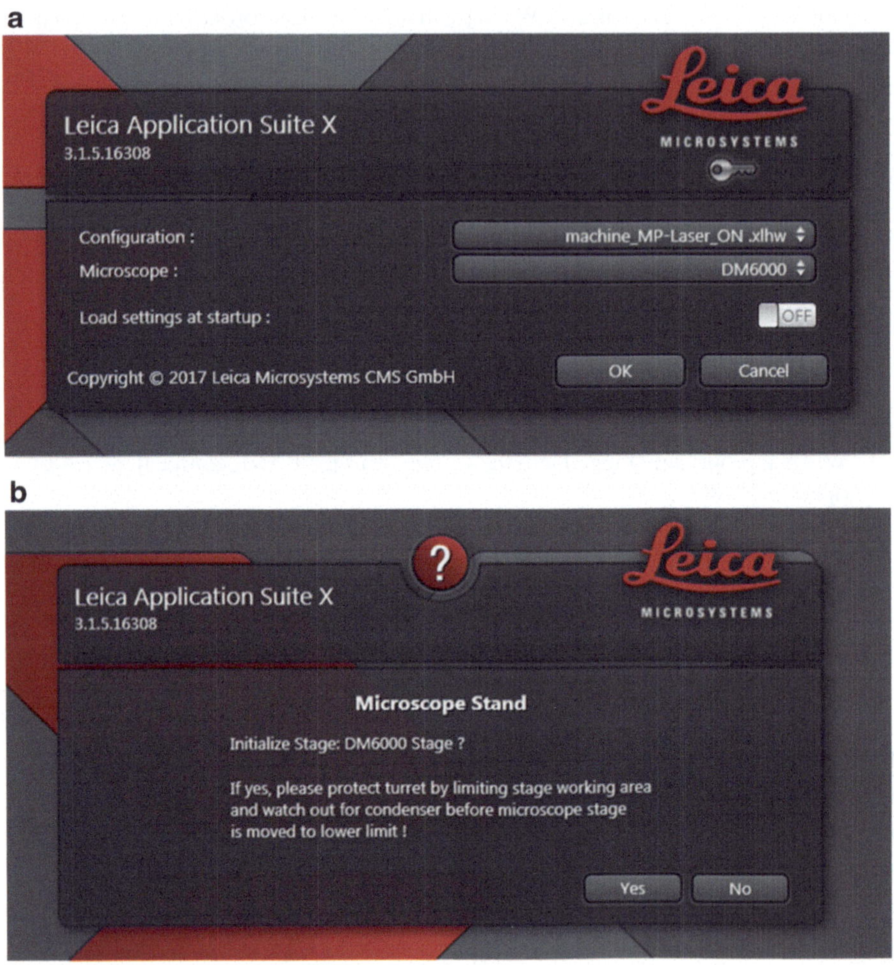

Fig. 9.3 Upon startup of the LAS X program, two pop-up screens will prompt the user for laser and stage configuration. (**a**) "Configuration" of the multiphoton laser: "machine_MP-Laser_ON" or "machine_MP-Laser_OFF." (**b**) Initialization of the automated XY microscope stage if the system is equipped with a motorized XY stage. The stage should only be initialized if a tiling experiment will be performed. If initializing the stage, the condenser lens should be set to its lowest position to prevent contact with the stage during initialization

This prevents the stage from striking the edge of the condenser when it is initialized. Stage initialization is only required when tiling with a XY motorized stage. If tiling is not required, select "no" (Fig. 9.3b).

7. The LASX software will open in the "Acquire" window (Fig. 9.4a). Visible lasers can be turned on in this window by clicking on the "+" which will display the available lasers (Fig. 9.4b). Each laser can then be activated and the percent transmission set with the vertical slider. Note that this system has 488 nm, 552 nm, and 638 nm lasers. Lasers may also be activated through the

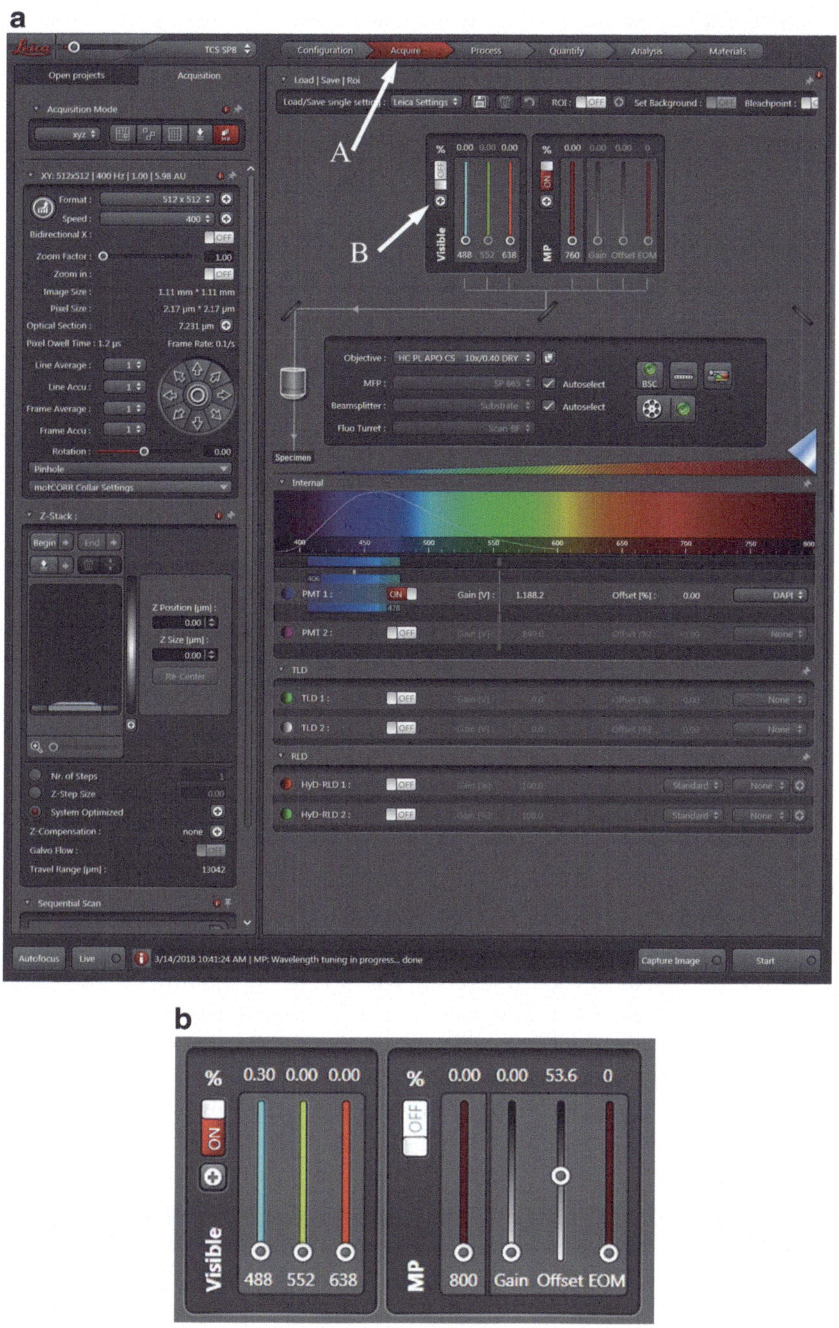

Fig. 9.4 Opening screen of the LAS X software. Overview of the software in the Acquire menu (**a**). From this window it is possible to access the Configuration, Process, Quantify, Analysis, and Materials windows. (**b**) The visible lasers are accessed by selecting the "+" button indicated by arrow B

"Configuration" window as discussed below. Only the lasers needed should be activated to preserve hours on the unused lasers.

8. Alternatively, in the "acquire" window, a saved project can be selected as described in Sect. 9.7.1. The header information in the saved file contains all settings including laser and detector settings; scanning parameters, such as speed, resolution, bit depth, etc.; the collection parameters such as pinhole diameter; and Z-series parameters.
9. Obtain a wide-field fluorescent image by setting up the light microscope operating parameters.
10. Proceed with setting up the system for confocal imaging as discussed in Sect. 9.5 and subsequent sections below.

9.2 The Leica SP8 STP8000 Controller and USB Control Panel

As mentioned above there are often several ways to change the setting of an operating parameter in the SP8 software. An additional feature of the Leica Sp8 Multiphoton system is the USB Control Panel (Fig. 9.5) that allows increased ease of use when adjusting the detector gain and offset, field rotation, pinhole diameter, image zoom, and specimen Z position with control knobs rather than the software. These can be reconfigured as desired in the configuration panel as discussed in Sect. 9.6.5.

Similarly, the STP8000 Controller of the DM6000 CFS upright microscope (Fig. 9.6a) provides control of all microscope parameters. The X, Y, and Z positions of the microscope stage can be adjusted with the controls on the right-hand side of the Controller. Each icon on the left of the touch pad window provides separate controls for operation of the microscope platform. For example, when the microscope icon on the top left of the window is selected (Fig. 9.6b), adjustments can be made in light intensity, condenser aperture settings, field of view, and the shutter can be opened or closed as needed. The color wheel (Fig. 9.6c) allows viewing of the specimen in brightfield (BF), fluorescence (FLUO), or a combination of the two (CS) modes. The transmitted light (TL) shutter is used for imaging with TL in the TL mode and the IL shutter in fluorescence mode. The Objective Icon (Fig. 9.6d) allows selection of the available objectives on the system by simply touching the

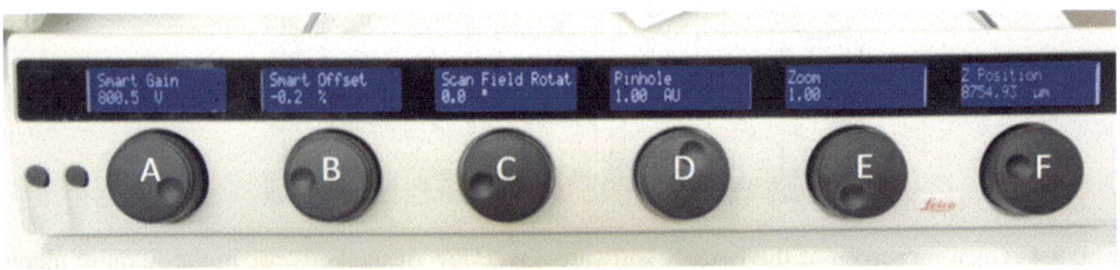

Fig. 9.5 The Leica USB control panel easily provides user-friendly access to the detector smart gain (**a**) and smart offset (**b**), scan field rotation (**c**), pinhole diameter (**d**), image zoom (**e**), and Z position (**f**)

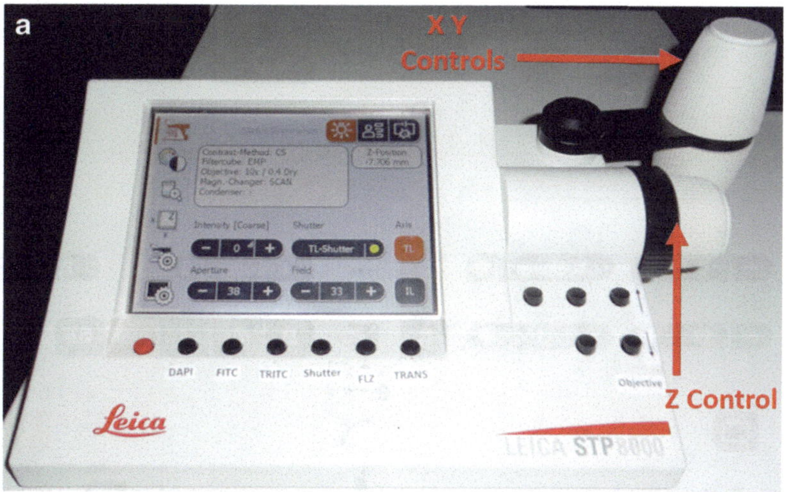

Fig. 9.6 STP8000 Microscope Control Panel. (**a**) Overview of the control panel indicating XYZ stage and imaging controls. As each icon on the far left of the window is selected, additional microscope functions can be accessed. (**b**) Selection of "microscope" image on top left allows adjustment of light intensity, Condenser Aperture settings, and field of view and the shutter. (**c**) The "Color Wheel Icon" allows viewing of the specimen in brightfield (BF), fluorescence (FLUO), or a combination of the two (CS) modes. (**d**) The "Objective Icon" allows selection of the available objectives on the system by simply touching the objective of choice. (**e**) The x, y, z icon provides stored parfocality positions for the objective focus (Z) and the X, Y stage position. (**f**) The "Microscope/Wheel" icon allows setting of various functions for the black buttons on the bottom of the Microscope Control Panel as shown by the DAPI, FITC, etc. labels. These functions can be changed in this window to customize a microscope. (**g**) The "Monitor/Wheel" icon allows calibration of the Control Panel

objective of choice. While it is still possible to manually change the objectives, this increases the ease of use as well as convenience. The x, y, z icon (Fig. 9.6e) provides stored parfocality positions for the objective focus (Z) and the X, Y stage position. The two bottom icons (Fig. 9.6f and g) provide settings for each of the black buttons on the bottom of the unit including language of use, opening of specific shutters, etc. for ease of use. The red button on the bottom provides information for setting function key options. These are adjustable settings and will not be used often, if at all, by the everyday user. One such setting icon can be seen in Fig. 9.6e.

9.3 Test Specimens and System Performance

As discussed in Chap. 1, Pawley (2000) posed the question, does a fluorescent micrograph reveal the actual location and number of labeled molecules in a cell or tissue, to members of his well-known confocal microscopy workshop. Based on the several

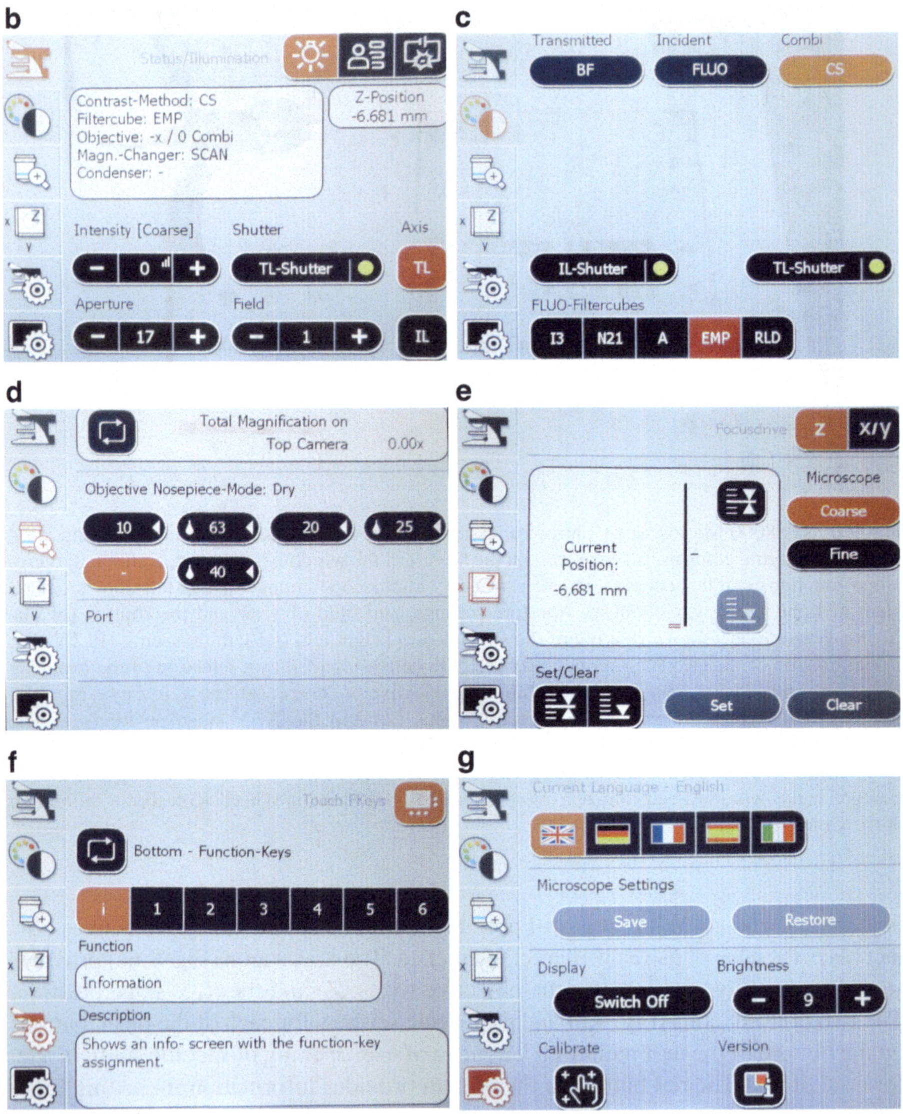

factors that can affect the numerical values stored in a computer that represent a fluorescent micrograph (see Chap. 6), the conclusion was that all we "can really be sure of measuring with most confocal microscopes in the fluorescence mode is some feature of the number of photons collected at a particular time." In the previous chapters, we have addressed many of the factors, including fluorescence, the structure of digital images, specimen preparation, and system hardware that can affect these numerical values and the resulting quality and interpretation of the image.

In this chapter, we address the setup of the microscope operating parameters for image acquisition and how each affects the final image. Primary considerations when determining the setting of each of the adjustable parameters on an instrument include X, Y, and Z resolution, signal-to-noise ratio (SNR), appropriate spectral resolution so no bleed through or spectral overlap exists, specimen damage, bit depth, and file size. Even with all of the variables present while imaging with a confocal microscope, it is possible for a user to obtain images on confocal instruments that are superior to wide-field fluorescent images with relatively little training. However, minimally trained users seldom take advantage of the superior imaging capabilities of confocal microscopes and rarely do they realize the compromises (Table 9.1) required for minimization of specimen damage while optimizing image quality when each instrument setting is altered. Improper setup of the system may also result in the collection of data artifacts that may not be recognized. It is only through a thorough understanding of the operation of each component of the confocal system that informed decisions can be made and optimal images obtained on a routine basis.

Zucker and Price (1999) published an excellent article on methods of instrument maintenance and quality control to ensure an instrument is operating at optimal specifications. While these topics are important and it is essential that attention is paid to these details, we will not go into extensive discussions concerning instrument maintenance and specifications here. Rather, this discussion focuses on the

Table 9.1 Advantages, disadvantages and imaging compromises associated with adjustment of confocal microscope operating parameters

Parameter	Advantage	Disadvantage
Increased laser output	Improved S/N ratio	Increased specimen damage
Decreased scan speed	Improved S/N ratio	Increased scan time and specimen damage
Line or frame averaging	Improved S/N ratio	Increased scan time and specimen damage
Increased pinhole size	Improved S/N ratio	Decreased contrast and resolution in x, y, and z
Increased number of pixels (smaller pixel size)	Improved resolution	Increased scan time; large file size
Increased amplifier gain	Fewer photons required	Poor S/N ratio
Increased amplifier offset	Improved gamma	Decreased contrast
Narrow band pass filter	Minimal bleed through	Increased scan speed time and specimen damage
Sequential imaging	Eliminates bleed through	Increased scan time and specimen damage
Simultaneous imaging	Decreased scan time and specimen damage	Increased bleed through of fluorochromes with overlapping emission spectra
Long wavelength fluorochrome	Improved depth of imaging	Decreased resolution
Short wavelength fluorochrome	Improved resolution	Decreased depth of imaging

proper setup of an optimally functioning instrument for everyday imaging. Inherent in these discussions is the assumption that a core director or other personnel extensively trained in instrument maintenance are routinely checking and correcting instrument specifications and alignment.

For discussions below concerning parameters which affect image quality and proper setup of the confocal microscope, test specimens that consist of 4 um beads, filter paper stained with various fluorochromes, or *Convallaria* (lily of the valley) are used as examples. Standard specimens are essential tools for training and as quality control assays to ensure that an instrument is working properly. Often, investigators blame poor image quality on the instrumentation when the actual problem is poor specimen preparation or operator error. The use of test specimens with known fluorescence characteristics to definitively show an instrument is working properly is the only factor that convinces an investigator that a problem may not be with the instrument, but instead is a problem with the specimen preparation or fluorophore specificity.

9.4 Definition of a Good Confocal Image

Even though a number of operating parameters on a confocal microscope can be adjusted, the goal should always be to set these operating parameters so that the digital image which is collected accurately represents the image as seen through the microscope.

A large number of operator decisions and compromises affect image resolution, SNR, and qualitative and quantitative assessment of colocalization and structural information in a confocal image. Before discussing the changes that can be made in various operating parameters that affect specimen damage and image quality, reviewing the definition of a good confocal digital image is essential. As discussed in Chap. 6, an ideal confocal image is one that has appropriate resolution, a good SNR so that little or no noise is apparent, good balance between contrast and brightness, and use of the full dynamic range of the available pixel values. For an 8-bit image, this would represent pixel values between 0 and 255 and for a 12-bit image pixel values between 0 and 4095. It may be desirable to leave 5–10 pixel values open on each end of the range for the application of histogram stretch functions for contrast enhancement. However, if any type of quantitative or semiquantitative analysis is the goal of the study, this must be performed prior to the histogram stretch operation.

With the large number of specimen, hardware, software, and user-definable factors that determine the quality of a confocal micrograph, it is essential that a mechanism be in place to ensure routine collection of quality images that include all available data without the introduction of artifacts. Within the confocal software, lookup tables (LUTs) are available that take advantage of the ability of the human eye to detect colors more efficiently than gray tones. These LUTs set pixel vales of zero to a specific color, typically blue or green, and pixel values of 255 in an 8-bit image to a contrasting color. In the SP8 software, the LUT assigns pixels of low value green and pixels of high value blue. All intermediate pixels are assigned a shade of red (Fig. 9.7). This allows the operator to quickly determine that the full range of pixel values is being used and that no detail is being lost on low or high ends of the image histogram. Comparison of the gray tone images in Fig 9.7b, d, and f shows the loss of significant detail in both regions where large numbers of blue and green pixels are present. In the green region, details are lost since all pixels are saturated at the 255 range, and the entire region is shown as white. In the blue range, details are lost since all pixels are undersaturated, and the entire region is shown as black. In the center images, the fine details (arrows) inside the Convallaria cells are evident. Details on the use of LUTs to assist in the setup of confocal operating parameters are discussed below. Examples showing the effects on the range of pixel values, SNR, resolution, and other factors affecting image quality are illustrated for the proper collection of an image.

9.5 The Opening Screen

A daunting task for many inexperienced confocal microscopists, and often for those with considerable experience, is navigating the extensive software necessary to operate a confocal microscope. This task is often made more intimidating because many optional functions that are not available on a system are present in the software creating the need for additional pages and icons, and there is often extensive redundancy in various windows of the software which, while designed to make access to functions such as turning lasers on or selecting dyes simpler, may make initial use of a system seem much more complicated than it actually is. The goal of the rest of this chapter is to discuss the main software pages for the SP8 and the primary functions necessary for basic operation of the instrument.

Figure 9.8 shows the top portion of the opening screen from which it is possible to access multiple menus for operation of the system and data processing. For example, with the TCS SP8 pull-down menu, open options such as "Live Data Mode," "Electrophysiology," "FRAP" and "FRET" functions, and "HyVolution" super-resolution software, if available, can be accessed. These advanced imaging functions go beyond basic confocal microscope operation and will not be discussed in this chapter. The "Configuration" menu, discussed in detail in the next section, provides several separate windows for setting user profiles including laser settings, choice of fluorochromes, objective choices, etc. The "Acquire" menu provides user access to all functions necessary for collection of images. Each of the functions will be discussed in detail in Sect. 9.7. The "Process," "Quantify," "Analysis," and "Materials" tabs are all image processing functions that go beyond the image acquisition scope of this chapter. Some basics of image analysis functions are discussed in Chaps. 10 and 11.

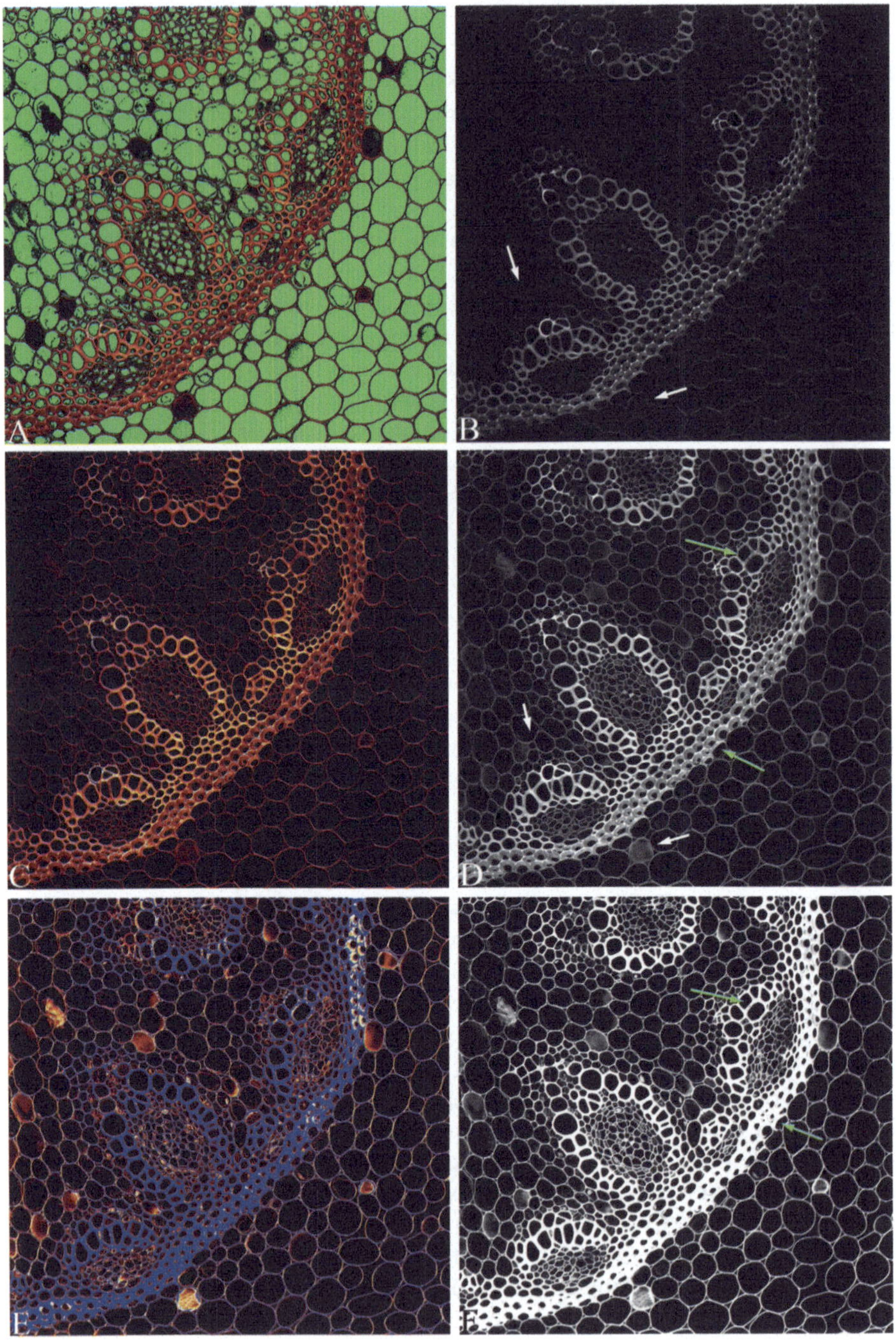

Fig. 9.7 Images of Convallaria to demonstrate the use of LUTs for setting the appropriate dynamic range in an 8-bit image. In images **a**, **c**, and **e**, the LUT is applied, and pixel values of 0 are shown in green, pixel values of 255 in blue, and all other pixel values as a shade of red. The goal is to set the pixel values in the image so that few green and blue pixels are present as shown in panel C. In **b**, **d**, and **f**, the corresponding grayscale images are shown demonstrating loss of fine detail in areas that were undersaturated (white arrows) or oversaturated (green arrows). The appropriate use of LUTs is required to avoid loss of fine detail in samples

Fig. 9.8 The top portion of the opening LASX screen with the TCS SP8 pull-down menu selected to show various imaging modes that might be available on a system including "Live Data Mode," "Electrophysiology," "FRAP" and "FRET" functions, and "HyVolution"

9.6 The Configuration Screen

Selection of the "Configuration" button (Fig. 9.9) provides access to several icons explained in detail below. As noted above, there is redundancy in the software, and many operating parameters that can be set in the "Configuration" menu can also be accessed, set, and adjusted in the "Acquire" menu. Areas of redundancy will be mentioned in both sections since they often provide user-friendly access to various imaging functions. Several icons such as "USB Panel," "Hardware," and "Memory" pages are specific to the instrument setup and are not used in routine confocal imaging. These icons are infrequently accessed by individual users and are not discussed here.

9.6.1 Stage Configuration

The "Stage Configuration" icon (Fig. 9.10) adjusts the settings for the motorized XY stage if available and should only be adjusted by the manager of the system. Adjustment of the stage settings affects image orientation during tiling and stitching experiments, and if not adjusted correctly, the final stitched image will have artifacts.

9.6.2 Instrument Parameter Settings (IPS)

The "IPS" icon (Fig. 9.11) provides access to specific software masks that will be seen in the "Acquire" window and saved with a file. For example, in the "IPS Mask" shown the "Lambda Configuration" (wavelength), Look up Tables ("LUTs") and the "Shutter" settings will be applied when a file is opened. These are typically set upon instrument installation by Leica but can be adjusted depending on the file parameters an individual user may need. Typically most users will use the default settings.

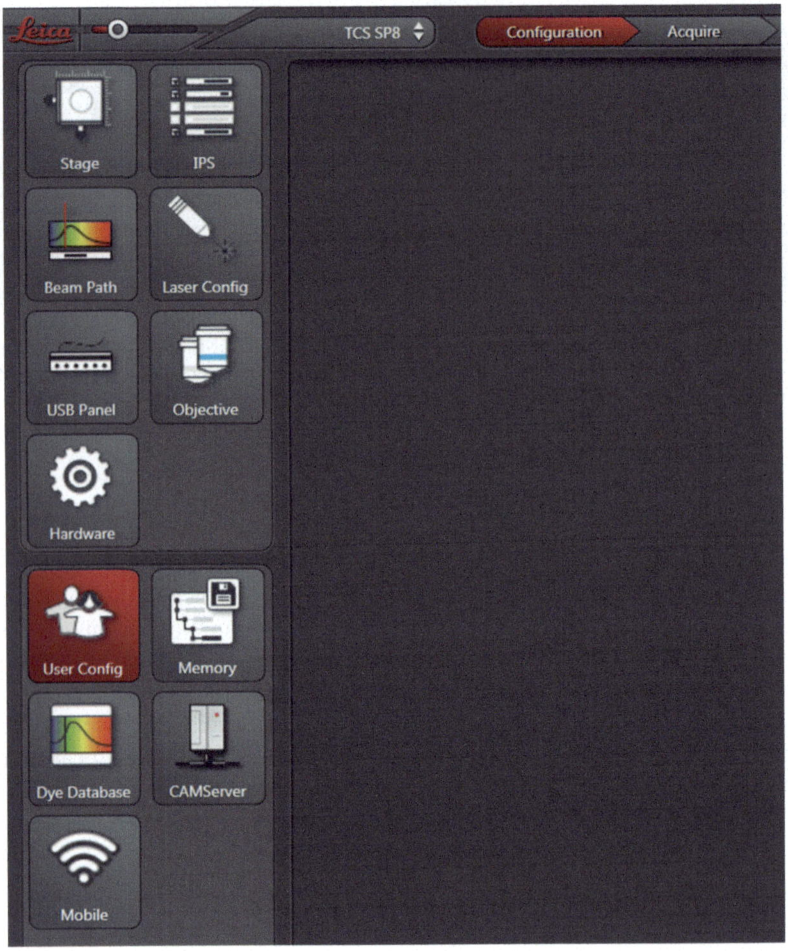

Fig. 9.9 An overview of the "Configuration" window. Various functions in this window are discussed in Figs. 9.10 through 9.18

9.6.3 Beam Path

The "Beam Path" icon, specifically the "Fluorifier Disc Settings" (Fig. 9.12), are used for setting of barrier or notch filters for use with stimulated emission depletion microscopy (STED), fluorescence correlation spectroscopy (FCS), or white tunable lasers. Since this SP8 system has none of these functions, there are no available filters seen in Fig. 9.12. However, it is possible to image "polarized" light with this system, and at the bottom of the screen, the polarized light settings can be adjusted by using the slide bar, if applicable in an experiment.

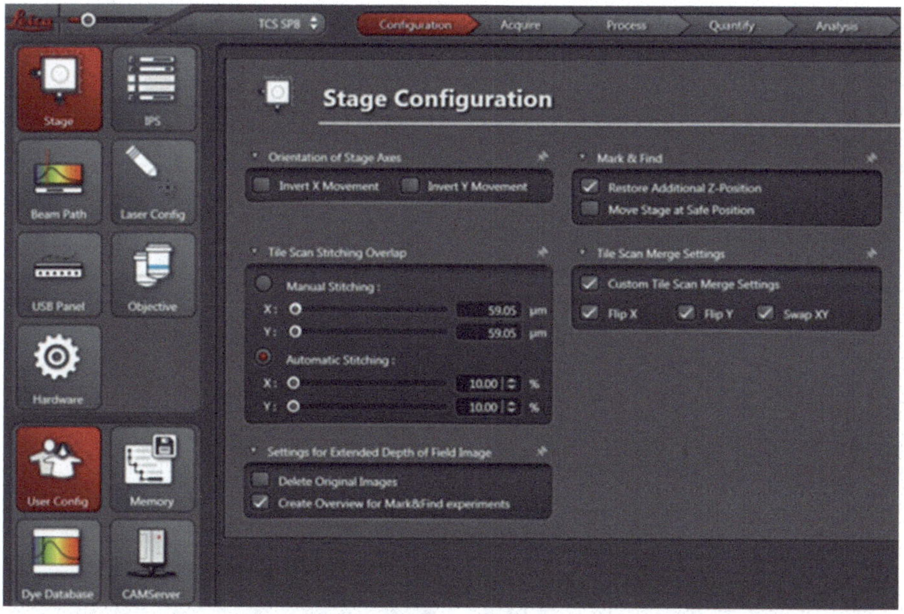

Fig. 9.10 The "Stage Configuration" window for control of the motorized microscope stage functions

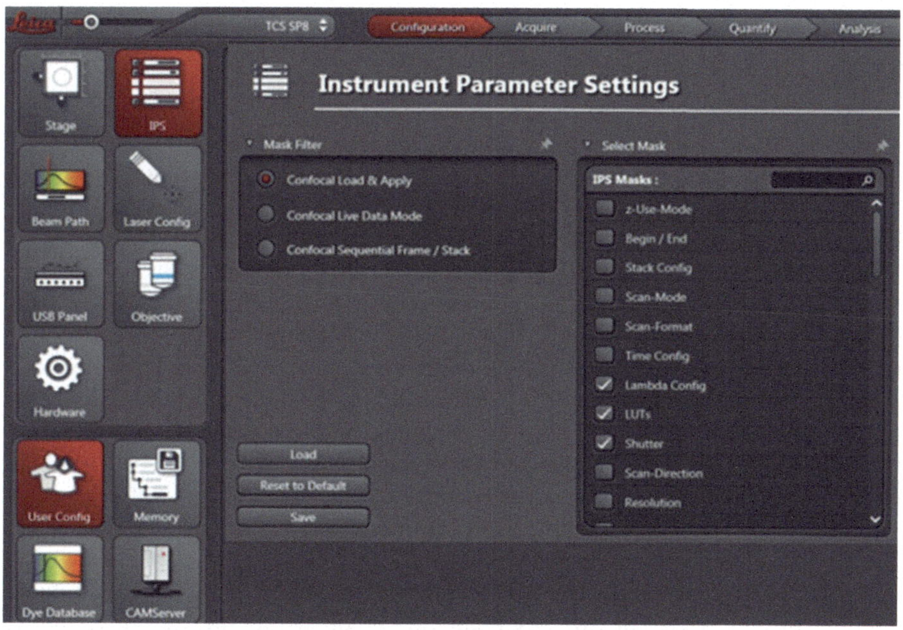

Fig. 9.11 "Instrument Parameter Settings" window for setting the information that will be seen in the "Acquire" window and saved with a file

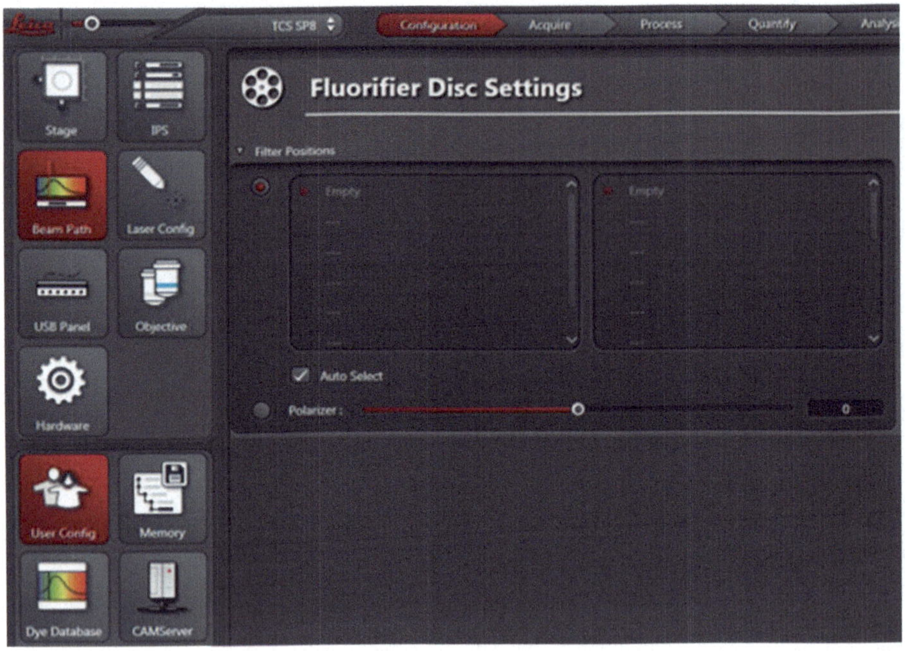

Fig. 9.12 "Fluorifier Disc Setting" controls for determining filters used with various imaging modes

9.6.4 *Laser Configuration*

Activation of the lasers is a primary example of software redundancy as they can be turned on in the "Configuration" menu as shown here (Fig. 9.13) and also in the "Acquire" menu as discussed in Sect. 9.8. By selecting the "Laser Configuration" icon (Fig. 9.13), it is possible to access the on/off controls for the OPSL 488, OPSL 552, and Diode 638 lasers available on this instrument. One should always be aware of the lasers present on an instrument prior to designing an experiment. There have been many cases where an investigator, without consultation with a core director or manager of an instrument, designs an experiment, purchases fluorochromes, and processes samples only to realize too late that an optimum laser is not available for the fluorochrome used and valuable tissue and expensive reagents are wasted. With the increased availability of tunable multiphoton/IR lasers such as the one available on this instrument and the increasing use of wide-spectrum white light lasers, this should become less of a problem. However, it is always wise to know the system hardware prior to design of an experiment and purchase of expensive reagents.

While the multiphoton laser (MP) can also be turned on in the "Configuration" window, the shutter for the MP laser can only be opened in the "Acquire" menu. In practice, only the lasers needed for a specific operating session should be turned on. This saves life time on the lasers not needed for a specific experiment.

Fig. 9.13 "Laser Configuration" window where one option for turning lasers on is provided

Fig. 9.14 "USB Control Panel" window. Each knob is assigned a parameter by selecting from the top dropdown menu, and the sensitivity of each turn is set with the bottom dropdown menu

9.6.5 USB Control Panel

The "USB Control Panel" icon (Fig. 9.14) provides adjustment controls for the functions on the "USB Control Panel" shown in Fig. 9.5. Each knob is assigned a parameter by selecting a function from the dropdown menu directly above each knob. The sensitivity of each turn is then determined by the dropdown menu directly below each knob. For example, the second knob controls the "Smart Offset" and is set to increase or decrease in 1% increments when turned to the right or left. As previously mentioned, in heavily used facilities, it is often advisable to allow access to these settings only by the manager of the equipment or qualified users with the understanding that all settings must be returned to those commonly used in the facility. This avoids problems when new or relatively inexperienced users realize that knobs no longer control the imaging functions that were set during training.

9.6.6 Objective Configuration

The "Objective Configuration" icon lists the specific objectives and their specifications available on the system. As seen in Fig. 9.15, the objectives available on this system are a 10×/0.40 DRY, 63X/1.40 OIL, 20x/0.75 DRY, 25X/0.95 WATER, and a 40x/1.10 WATER. The Clarity objective available on this SP8 is available through a separate nosepiece and is not shown here. By selecting each objective,

Fig. 9.15 "Objective Configuration" window showing available objectives and specifications for each

the attributes including type, magnification, numerical aperture, immersion type, and resolution can be seen. For example, the 25X water immersion objective is selected here and indicates that it is a Fluotar L objective with a NA of 0.95.

9.6.7 *Hardware Settings*

The "Hardware Settings" window (Fig. 9.16) provides access to many of the functions active during collection of an image such as "Line Averaging During Live Acquisition"; "Image Orientation," in which detector combinations are used; and "Bit Depth" of the image (see Chap. 6, Sect. .6.4). While many of these settings as shown here are default settings established when the instrument was installed, the advanced user may make changes in the way live images are collected and data transferred. However, prior to making changes and in a heavily used core facility, it is wise to understand ramifications of each change and to always reset the operating parameters to the default settings for ease of use by less experienced users. As many of these changes such as "Data Transfer Modes" and "Projection" of pixel values during acquisition go beyond basic operating practice, they will not be discussed here.

Fig. 9.16 "Hardware Settings" window for setting image collection parameters

As discussed in Chap. 6, pixel "Bit Depth" is an important consideration when scanning a specimen. Many systems have the option of scanning at 8 or 12 bit levels of information (Fig. 9.16). While intuitively one might always want to select the option to scan for the collection of 12 bits to increase the information available in an image, many of the above considerations on file size should be considered. File size of a single optical section can double in size when scanning at bit depths of 8 and 12, respectively. For example, when exported as a Tif file, file size for an 8-bit image collected at 1024 × 1024 was 3077 kb. With the same imaging parameters, but collected at 12 bits, the exported Tif file was 6,153 kb in size. When imaging multiple channels in a large Z-series, this may become a major factor in being able to process images in many enhancement and analysis programs. The major advantage of scanning at a bit depth of 12 is the increased information available in the data set for quantitative experiments. However, if the end point of the experiment is a qualitative assessment of the data, scanning at a bit depth of 8 typically provides more information than the human visual system can interpret and is an option to reduce file size.

9.6.8 User Configuration and Memory Management

As with many functions in the "Configuration" menu, the "User Configuration" (Fig. 9.17) icon and window are best used only by experienced users and then reset to system-defined (default) functions following an operating session. Functions in this window such as "Use Last System Settings" and working in" Folders" or "Projects" when storing data are self-explanatory, but for novice users or those less familiar with these software pages, selection of various icons can drastically affect the appearance of the viewing window, how data is stored, and even the unit of measurement. To avoid potential confusion it is best to encourage operators to use default functions in this case.

The "Memory Management" window (not shown) is a standard browser for setting the storage tree for auto saved and avilable system memory.

9.6.9 Dye Database

The "Dye Database" (Fig. 9.18) provides a list of excitation and emission spectra for many fluorochromes that can be imaged with the SP8. By selecting a fluorochrome from the pull-down menu, the absorbance ("Excitation") (red) and "Emission" (white) curves and the peak values are shown. For example, in Fig. 9.18 Alexa 532 is selected showing the maximum excitation wavelength as 534 nm and emission wavelength as 553 nm. Excitation and emission curves are also shown which is important when selecting fluorochromes to minimize bleed through as discussed in Chaps. 2 and 3. While data for specific fluorophores are available in the "Configuration" window, for convenience there is redundancy in the "Acquire" menu. The "Dye Database" and "Dye Assistant" will be discussed in detail in Sect. 9.9.

Fig. 9.17 "User Configuration" window for selection of how the system will be set up and data stored

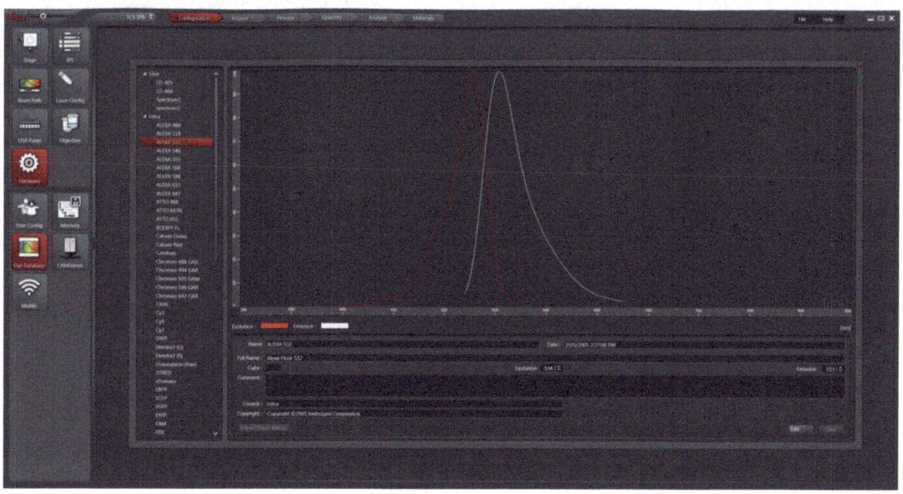

Fig. 9.18 The "Dye Database" provides a library of information on the excitation and emission spectra of a large number of fluorochromes

9.7 The Acquire Menu

Most user interaction with the software will occur in the "Acquire" window (Fig. 9.19) where sample fluorochrome excitation and emission spectra are used to establish laser and acousto-optical beam splitter (AOBS) settings, pinhole and detector settings, scan speed, Z-series and tiling considerations, and acquisition parameters. At this point in the decision-making process, it is advantageous to collect images with knowledge concerning the end point of the experiment and how the data will be presented. It will save considerable time and effort to collect high-quality images using the full dynamic range of the microscope rather than working in Photoshop, ImageJ, LASX 3D software, or other post-acquisition image enhancement programs to adjust image orientation, gamma, contrast and brightness, etc. For example, is the experimental end point a qualitative assessment of structural components at a relatively low magnification, or is the end point a high-resolution quantitative colocalization experiment? If recurring components such as myofibrils are present in the cell, is it possible to compose groups of images prior to collection so that orientation of the recurring components will all be collected in the same direction? It is much easier and more time efficient to address image quality and composition concerns during data acquisition rather than trying to enhance images post-collection.

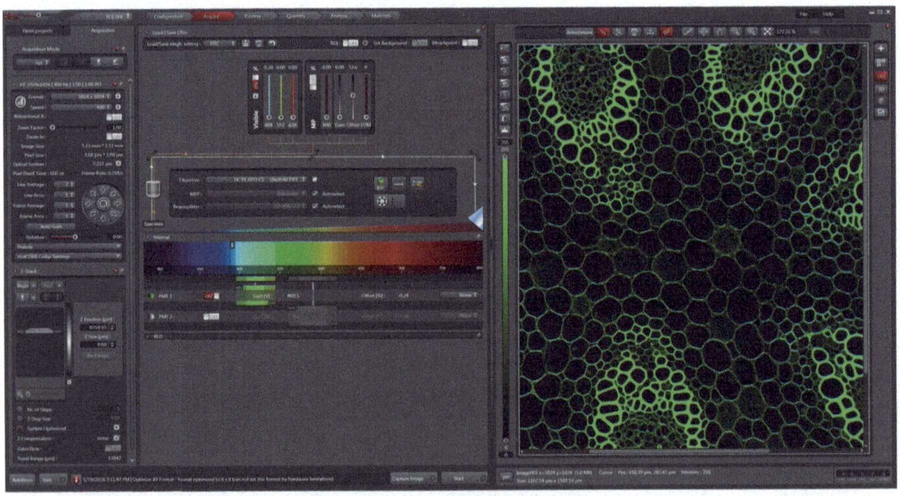

Fig. 9.19 An overview image of the "Acquire" window. The left panel provides access for setting imaging parameters including scan speed, resolution, averaging and summing functions, Z-series parameters etc. The superimposed white arrow on the left of the figure indicates a small arrowhead next to the "Z-Stack" window. Each window in the left panel has a similar arrowhead that allows contraction of the window if that particular function will not be used in the imaging session. The center panel provides access to laser controls, the Dye Assistant, and detector settings. The right panel shows the collected image and image processing tools. Each of these windows and functions is discussed in detail in the subsequent figures

If decisions are not possible concerning the end point of the experiment at the time of image collection, then a general rule of thumb is collect big and work small (Commandment #8: resolution is a one-way street). Once an image is collected, it is not possible to improve the resolution or increase the bit depth. However, if images are collected at high resolution and bit depth, it is possible to selectively reduce the data (i.e., number of Z-sections, pixels, or voxels) while working with images in enhancement and analysis programs. It is essential that the original files be archived if data reduction processes are used to ensure no significant loss of information occurs. It is also important to consider compromises related to file size and collection time when collecting high-resolution images or scanning slowly as discussed in Chaps. 6 and 7.

All settings for components in the optical path, including laser and detector settings, are user definable and can be easily stored for future use. This provides rapid access to established protocols for commonly used fluorochromes. Individualized protocols for imaging routines can also be written, stored, and accessed through the "Acquire" menu or by opening an existing file and applying the settings used to image that specific data set. A precaution here is that in multiuser situations, saved protocols should be routinely checked to ensure no changes have been made and saved by previous users. Settings should be stored in a separate area limited to Administrator access for future use. Investigators should also realize that since no two specimens are identical, fine adjustments in settings are usually required. Caution must also be used in comparison of images collected during different operating sessions as laser conditions and other operating parameters can vary over time and as components in the system age or are replaced. However, established configuration protocols and reuse functions are very useful in rapidly establishing instrument settings to obtain an image when starting a session (see Fig. 9.3a).

Figure 9.19 shows an overview image of the "Acquire" window in the Leica LAS X software. The left panel provides access to settings necessary to set specimen scanning parameters, the center panel access to laser and detector settings with regard to wavelengths and the Dye Assistant, and the right panel shows the collected image (in this case Convallaria) and image annotation functions. The subsequent sections discuss each of the parameters in these three windows. Note that in many cases there is a small arrowhead next to the window title (left panel "Acquisition Mode," XY, Z-Stack, etc.). These arrows allow windows to be expanded to provide the information as shown here or closed to allow space for other windows to be opened. For example, if a Z-Stack will not be collected, it is possible to click on the arrowhead to close this window and provide space for other expanded windows to be shown.

Along the bottom of the Acquire window are "Autofocus," "Live," "Capture Image," and "Start" buttons. The "Autofocus" button is useful in obtaining a ballpark focused image, but many experienced users avoid "Auto" functions and prefer setting operating parameters based on experience rather than allowing the instrument to do this. The "Live" button will start a continuous scan for viewing an image in the right panel while establishing operating parameters to obtain a good SNR image. The "Capture Image" button will scan only the channel that is currently active (discussed below). The "Start" button should be used to collect the final image that will be saved. To begin collection of multichannel images, Z-series, or tiled images, "Start" must be selected.

9.7.1 Acquire Window Left Panel: Open Projects

The "Open Projects" window (Fig. 9.20) provides access to the various drives and files present in the system. The top icons in the "Open Projects" window allow opening, saving, and browsing of files. From the left the icons provide access to the "Explorer" folder tree, "Open New Projects," "Open an Existing Project," "Save All," "Apply," "Browse," and "Gallery View." When a project is open and the file is highlighted by selecting the "Apply" icon, it is possible to reuse the parameters set in the data collection of the image displayed in the right window on the monitor (not shown here).

9.7.2 Acquire Window Acquisition Mode: Resolution

As shown in the overview image of the "Acquire" window (Fig. 9.19), several image acquisition functions can be accessed through separate regions of the "Acquisition Mode" panel. These include various scanning methods such as the default "XYZ"

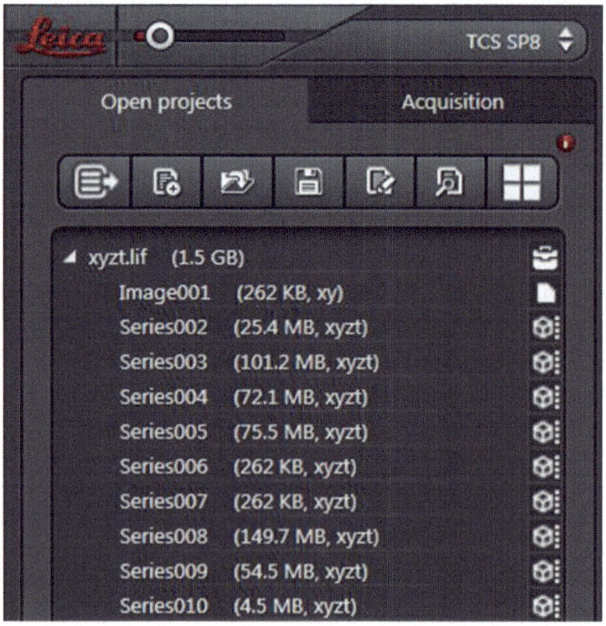

Fig. 9.20 The "Open Projects" window allows opening, saving, and browsing of files. From the left the icons provide access to the "Explorer" folder tree, "Open New Projects," "Open an Existing Project," "Save All," "Apply," "Browse," and "Gallery View." When a project is open and the file is highlighted by selecting the "Apply" icon , it is possible to reuse the parameters set in the data collection of the image displayed in the right window on the monitor

scan mode, "XYt" for time mode scanning, and "XYλ" mode for spectral imaging. Each of the "XYZ Acquisition" functions present on the SP8 system will be described in the subsequent paragraphs. Systems configured differently may have other sub-windows. The "XYt" and "XYλ" modes include more advanced techniques and will not be discussed.

Image "Format" (Fig. 9.21) defines the resolution and shape of the image that will be collected by determining the number of pixels present in a set frame size. As an example, Figure 9.21a shows an image format of 2488 pixels × 2488 pixels (arrow) that would represent a square scanning format with an "Image Size" of 175.74 μm × 175.74 μm and XY "Pixel Size" of 70.66 nm × 70.66 nm (boxed information). In Figure 9.21b the "Format" has been changed to a higher resolution of 4472 × 4472 (arrow). While the "Image Size" remains the same, the number of pixels has doubled, and XY "Pixel Size" is now 39.31nm^2 (boxed information). By selecting the "Format" dropdown menu, a range of preset format sizes can

Fig. 9.21 Use of the image "Format" allows selection of resolution and shape of the region of interest. In (**a**) a "Format" of 2488 × 2488 is selected (arrow) resulting in an "Image" and "Pixel" size shown in the boxed area. In (**b**) a larger number of pixels will be collected resulting in the same image size but smaller pixels (boxed area)

Fig. 9.22 "Format" window from which preset pixel arrays can be selected

be chosen (Fig. 9.22). It is often advantageous to select a rectangular format such as 1024 × 256 for framing a region of interest. If the desired format is not listed, the "+" arrow to the right of the "Format" information allows for custom selection of image resolution (Fig. 9.23).

Frame size and the number of pixels per frame in the X and Y directions are selected by default based on the objective lens selected but can be manually changed. As discussed in Chap. 7, the final resolution of the image will be determined at this point. With respect to confocal imaging, several choices must be made in selection of the appropriate resolution to minimize specimen damage and optimize image quality. The Leica LASX software provides several choices of pixel resolution that range from low-resolution images that include relatively few pixels (128 × 128) to high-resolution images (8192 × 8192). The effect of pixel size on image quality was shown in Fig. 6.3, and ideally selection of pixel resolution should be based on the Nyquist-Shannon theorem as discussed in Chap. 6.

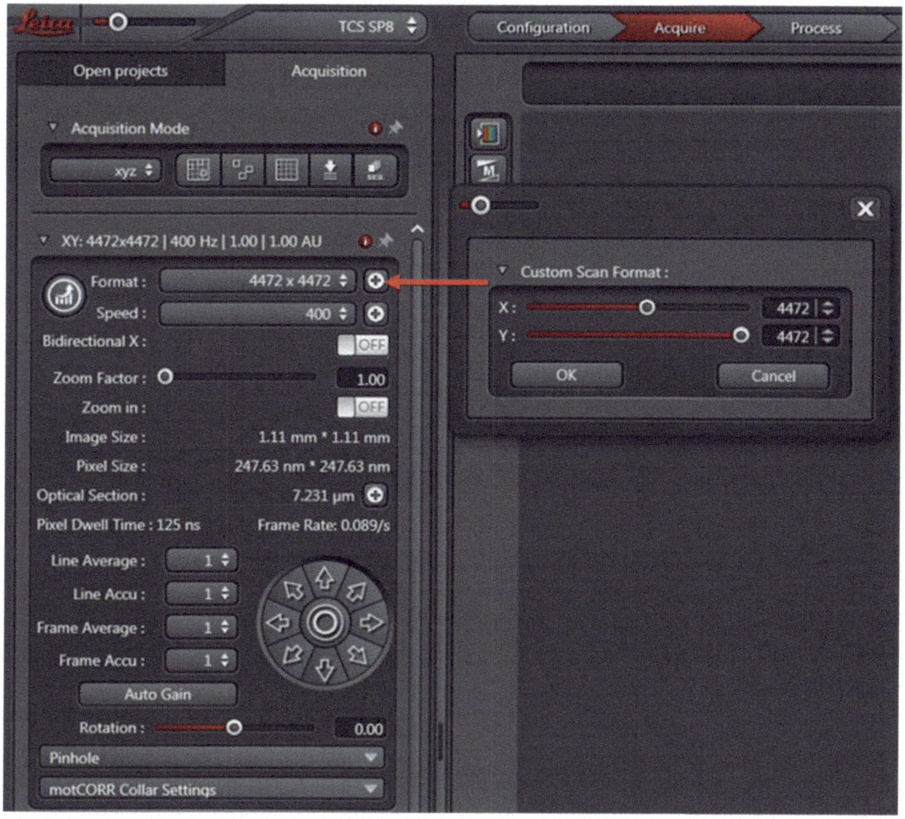

Fig. 9.23 By selecting the "+" button (arrow), a "Custom Scan Format" can be created

While increasing the number of pixels in an image will improve resolution beyond that required by the Nyquist theorem, the time required to scan a specimen will also increase which may result in increased specimen damage. The increased number of pixels will also significantly increase the file size. For example, in the Leica SP8 system, scanning at a resolution of 512 × 512 at a scan speed of 700 Hz results in a "Pixel Dwell Time" (the time the laser is in contact with that spot in the specimen) of 675 ns and a frame rate of 1.35 frames/s. At the same scan speed, an increase in image resolution to 2048 × 2048 results in the same "Pixel Dwell Time," but since many more pixels are present, the frame rate slows to 0.34 frames/s. Although resolution is improved significantly, the increased time may result in significant photobleaching.

File size and image processing capabilities must also be considered when selecting image resolution. While confocal systems are typically equipped with computers containing large data storage and processing capabilities, many offline processing systems do not have the storage and RAM capabilities to work with the large data sets collected with confocal microscopes. For example, a single channel image collected at a pixel resolution of 1024 × 1024 will be slightly over 1 Mb in size. In comparison, file size for an image collected at a resolution of 8192 × 8192 is 67.1 Mb for a single optical section of one channel. However, most confocal data sets are multichannel. A four-channel image collected at a resolution of 1024 × 1024 is over four megabytes in size, and a four-channel image at 8192 × 8192 would be approximately 269 Mb. It is easy to see that a large four-color Z-series collected at high resolution would exceed several gigabytes in size. This often challenges the image processing capability of standard computers, and it is now common for collection of data sets with some imaging systems to exceed a terabyte in size. Data collections of this magnitude quickly overwhelms standard computer processors demanding most imaging systems to house multiple hard drives exceeding terabytes in size or the use of cloud storage.

9.7.3 Acquire Window Acquisition Mode: Scan Speed

As noted above the "Pixel Dwell Time" represents the amount of time the laser remains in contact with the region of the specimen (pixel or voxel) being imaged. If the laser remains in contact with the region for a longer period of time, the potential for emission of a greater number of photons (signal) increases. Thus, the selection of scan "Speed" is a critical component in determining the SNR with a slow scan "Speed" resulting in the generation of an increased number of photons. The selection of scan "Speed" can be made directly below the image "Format" button (Fig. 9.24). Decreasing the scan rate will improve the SNR; however this is at the expense of increased photobleaching, which could be a problem with some samples. In Fig. 9.24a, a scan "Speed" of 400 Hz is selected with an" Image Size" of 1024 × 1024. With these operating conditions, the "Pixel Dwell Time" is 600 ns and the "Frame Rate" 0.388 frames/s. If the scan "Speed" is increased to 700 Hz (Fig. 9.24b), the "Pixel Dwell Time" is decreased to 325 ns, and the "Frame Rate" becomes 0.678 frames/s. While the difference in actual time to collect an image may seem small, when imaging events in live cells or preserved samples that are rapidly photobleaching, the increased scan speed will reduce the exposure of a sample to high-intensity lasers. This may mean the difference between being able to collect meaningful data, although at reduced SNR, and the inability to collect data necessary to finish an experiment.

It should be noted that as the scan "Speed" is increased passed 700 Hz, a "Zoom Factor" is forced due to limited movement of the galvanometer mirrors above a certain speed. This zoom can be seen in Fig. 9.25 which compares images collected at 400, 700, and 1400 Hz. When scan "Speed" is increased to 1400 Hz, "Pixel Dwell Time" is decreased to 150 ns, and the "Frame Rate" becomes 1.35 frames/s. Note the increased noise in Figs. 9.25b and c in comparison with the slower scan speed used for Fig. 9.25a.

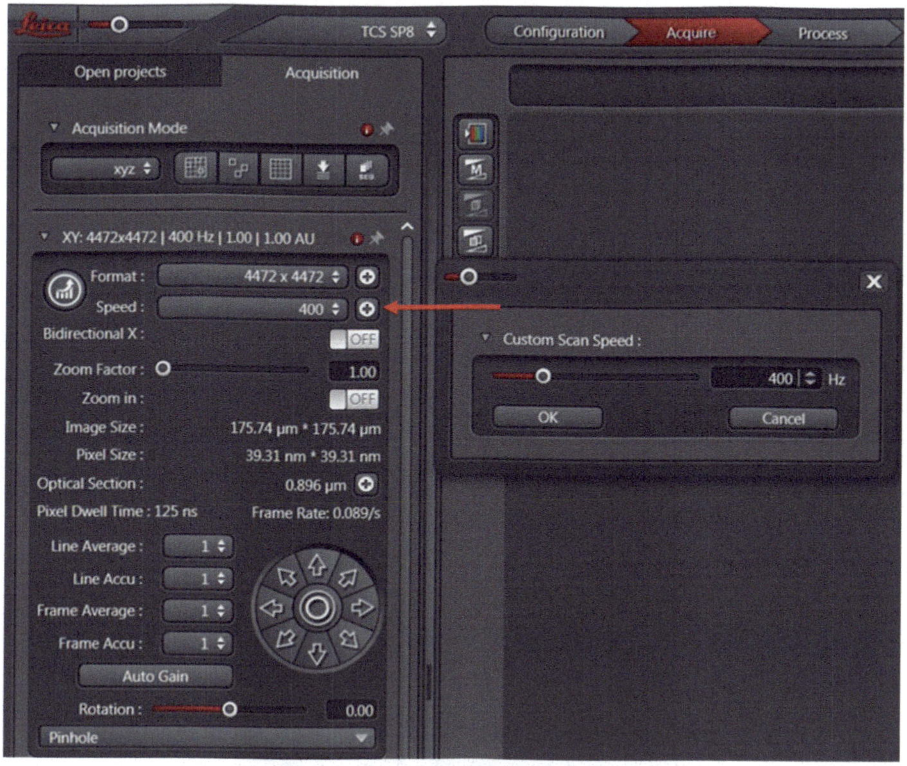

Fig. 9.24 As scan "Speed" is increased from 400 Hz (**a**) to 700 Hz (**b**), the "Pixel Dwell Time" and "Frame Rate" decrease. This may adversely affect the SNR but can protect a specimen from photobleaching

As described above, in single-photon confocal imaging, the laser interacts with a large volume of the specimen during each scan. This may result in excessive bleaching and loss of signal in optical planes above and below the focal plane of the specimen if the sample is not protected against photobleaching with mounting media, such as DABCO, as described in Chap. 4. While the time necessary to collect a data set should not be a major consideration in most imaging protocols, by knowing the options available on an instrument and compromises made in specimen damage and image quality, both instrument access and operator time can be used efficiently. For example, for a four-color 1024 × 1024 Z-series of 50 optical sections collected at 400 Hz, total scan time will be 13.5 m. If sufficient signal is present to scan the same data set at 700 Hz, total scan time can be reduced to 5.25 min. If photon emission from the sample is sufficient to collect an image with a good SNR at a fast scan rate, this option should be used saving both operator and instrument time. This allows optimal use of the specimen and instrument.

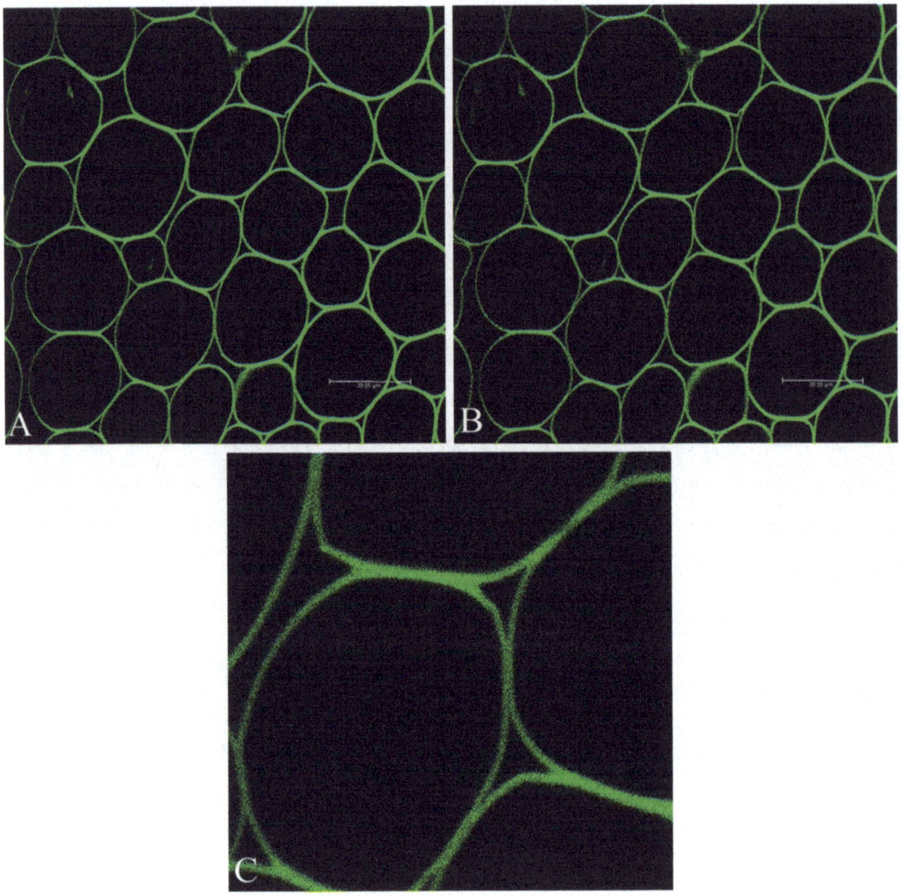

Fig. 9.25 As scan "Speed" is increased from 400 Hz (**a**) to 700 Hz (**b**) and 1400 Hz (**c**), the "Pixel Dwell Time" is decreased resulting in an increase in noise as shown in (**c**)

9.7.4 Acquire Window Acquisition Mode: Bidirectional Scanning

The default scan pattern for most confocal systems is unidirectional where the beam scans in the X-direction, is blanked on return, and again scans in the X-direction for the set number of lines. "Bidirectional X" scanning patterns (Fig. 9.26) do not blank the beam resulting in faster scan rates, but may introduce some scanning artifacts that must be corrected. For example, if a unidirectional scan "Speed" of 600 Hz at a resolution of 1024 × 1024 is chosen, the "Frame Scan Rate" is 0.15 frames/s. If the bidirectional X function is turned on, the frame rate doubles to 0.0.291 frames scanned per second (Fig. 9.26). The bidirectional X scan pattern can be turned on just below the scan" Speed" button (Fig. 9.26 arrow head). This also activates the

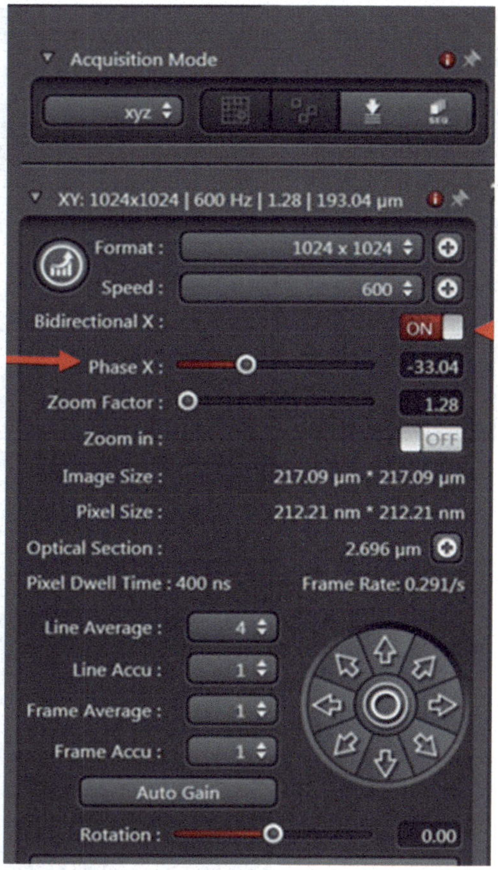

Fig. 9.26 "Bidirectional X" (arrow head) scanning increased the "Frame Rate" but may introduce artifacts as shown in Fig. 9.27 that need to be corrected with the "Phase X" control

"Phase X" control that can be used to correct the bidirectional scan artifacts seen in Fig. 9.27a–c by sliding the bar until the edges of the image do not show artifacts. Increased scan speed with the "Bidirectional X" function may reduce photobleaching and can increase scan rates during live cell imaging. Care must be taken when using "Bidirectional X" scanning that unintended artifacts are not introduced by overcorrection along the edges of structures in an image.

9.7.5 Acquire Window Acquisition Mode: Zoom and Rotation

"Zoom Factor" and "Rotation" (Fig. 9.28) provide a mechanism to frame an image and to digitally increase magnification by scanning a smaller region of the specimen and projecting it to the detector. For example, Fig. 9.29a shows an image of Convallaria at 10X with a zoom factor of 1. The pixel size at 512 × 512 resolution is $2.17\mu^2$. Without changing objectives and increasing the zoom factor to 2 (Fig. 9.29b),

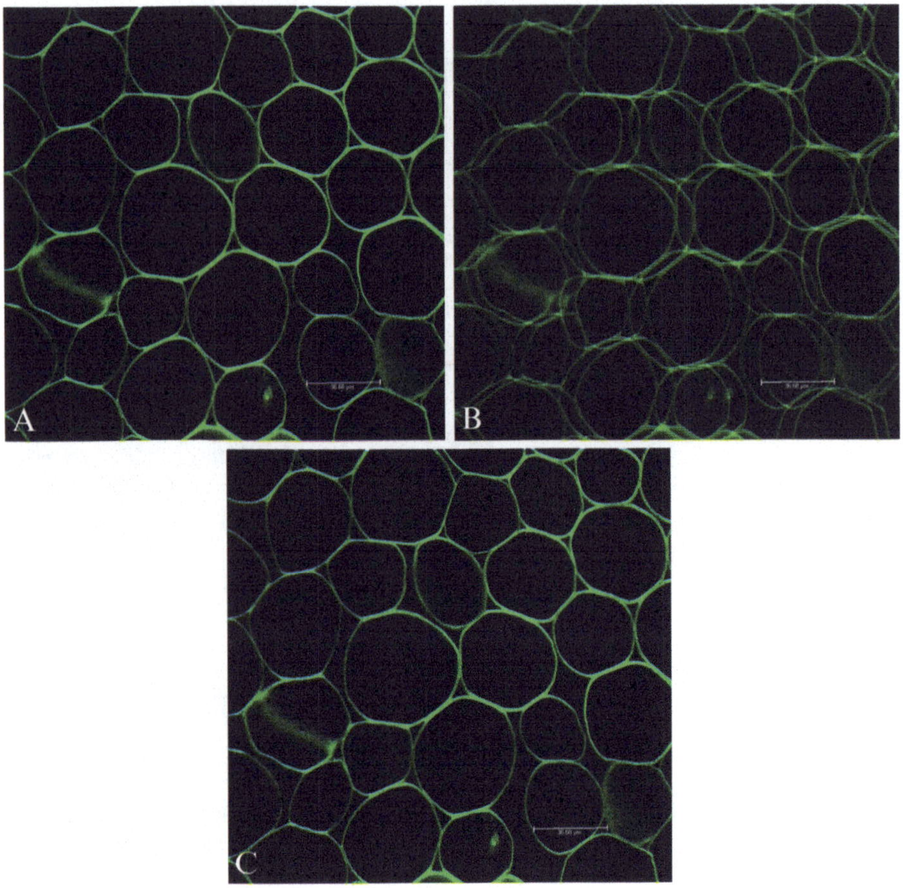

Fig. 9.27 Edge artifacts may be introduced when "Bidirectional X" scanning is used to increase scan speed. (**a**) Image collected with unidirectional scanning. (**b**) Uncorrected image collected with "Bidirectional X" scanning. (**c**) Image collected with "Bidirectional X" scanning and "Phase X" correction

the pixel dimensions are changed to $1.08\mu^2$, and the region is magnified without changing objectives. Similarly, if the Zoom Factor is decreased to 0.75 (Fig. 9.29c), the image magnification is reduced, and pixel dimensions are now $2.89\mu^2$. This is an effective mechanism to change magnification to a certain degree without changing the objective. For instance, it may be desirable to continue imaging with a 20X dry objective rather than changing to a 40X oil objective to avoid adding oil to a slide. However, remember that resolution is set by the objective and imaging with a lower NA 20X objective will not provide the resolution that a higher NA 40X oil objective will provide. A large increase in magnification will result in blurring of the image, and this should be performed by using an objective of higher magnification rather than doing so digitally.

Fig. 9.28 "Zoom Factor" and "Rotation" controls (arrow heads) for electronic adjustment to magnification framing of an image

Another mechanism to control the electronic increase in magnification of a region is to use the "Zoom in" function located just below the "Zoom Factor" control. When "Zoom in" is activated, a region of interest can be selected with the red highlighted rectangle in the image window on the right of the monitor. Only the region selected is then scanned at a magnification determined by the size of the rectangle drawn on the image.

Once the "Zoom Factor" is applied, images can also be moved laterally with the arrows and rotated with the "Rotation" function (Fig. 9.30) or with the scan field rotation on the USB panel box. These factors assist with framing an image and are very useful if repeating structures such as myofibrils are present in a series of images. With the rotation function, all images can be aligned identically at this stage rather than with image processing programs at a later time.

9.7.6 Acquire Window Acquisition Mode: Optical Section Thickness and Pinhole Setting

When using standard lasers, the optical section thickness is determined by the pinhole diameter, NA of the objective, and the emission wavelength as shown in the formula for the "Optical Section Thickness" in Fig. 9.31. For example, when using

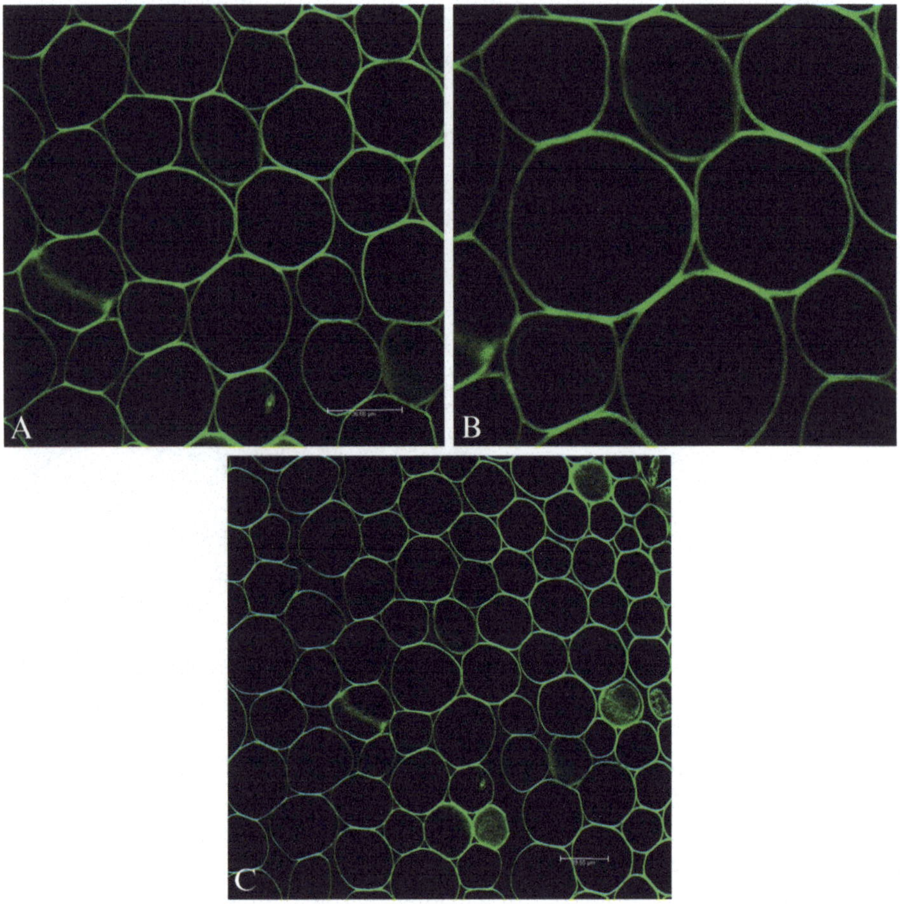

Fig. 9.29 Examples of electronically changing the magnification of an image by use of the "Zoom Factor" control. (**a**) "Zoom Factor" of 1. (**b**) "Zoom Factor" of 2. (**c**) "Zoom Factor" of 0.75

an emission wavelength of 580 nm, a 63X oil objective with a 1.4NA, and a pinhole diameter of 1 AU, the optical section thickness is 0.896 µm (Fig. 9.31a). Since the emission wavelength and NA of the objective are set when collecting an image, the optical section thickness can be changed by adjusting the pinhole diameter/Airy units. In Fig. 9.31b the pinhole diameter has been increased to 3 Airy units, and the optical slice thickness is increased to 2.368 µm. If using a multiphoton laser to obtain the correct calculation of section thickness, the "+" button (a) next to the "Optical Section" information and the "Multiphoton" box (b) must be checked as shown in Fig. 9.31c for the system to correctly calculate the optical slice thickness. Note that when using the multiphoton laser, the pinhole diameter is not a factor as only fluorochromes in the focal plane are excited as discussed in Sect. 2.4.

Setting the pinhole diameter is one of the most important decisions made when imaging a sample. Adjustments to the pinhole can directly affect image resolution,

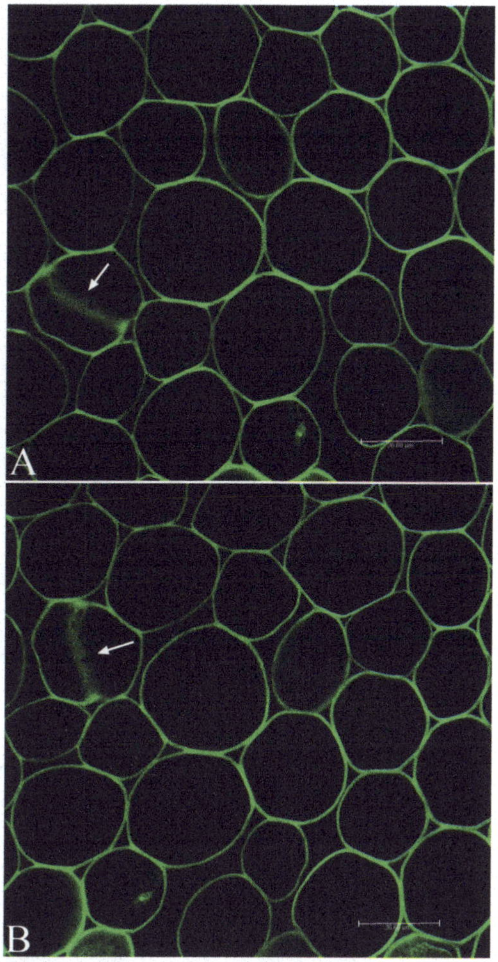

Fig. 9.30 Use of image "Rotation" to change the orientation of a structure (arrows)

SNR, quantitative analysis of structure size, and interpretation of colocalization data. Since the effects of pinhole diameter adjustment occur after interaction with the specimen, effects on specimen damage are indirect. For example, increasing the size of the pinhole improves the amount of signal at the detector which allows reduced laser intensity and/or scan time to saturate the detector. This can protect the specimen from laser damage since laser intensity and scan time can be decreased. If equipped with Leica HyD detectors the pinhole should also be set to a minimal value to protect the detectors from oversaturation damage.

If all other factors are held constant, opening the pinhole will increase the signal (brightness) of a confocal image as noted above, but this also decreases contrast and Z-resolution due to inclusion of information in the image from out-of-focus areas of the sample. This is illustrated in Fig. 9.32 which shows images collected at pinhole diameters of 1 AU (Fig. 9.32a) and 3 AU (Fig. 9.32b). Note that in this example all

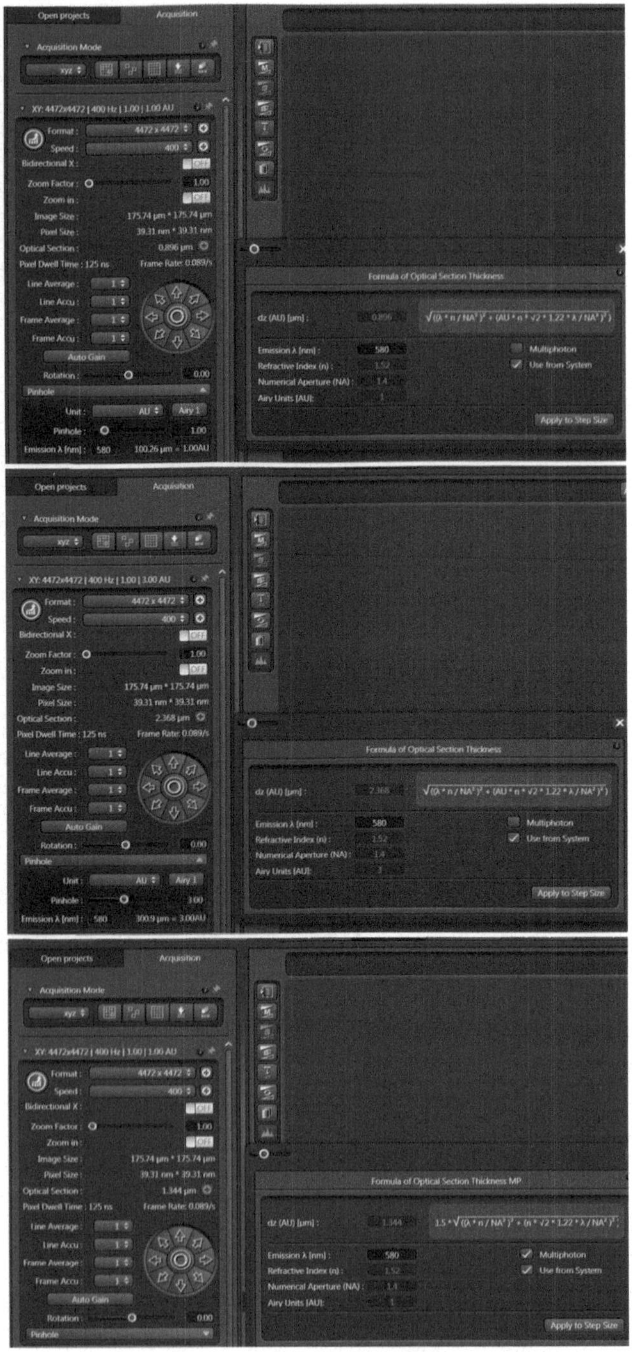

Fig. 9.31 Calculation of the "Optical Section Thickness" is based on pinhole dimeter and the number of Airy units (**a** and **b** arrows) when using single-photon confocal microscopy. As pinhole diameter increases the "Optical Slice Thickness" also increases. When using multiphoton microscopy (**c**), for correct calculation of the optical slice thickness, the "+" (arrow a) button should be selected and the "Multiphoton" box checked (arrow b) for correct calculation of the "Optical Slice Thickness"

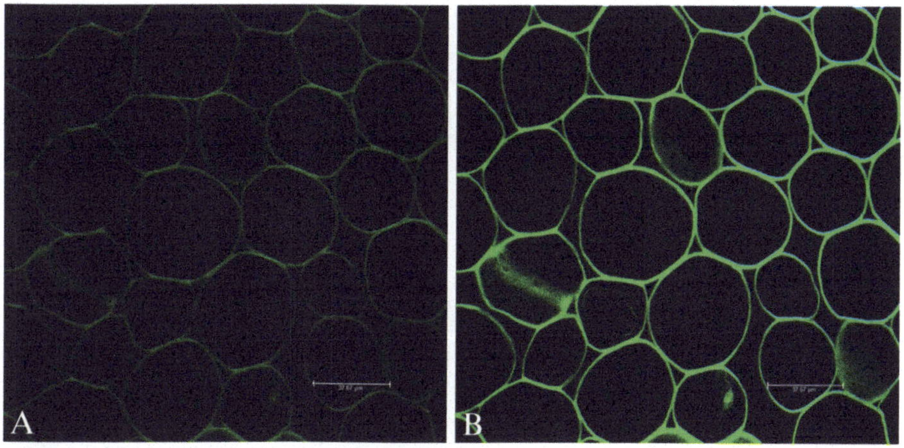

Fig. 9.32 Opening the pinhole from 1 AU (**a**) to 3 AU (**b**) increases the amount of signal present, but much of the increased signal is from out-of-focus planes which may decrease resolution and contrast in the image

detector and laser settings were held constant. At 1 AU, the signal is relatively low and the image appears dim. When the pinhole is opened to 3 AU, increased signal is present resulting in a brighter image. While increasing the optical slice thickness will increase the signal, the improved brightness is due to collection of photons from above and below the focal plane, and this option should be avoided if possible. By increasing the pinhole, the signal is improved, and it is possible to decrease laser intensity and scan speed to protect a specimen from photobleaching. These three factors, scan speed, laser intensity, and pinhole diameter, must always be considered when establishing image collection parameters as they will determine image quality versus specimen damage compromises as shown in Table 9.1.

9.7.7 Acquire Window Acquisition Mode: Line and Frame Average and Accumulation

Even with the best of specimens and operating conditions, many confocal images have the potential to contain noise in an image as a result of the electronics present in the hardware. This is most easily observed in the dark areas of an image that contain little or no fluorescence. In a single pass of the laser, some pixels in these areas may have a bright appearance that is above the zero value set for black. In subsequent scans the same pixel may have values that range from zero to high enough to show a bright spot in the image background. Shifting of the value of these pixels during subsequent scans indicates electronic noise in the image. If noise becomes a severe problem, it adversely affects image quality, and if a pixel is bright enough, it may affect quantitative assessment of the data.

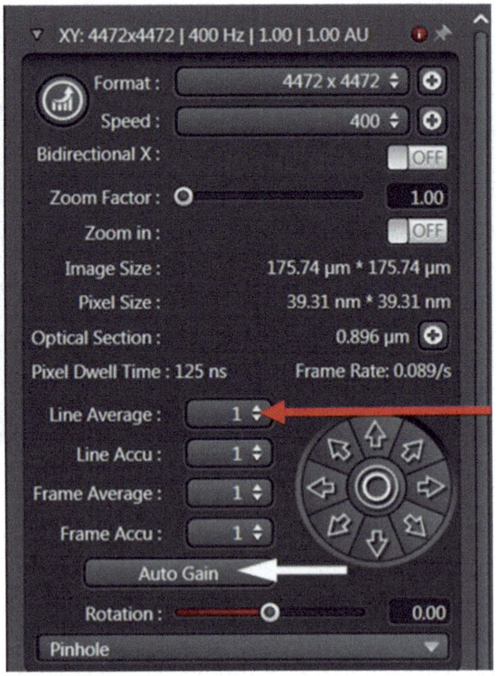

Fig. 9.33 To reduce noise in an image, "Line Average" (red arrow) or "Frame Average" may be used. "Line Accu" and "Frame Accu" can be used to add the value of each pixel on subsequent scans. "Auto Gain" (white arrow) provides a mechanism for the system to rapidly set the SNR of an image

Since electronic noise is random, it is unlikely that the same pixel experiences a noise event in subsequent scans. Imaging several lines or frames and averaging each pixel value reduces the background level by decreasing pixel values where noise has occurred. Two methods, "Line Average" or "Frame Average," are available for averaging during collection of an image (Fig. 9.33 red arrow). Note that each box (arrow) has a pull-down menu for selection of the number of scans averaged. During line averaging each line is scanned the set number of times and the average value obtained for each pixel position. This value is then assigned to the final image. As an example, if an averaging function of 8 is selected and one of the scans has a pixel noise value of 24 while the remaining scans have a pixel value of 0, the average pixel value assigned to that location in the image is 3. In the Frame Average mode, the averaging of pixel values occurs after the completion of each frame scan. Averaging should result in noise values falling at the end of the image histogram reserved for histogram stretch functions and result in a black image pixel not detectable by the eye. The effects of line averaging are shown in Fig. 9.34 by comparing an image collected without averaging (A) with those collected by averaging four (B) and eight (C) times.

A common question is if averaging improves image quality, why shouldn't several more frames be included while collecting images? In the above example, an averaging function of eight scans was selected which reduced the noise in the pixel from a value of 24 to 3. If the number of scans was doubled to 16 under the same operating conditions, the average pixel value for the location would have only dropped to 1.5. Both values are at the low end of the acceptable pixel range in the

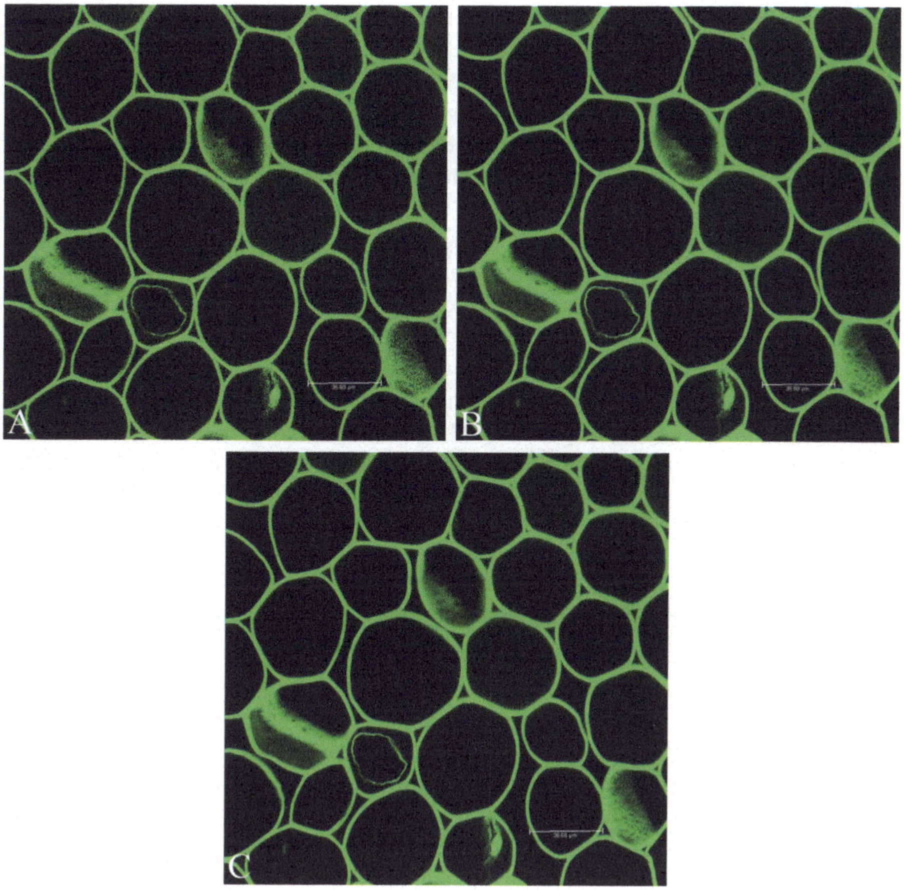

Fig. 9.34 Effects of "Line Average" to improve the SNR. (**a**) No averaging. (**b**) Average 4 times. (**c**) Averaged 8 times

image and would be detected by the eye as a black pixel. Several factors enter into the decision on the number of scans to average. The improvement in an image is directly proportional to the square root of the number of scans averaged. Thus, if an image is averaged 16 times, the image is improved by a factor of 4. If the image is scanned four times, it is improved by a factor of 2. However, to get the improvement in SNR ratio from 4 to 2, it takes four times as long to scan the sample which significantly adds to the total number of scans and time the specimen is exposed to the laser, which may result in photobleaching. For example, if a sample is scanned with a line average of 1 and a speed of 400 Hz as shown in Fig. 9.33, the pixel dwell time is 125 ns. If the same sample is averaged 4 times, the cumulative pixel dwell time would be 600 ns. While this difference may seem small when imaging specimens that are rapidly photobleaching, this may result in loss of signal after only a few scans or the death of cells during live cell imaging.

As a rule of thumb, if possible images should be averaged at least by a factor of 2 to reduce noise that may not be apparent to the eye, but will be present if an image is analyzed by computer. In samples where SNR is a significant problem and the fluorochromes are stable, averaging of more frames improves the image quality, but the compromise of SNR versus sample stability and time should always be a primary consideration when establishing the number of lines or frames averaged (Table 9.1).

It is also possible to accumulate the signal present at each pixel in an image (Fig. 9.33 – "Line Accu" and "Frame Accu"). In this case if a pixel value was 20 on each scan and a sample was scanned 5 times, a pixel value of 100 would be assigned to that location in the image. Use of the accumulate function is an excellent way to increase the brightness of an image while keeping laser intensity very low, but it is also possible to introduce significant artefactual data by accumulating the signal and creating an appearance of a bright fluorochrome when in reality little signal is present. It is essential that the final image mirror that seen in the microscope and the accumulate function not be used to create data.

9.7.8 Acquire Window Acquisition Mode: Auto Gain

The "Auto Gain" function (Fig. 9.33 white arrow) is a mechanism to allow the system to rapidly determine the optimum gain for a sample region. When auto gain is selected, 32 lines in the middle of the sample are scanned and the gain then set for the entire sample. While this represents a rapid method to reach a ballpark setting, it may be necessary to manually adjust the detector gain to optimize the image for regions that may not fall within the 32 lines initially scanned. Many users may be hesitant to allow the instrument to set critical parameters automatically.

9.7.9 Acquire Window Acquisition Mode: Z- Series Collection

One of the most common uses and largest advantages of confocal imaging is the collection of Z-series data stacks in which every optical section through a sample is in focus and the data set can be reconstructed for presentation in 3D and animated images. This is accomplished by collecting a section, moving the stage an established distance in the Z-direction, and collecting another section. This process is repeated until the required number of sections are collected. Several considerations such as optical slice thickness when collecting sections and the depth from which images can be collected are discussed in Chaps. 3, 4, and 10.

To correctly set up a "Z-Stack," all of the above considerations such as resolution (file size), scan speed, optical section thickness, and averaging or summing functions must be taken into consideration. Compromises based on the desire to collect as much data as possible versus specimen stability, time, and file size are often necessary. When operating in single-photon mode, one should also remember that photobleaching occurs in optical planes above and below the focal plane. When collecting data sets of several microns, the final sections may have suffered photobleaching damage by the time they are imaged. By understanding the compromises made between image quality and specimen damage, informed decisions can be made to avoid oversampling of data, while making sure adequate amounts of data are collected to complete an analysis. Once an operator understands all of the considerations and compromises necessary for collection of a single optical section, setting up the instrument for "Z-Stack" imaging is relatively simple.

Figure 9.35 shows the "Z-Stack" software. With most modern instruments, the process of collecting a Z-series is fairly easy. To begin collection a single channel

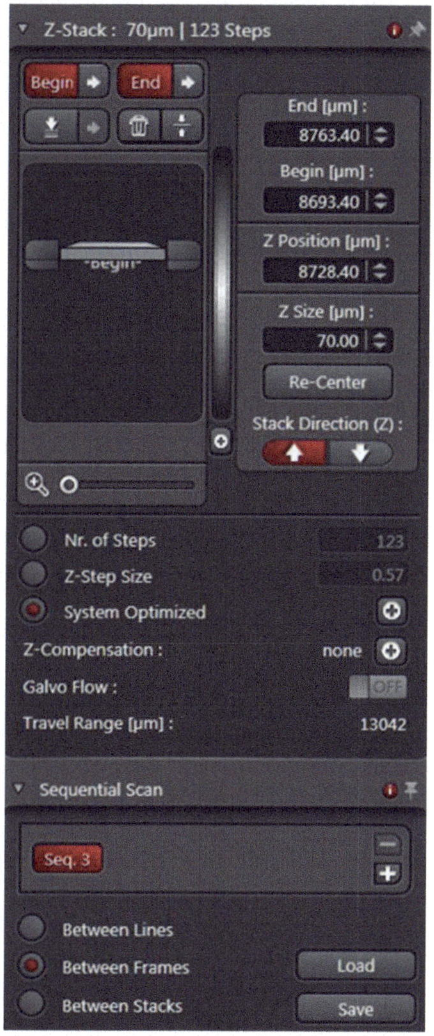

Fig. 9.35 "Z-Stack" window showing "Begin," "End," and information concerning Z-stack parameters. As set up the "Z-Stack" would be 70 µm thick collected at 0.57 µm intervals for a total of 123 sections that would be collected sequentially between frames

should be selected for setting the top and bottom of the Z-Stack. This allows rapidly setting the depth parameters without waiting for switching of channels to occur. The specimen should then be focused to either the top or bottom of the region of interest and "Begin" selected. Then focus to the last optical plane in the region of interest and select "End." This will establish the region that will be included in the "Z-Stack." It should be noted that an entire sample, especially when working with thick samples, does not need to be imaged. The "End" (8763 μm) and "Begin" (8693 μm) values shown in the boxes to the right in Fig. 9.35 represent the stage position at each point in the Z-series, and in this case a "Z-Stack" of 70 μm will be collected. Based on the pinhole diameter selected earlier, the "Z-Step Size" between each section will be 0.57 μm when the system is optimized to meet the Nyquist-Shannon considerations discussed in Chap. 7. In the example shown, to collect 70 μm of tissue at 0.57 μm intervals, the "Number of Steps" in the "Z-Stack" will be 123 sections. Often when sampling thick specimens with the system optimized, it may require several hundred sections to meet the Nyquist-Shannon requirements. This may result in very long data collection times, photobleaching, and excessively large data files. It is possible to reduce the number of sections (i.e., under sample by collecting sections at 1 μm intervals), but there are ramifications related to 3D reconstructions as discussed in Chap. 10, and factors such as missing data and/or data interpolation must be considered when compromises in data collection are made.

When imaging thick specimens, a choice is often necessary on where to set the SNR for optimum image quality throughout the Z-series. If the optimal SNR is set at the midpoint of the series, signal from the first section is often too bright, and signal from the last section is often minimal making reconstructions difficult. The low signal from the deepest sections may be a result of a phenomenon termed Z-drop which involves penetration of the laser into the sample and the ability of emitted photons to exit the sample and be collected. Many factors including excitation and emission wavelengths, photobleaching and refractive index mismatch, as discussed in several places throughout this book, may affect Z-drop. Selection of the "Z-Compensation" button provides an adjustment mechanism to minimize the effects of Z-drop when reconstructing data sets as discussed further in Chap. 10. If Z-drop is a problem, the optimal SNR should be set using an optical section near the top of the sample, and the system will compensate for Z-drop as deeper images are collected.

A final important choice to make in setting up a Z-series is how the scan will occur. At the bottom of Fig. 9.35, the Scan window is open showing "Sequential Scan." This window will be discussed in detail in the next section in the context of writing protocols, but the method of scanning a Z-series can also be selected here. A choice must be made to scan" Between Lines," "Between Frames," or "Between Stacks." If "Between Lines" is chosen, lasers will be switched after each line is scanned. "Between Frames" results in less switching of lasers as all channels will be scanned prior to the stage being moved to the next position. "Between Stacks" results in minimal switching of lasers since one channel is scanned through the entire region of interest, and then the stage is moved back to the "Begin" position for scanning of the second channel. In a four-channel sequence, this results in the lasers being switched only four times which decreases the total amount of time required to collect the "Z-Stack" but relies on the stage being returned to the exact "Begin" position following collection of each channel. In our laboratory "Between Frames" or "Between Stacks" is routinely used for collection of "Z-Stacks."

9.7.10 Acquire Window Acquisition Mode: Scan Window Simultaneous and Sequential Modes

The "Scan" window provides information for the protocol that is currently in use. For example, the protocol shown in Fig. 9.36 is a three-channel "Sequential Scan" set up to change lasers "Between Frames." As shown the windows showing the "XY" scan (resolution, speed, zoom, and pinhole) and the "Z-Stack" information are minimized. Sequence 3 is highlighted in red to indicate it is the currently active sequence for which the above information is displayed. If an additional sequence needs to be added, the "+" button should be selected, and to remove a sequence, the "-" button should be selected when the "Seq" to be removed is highlighted red. It is also possible to "Load" and "Save" protocols in this window. Creating simultaneous and sequential imaging protocols is discussed in the next section.

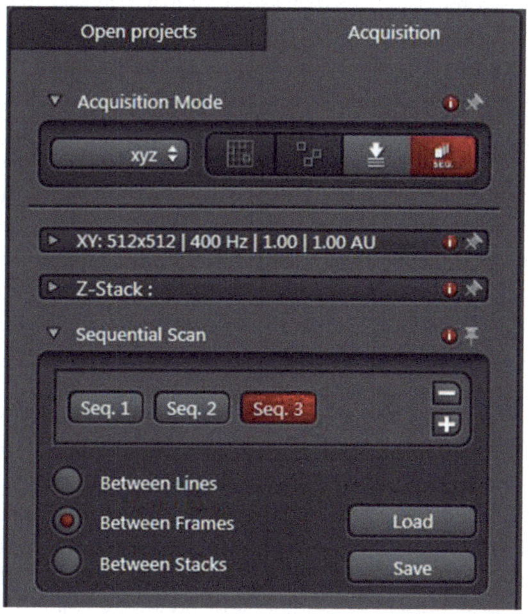

Fig. 9.36 The scan window indicating the third channel ("Seq 3") is active in this protocol set up to scan "Between Frames." The "XY" and "Z-Stack" windows are minimized as shown. To expand these windows, the arrowhead next to each should be selected

9.8 Acquire Menu: Center Window

The center window (Fig. 9.37) in the "Acquire" mode provides access to stored protocols and information for creating new protocols, control of the lasers, objectives, dye assistant, and the detector settings. Each component will be discussed in detail in the following sections. At the top of the window, a row of icons is provided to "Load" existing protocols," Save" newly written protocols (described below), select an XY region of interest ("ROI") for scanning, and setting the "Background."

Turning on the visible lasers (Fig. 9.37b) is a multistep process. First they must be activated by selecting the "+" button in the "Visible" laser window. This will open the Currently Available Lasers window. In this window select "ON" which will activate lasers but they are not yet turned on. It is then necessary to select the "ON" button in the Visible Laser window. During "Live" imaging the % transmission can be set to optimize the signal as discussed later, but this should be as low as possible to prevent photobleaching. Only the lasers used in an experiment should be turned on to preserve life on the other lasers.

If a multiphoton laser is available, a separate window will be present (Fig. 9.37c). Turning on the multiphoton laser is also a multistep process. First, select the "+" button in the MP panel which will open the MP laser control window (Fig. 9.37c). Then click and hold the dot in the MP Shutter Status window until the "Power" is on and the "Mode" indicates "Pulsing" this may take up to 5 s. Once the MP laser is on, select the wavelength of choice in the "Adjust Laser window" and select "Tune." It is then necessary to adjust the GDD (Group Display Dispersion) which is objective specific. As shown the GDD is adjusted for the 25× water objective on this instrument. The MP Laser Control window can then be closed. At this point the "ON" button in the MP window should be selected, and the "EOM" (electro-optical modulator) set to "1." During live acquisition the % transmission, gain, and offset can be adjusted to optimize signal.

In Fig. 9.37a the orange line (arrow D) exiting from below the lasers, going to the left through the image of the objective, and reaching the specimen represents the excitation photons. The spectrum of light represents the emitted photons shown as the white line around the boxed area with the objective, multifunction port (MFP) used to move filters into the microscope beam path, beam splitter, Dye Assistant (discussed below in the context of protocol creation), USB Control Panel, and Fluorifier Disc. While the objectives can be changed with the microscope STP8000 Controller, they can also be accessed by using the objective pull-down menu in the software (Fig. 9.37a, arrow E).

In the example shown in Fig. 9.37a, the 488 nm laser is activated, and 0.73% of the available laser light is allowed to interact with the specimen. The "%" transmission can be adjusted by raising or lowering the scale. The emission spectrum of the selected fluorochrome is shown as the white line superimposed on the color spectrum (Fig. 9.37, arrow F). When using multiple fluorochromes, the "Dye Assistant" (Fig. 9.37, arrow G) is a very useful tool. By selecting the Dye Assistant button, a new window (Fig 9.38a) will open. By selecting the three dots (arrow A) a list of

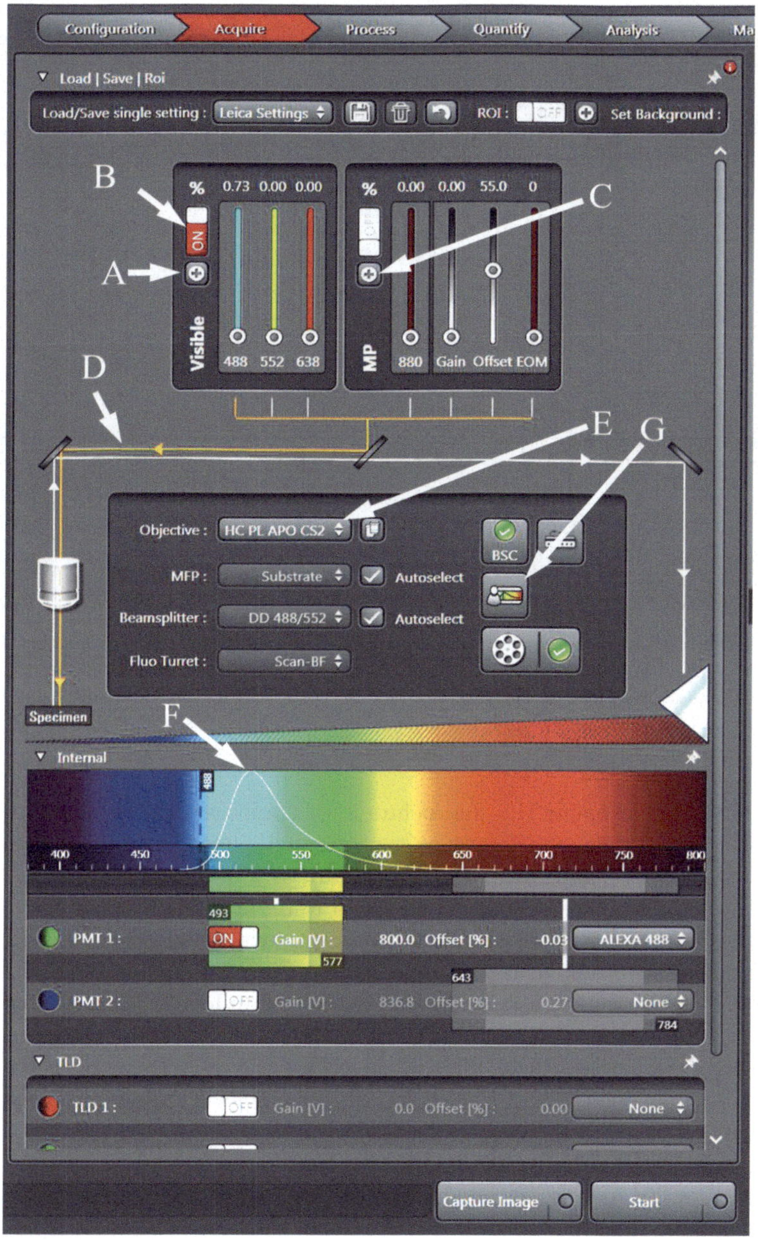

Fig. 9.37 (**a**) Overview of “Acquire” menu center panel showing the available lasers, objectives, Dye Assistant, and detectors. The path of the excitation photons is shown by the orange line (arrow D) below the lasers. Arrow E indicates the pull-down menu where the objective of choice can be selected, Arrow F indicates the emission spectrum of fluorochromes, and arrow G indicates the “Dye Assistant” that can be used to determine overlap of spectra from a large number of fluorochromes. (**b**) To turn on the “Visible” lasers, first select the “+” button (arrow A in **a**) to open the

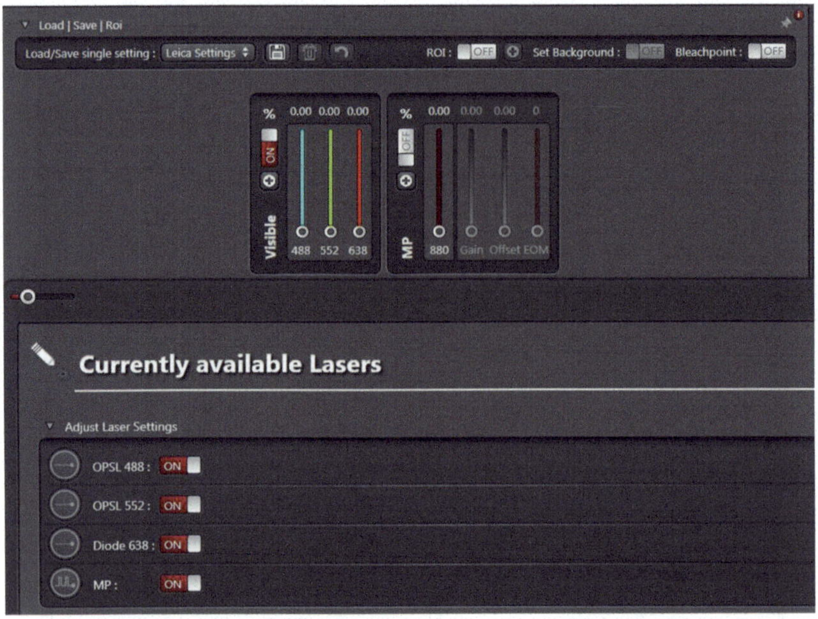

Fig. 37 (continued) window where lasers to be turned on can be selected. Then select the "on" (arrow B in **A**) to turn on the selected "Visible" lasers. (**c**) To turn on the multiphoton laser, select the "+" button (arrow C in **a**) to open the window shown. To start the MP laser, click and hold the red button next to "MP Shutter" until the "Power" is "On," "Mode" is "Pulsing," and "Shutter" is "Open." Then select the appropriate objective in the "GDD" window

fluorochromes will open. The fluorochrome of choice should be selected, and a grayscale image showing the emission spectrum will open. As shown in Fig. 9.38a, Alexa 488 and Cy3 have been added to the fluorochrome list. If a third fluorochrome were to be added, the three dots next to the white box would be selected, and an additional dye could be added to the protocol. Additional fluorochromes may be added accordingly to suit experimental design. By selecting the adjacent color box a LUT can be assigned to designate the color that will appear in the image. By convention, the color should be chosen based on the emission wavelength. For example, for Alexa 488 a green LUT was chosen since this is the wavelength for green light. The boxes below the dye selection window indicate the amount of "Crosstalk" if "None Sequential" (simultaneous), "Line Sequential," or "Frame Sequential" imaging modes are selected.

If "Edit" is selected, the actual "Excitation" and "Emission" values are shown on a graph, and the amount of overlap or "Crosstalk" between the excitation and emission spectra can be determined (Fig. 9.38b and c). In Fig. 9.38b "None Sequential" or simultaneous detection of the Alexa 488 and Cy3 was selected, and significant "Crosstalk" in the emission region near 550 nm (Fig. 9.38b arrow) is shown. This would not be a good choice for imaging these fluorochromes. In Fig. 9.38c "Frame or stack sequential" imaging has been selected and the detectors set to avoid "Crosstalk" between the channels. Setting detector parameters will be discussed in the next paragraph. Once the fluorochromes have been selected and the imaging mode determined, selecting "Apply" will load these settings into the protocol.

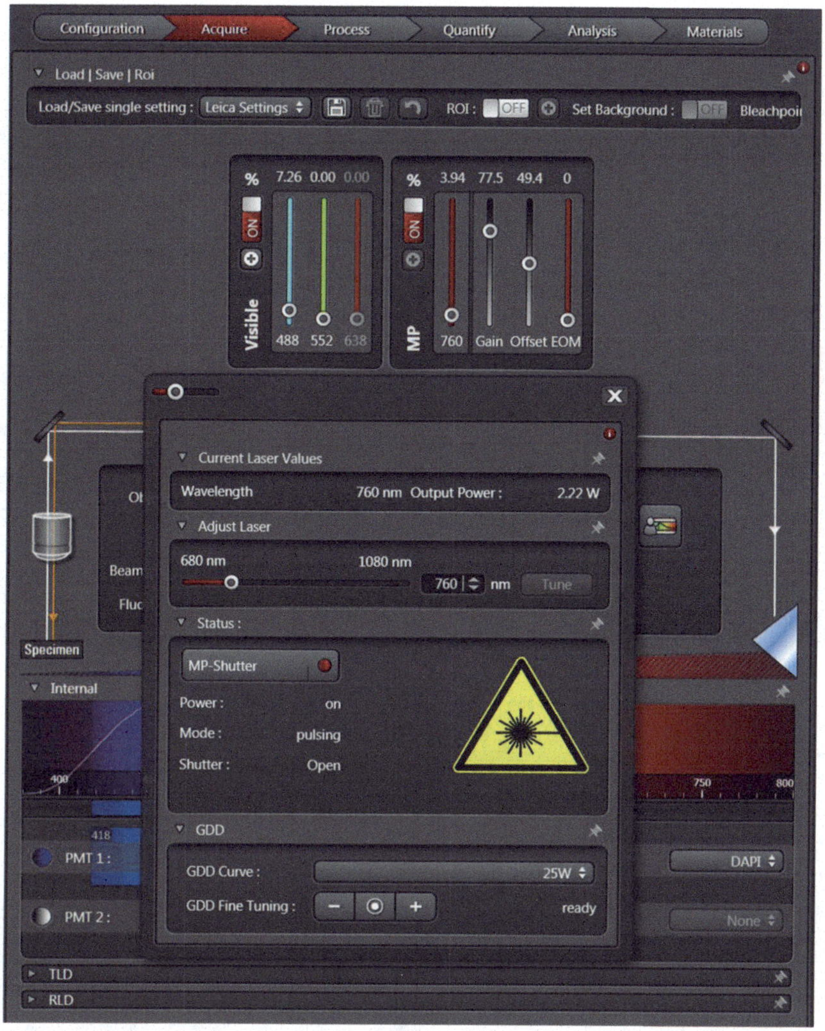

Fig. 37 (continued)

The detectors can be used to finely tune the detected emission spectrum for each channel. In the example shown here (Fig. 9.39) with Alexa 488, the laser excitation wavelength is shown as a vertical black dashed line at 488 nm (arrow A), and the emission spectrum is shown superimposed as a white line over the spectrum (arrow B). The emission range for Alexa 488 is approximately 490 nm through 600 nm, but with the AOBS the detection range is set for 493 nm through 550 nm. This results in emitted photons outside of this range being blocked from detection. In this case, if bleed through with the red channel became a problem, it would be possible to further decrease the emission range further to eliminate interaction with photons in the red wavelengths. In this example only a single emission channel is shown. As additional

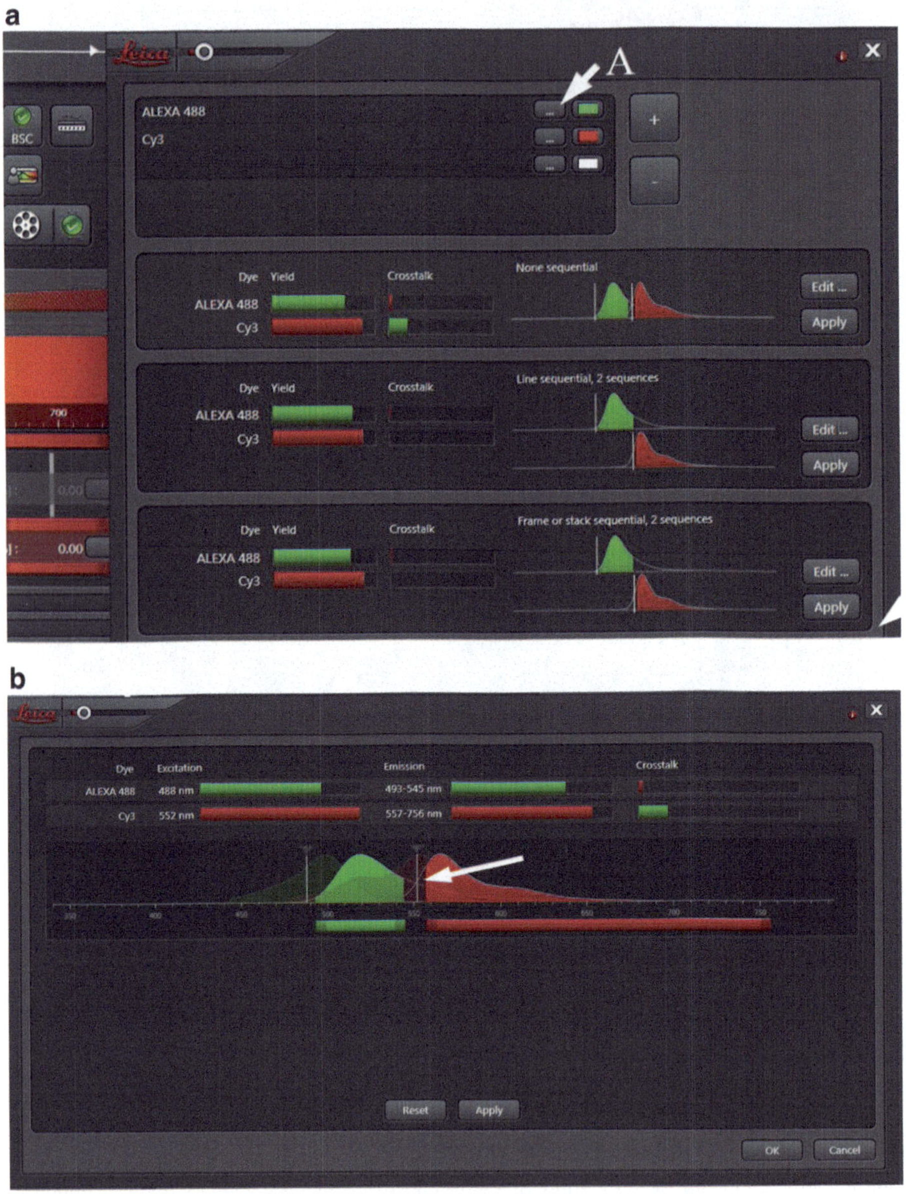

Fig. 9.38 (**a**) The "Dye Assistant" can be used to add fluorochromes to a protocol by selecting the three dots (arrow A) to open a list of dyes. Each dye can then be assigned a color for an image by selecting the box to the right of the arrow and using a lookup table. When more than one fluorochrome is selected, the "Crosstalk" between the emission spectra will be displayed in the windows under the dye list. The top window shows overlap if "None Sequential" (simultaneous) imaging is chosen and the bottom windows the "Crosstalk" for "Line Sequential" or "Frame or stack sequential." If "Edit" is selected, a window will open showing the emission spectra superimposed on a wavelength scale. (**b**) Significant overlap (arrow) exists if "None Sequential" is chosen as the imaging mode. (**c**) Overlap can be eliminated if "Line Sequential" or "Frame or stack sequential" is chosen

c

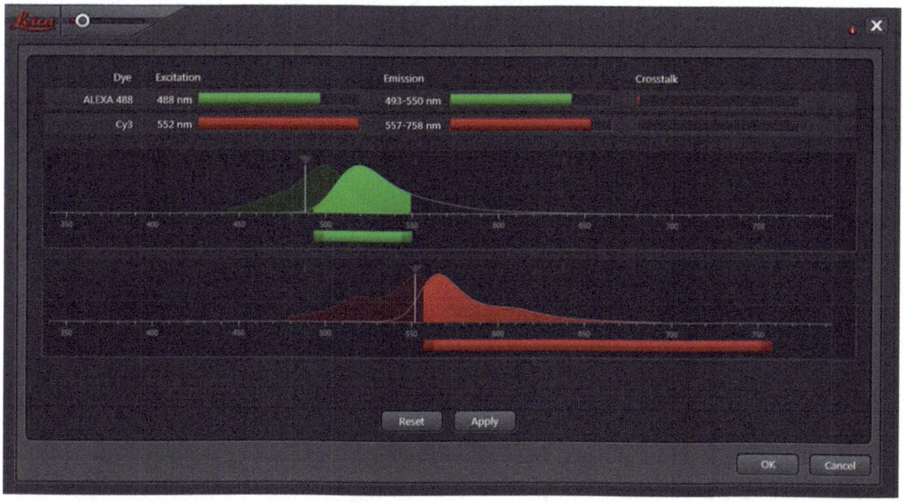

Fig. 38 (continued)

channels are added, multiple white lines representing the emission ranges will be present and overlap or bleed through easily recognized.

Just below the image of the light spectrum are controls to adjust the PMT gain and offset (Fig. 9.39 arrows C and D). This is again an area of redundancy since most often these are adjusted with the USB Control Panel. Increasing PMT gain makes the detector more sensitive resulting in fewer photons being required for saturation. If the PMT is made too sensitive, this may result in an image with a poor SNR. This can be compensated for by averaging the image as discussed above, but if possible detector gain should be decreased to avoid extensive averaging of the data. In combination with the PMT offset, the gain should be used to obtain images that use the full dynamic range available as shown in Fig. 9.7. These controls, along with laser intensity, pinhole diameter, and scan speed, all represent compromises in obtaining a high-quality image versus specimen damage (Table 9.1).

In addition to traditional PMTs, many systems also include transmitted light detectors (TLDs), GaAsP detectors, or if working with a Leica system their proprietary hybrid (HyD) detectors (Fig. 9.39). Combinations of various detectors can be used in protocols depending on the requirements of the experiment. The various types of detectors were discussed in Chap. 7.

9.9 Creating Protocols

In well-established systems, a number of standard protocols that cover a wide range of applications and fluorochromes are likely in place. In cases where new protocols are required, it is possible to use the Dye Database and Dye Assistant to create

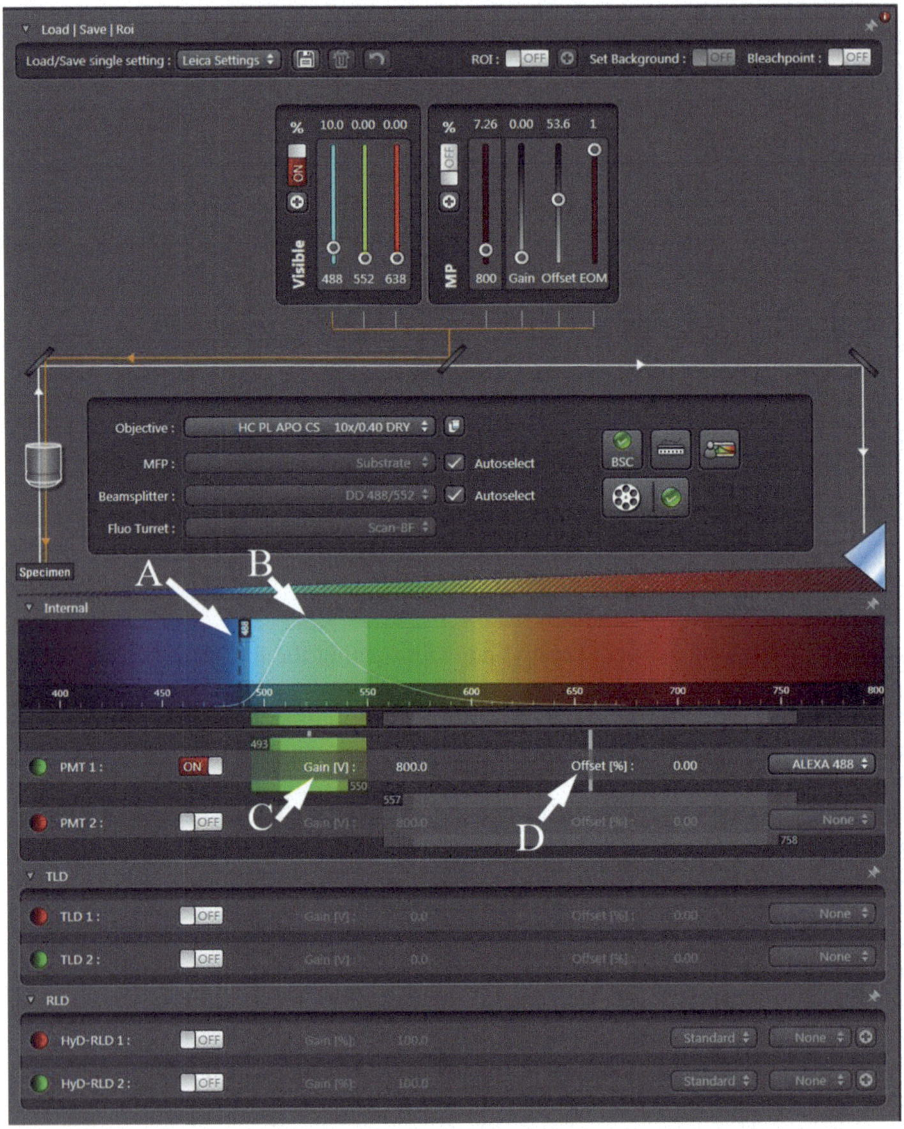

Fig. 9.39 Excitation (arrow A) and emission (arrow B) values are superimposed over the light spectrum to assist in setting the detectors to avoid overlap. Detector gain (arrow C) and offset (arrow D) to adjust image SNR can be adjusted here or with the USB Control Panel. In addition to traditional photomultiplier tubes or PMTs, many systems are also equipped with transmitted light (TLD) or highly sensitive detectors such as the Leica hybrid (HyD) detectors

simultaneous (multiple channels in a single scan) or sequential (multiple channels in separate scans) methods. Often, new laboratories using a core instrument will also create a standard protocol for their experiments and give it a specific name for easy access to standardized imaging parameters.

9.9.1 A Simultaneous Protocol for Alexa 488 and Cy5 Fluorochromes

As an example of a simultaneous two-channel protocol, the following steps can be used to create a method to image Alexa 488 and Cy5 fluorochromes:

1. Turn on the system and lasers needed to image the required fluorochromes.
2. Turn PMT 1 on and set detector gain to 800 as a starting point. Note you may also change the pseudo color by clicking on the circle to the left of PMT 1 (Fig. 9.40). Traditionally colors are assigned based on the wavelength of light detected as shown in the spectrum.
3. Adjust the emission spectrum captured by PMT 1 to the desired bandwidth (493–577 nm) for Alexa 488. As an aid for setting the emission spectrum to be detected, if the dropdown box on the right of the detector window is selected, the emission curve for the dye of choice can be opened.
4. Turn on (488) laser starting at lowest setting (less than 1%) which adds the black laser dotted line at 488 and the white emission line in the spectrum window (Fig. 9.40).
5. At this point it is possible to obtain a live image to adjust the detector gain and offset for the 488 nm channel as shown in the image window to the right. This can also be done at a later step.
6. Turn on PMT 2 and set detector gain to 800 to start (Fig. 9.40). The Cy5 window at the right of PMT 2 can be selected to show the Cy5 emission spectrum to set the appropriate detection wavelength (Fig. 9.40).
7. Adjust the emission spectrum captured by PMT 2 to the desired bandwidth (642–730 nm). Note that two white emission lines now appear on the spectrum. Since no "crosstalk" exists between the white lines, these fluorochromes can be collected simultaneously.
8. Turn on the 638 nm laser starting a low power (5%). Note the black 638 laser line is added to the spectrum window in Fig. 9.40.
9. Obtain a live two-channel image (Fig. 9.40) in the right window to make sure the protocol is correctly established. If necessary, detector settings can be adjusted at this time.
10. Select the save icon to make the protocol available for recall under the "load/save" single setting button.

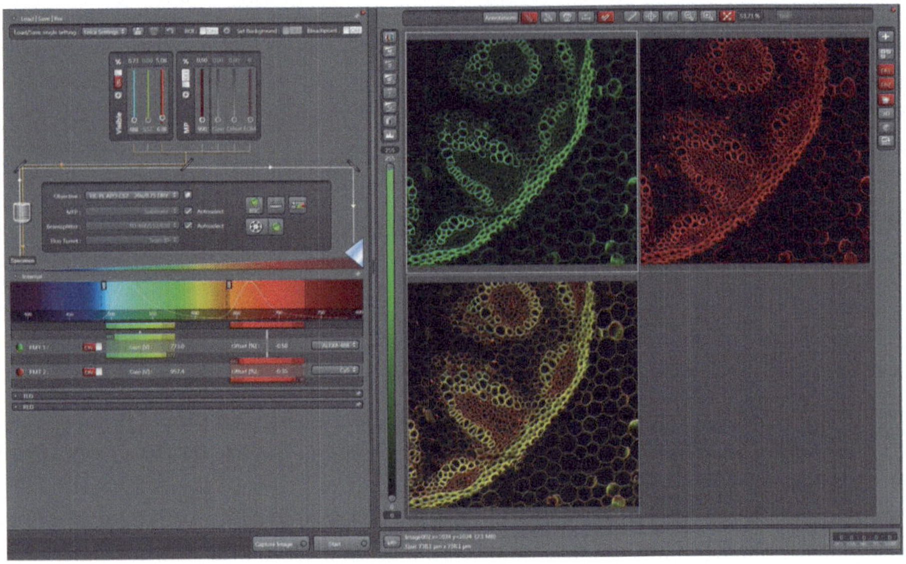

Fig. 9.40 Procedure for setting up a simultaneous imaging protocol for Alexa 488 and Cy5. Note that excitation lines are present at both 488 and 638 nm and that no overlap exists in the white emission lines. Clear separation in the emission spectra indicates simultaneous collection of these fluorochromes is an acceptable protocol. Images to the right show the Alexa 488 (green), Cy5 (red), and merged images

9.9.2 Sequential Protocol for Alexa 488, Cy3, and Cy5

When overlap of the emission spectra occurs, it is necessary to create a protocol where each channel is collected individually or sequentially. If there is concern about overlap of the emission spectra, the Dye Assistant can again be used to determine if overlap exists. Figure 9.41 shows the Dye Assistant for the three fluorochromes used in this protocol. Several options exist for establishing this protocol. The top panel shows the three fluorochromes selected and the color that will be assigned to each. Since both red (Cy3) and far red (Cy5) channels will be used, Cy5 is assigned blue. In the second and third panels, it shows that no emission spectra overlap exists between Alexa 488 and Cy5, but significant overlap between Alexa 488 and Cy3 exists. Minimal cross talk exists between Cy3 and Cy5. In the first and second windows, "Line Sequential, 2 Sequences" and "Frame or Stack Sequential, 2 Sequences" collection strategies are shown. In these cases Alexa 488 and Cy5 would be imaged simultaneously, and a second sequence to image Cy3 alone would be used. This would save time since each frame would only be scanned two times rather than three as described below.

It is also possible to image all three channels sequentially as shown in the bottom two panels, and this is the sequential protocol described below. The basics of this protocol can also be used for a number of other fluorochromes that have closely related spectra such as substituting Cy2 for Alexa 488 or Alexa 543 for Cy3.

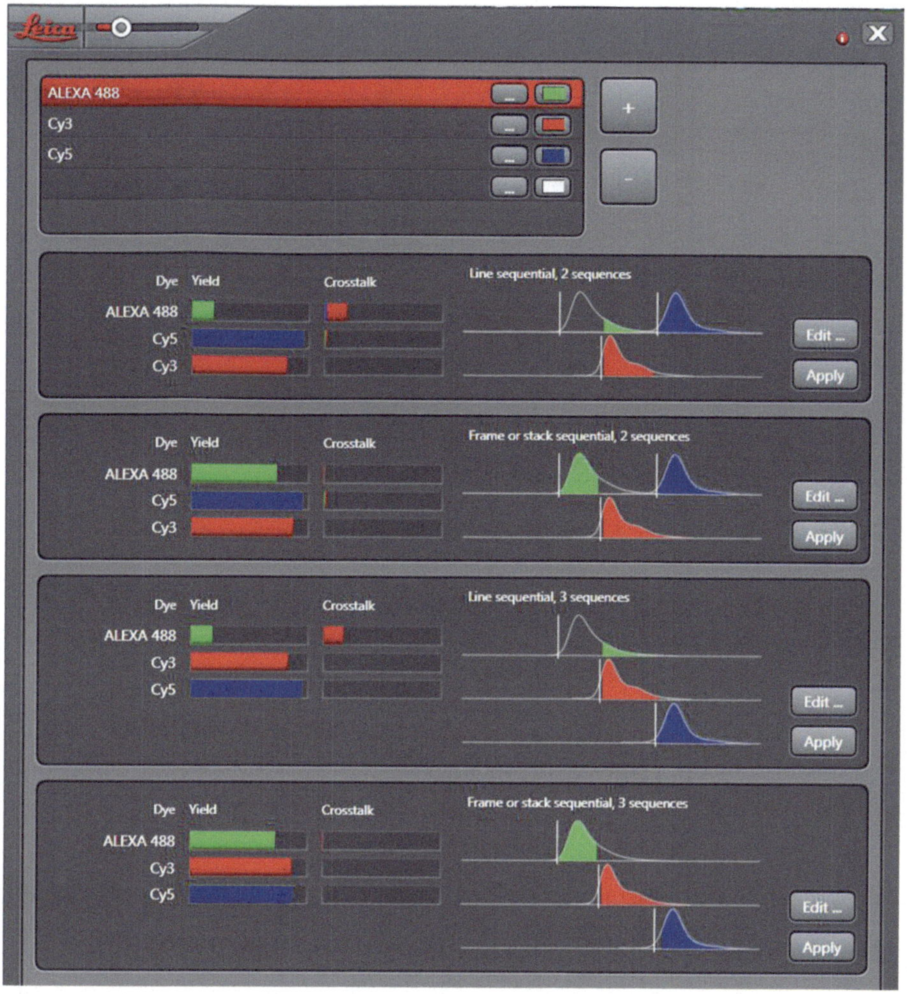

Fig. 9.41 Use of the "Dye Assistant" to establish a sequential imaging protocol for Alexa 488, Cy3, and Cy5

Setting up the Alexa 488 channel (Sequence 1):

1. In the left window Acquisition panel click on the "SEQ" button (Fig. 9.42 arrow A) which opens up the Sequential Scan panel.
2. From the Sequential Scan window protocols for each channel to be imaged can be added by clicking on the "+" button (Fig. 9.42 arrow B). In this protocol "Seq 1" will be Alexa 488, "Seq 2" Cy3, and "Seq 3" Cy5. Changes can only be made for the sequence highlighted in red so the first sequence established below will set the parameters for Alexa 488.
3. Select "Between Lines" or "Between Frames." Scanning "Between Stacks" is only relevant if collecting a Z-series as described above. This protocol will be set up to change the channel "Between Frames."

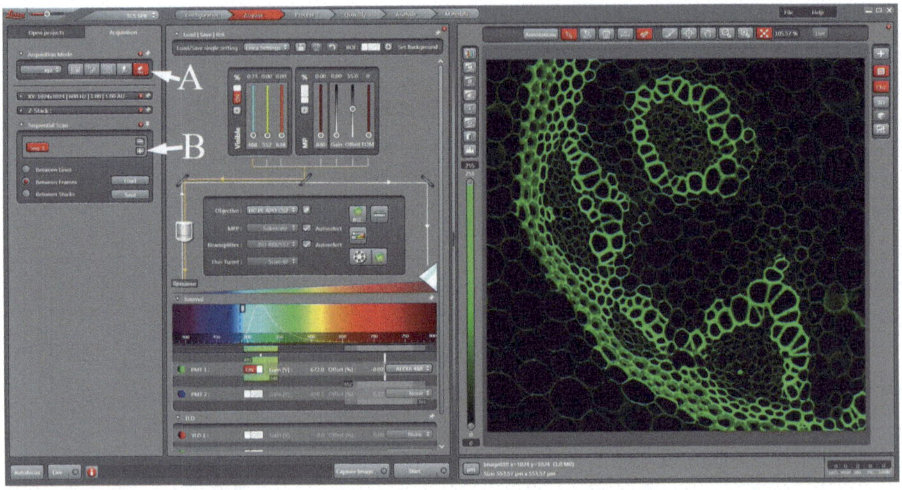

Fig. 9.42 Adding Sequence 1 (Alexa 488) to a three-channel sequential protocol

4. Turn on PMT 1 and set detector gain to 800 as a starting point.
5. Based on the "Dye Assistant" set the Alexa 488 emission spectrum to capture wavelengths between 493 and 548 nm (Fig. 9.42).
6. Set the 488 laser to less than 1% as a starting point. All other lasers should be at 0% at this time. Laser intensity, detector gain and offset, and other scanning parameters in the Acquisition window can be adjusted as needed once a live image is collected.

Setting up the Cy3 channel (sequence 2):

7. Select the "+" button in the Sequential Scan Window (left panel) to add the next sequence (Fig. 9.43).
8. In the Acquisition panel highlight Seq 2 and select "Between Frames."
9. Turn off PMT 1 and turn on PMT 2. Choose the color for this channel using the circle to the left of PMT 2.
10. Set the emission spectrum to the desired bandwidth for Cy3 (562–640 nm). It is also possible to choose Cy3 in the dropdown box to set the Cy3 spectrum.
11. Set the 552 nm laser to a relatively low power as a starting point. All other lasers should be at 0%. As above imaging parameters such as laser intensity, detector gain and offset, etc. can be tuned once imaging has started.

Setting up the Cy5 channel (Sequence 3):

12. Go back to the Sequential Scan window; click the "+" button to open Seq. 3 (Fig. 9.44).
13. Select Between Frames.
14. Seq 3 will again use PMT2 so this should be on.
15. Even though this is a far red channel, the color red was used for Sequence 2 so a different color needs to be used. In this case blue is chosen but any color in

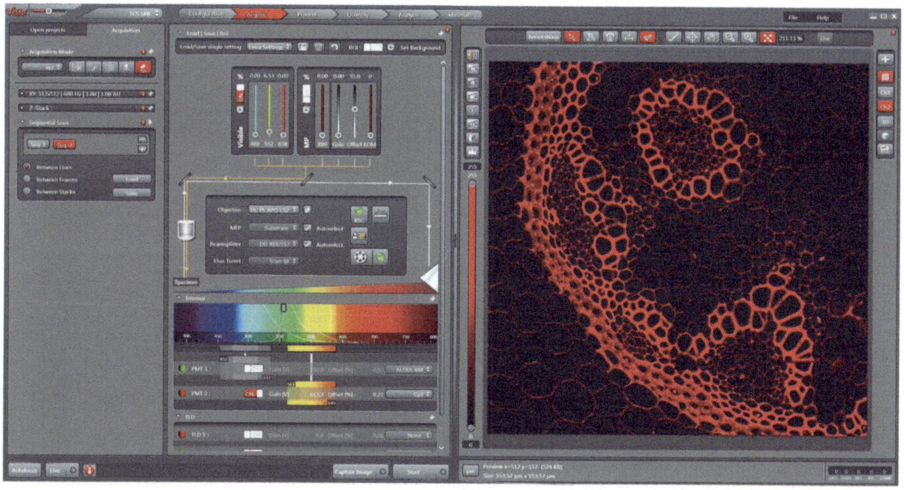

Fig. 9.43 Adding Sequence 2 (Cy3) to a three-channel sequential protocol

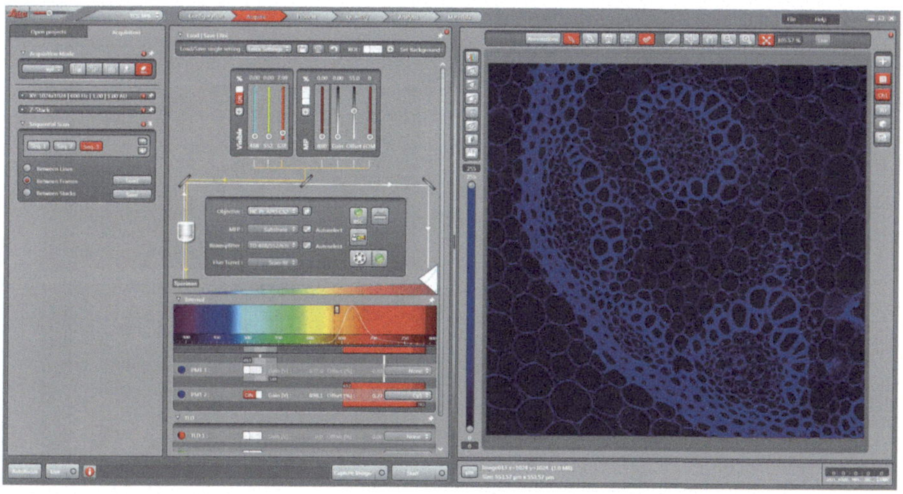

Fig. 9.44 Adding Sequence 3 (Cy5) to a three-channel sequential protocol

the spectrum except green or red could have been chosen. However, if colocalization between Alexa 488 (green) and Cy3 (red) occurs, this would show as yellow in a merged image so yellow should not be chosen for channel 3.

16. Open the emission spectrum to capture 652–783 nm. Again, the dropdown menu for Cy5 can be used to assist in setting up the emission range to be detected.
17. Turn on the 638 nm laser to a low percentage and set all other lasers to 0%.
18. As above imaging parameters such as laser intensity, detector gain and offset, etc. can be tuned once Live imaging is started.
19. Select "Capture Image" to collect a three-channel image as shown in the right window (Fig. 9.45 arrow).

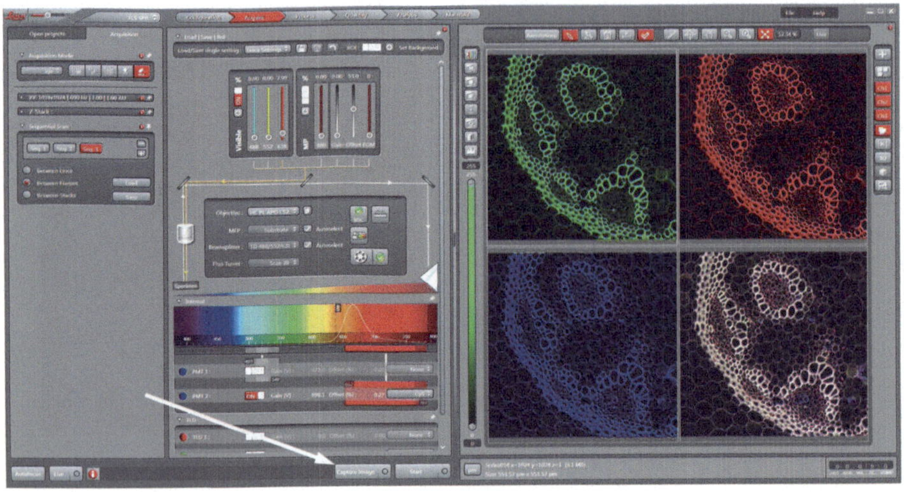

Fig. 9.45 Capturing the three-channel sequential protocol. The four windows show each channel plus a merged image

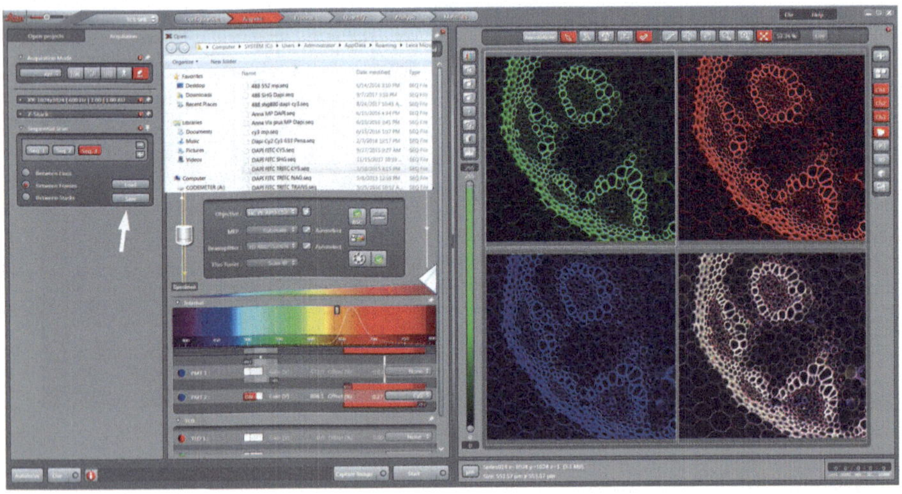

Fig. 9.46 "Saving" and "Loading" the protocols

To save protocols select the "Save" button, and to subsequently load saved protocols, select the "Load" button (Fig. 9.46).

9.10 Image Display Window

As with other windows in the LASX software, there are a large number of functions that can be performed in the "Display Window." Since separate chapters deal with digital images and two-dimensional and three-dimensional analysis, only a few of

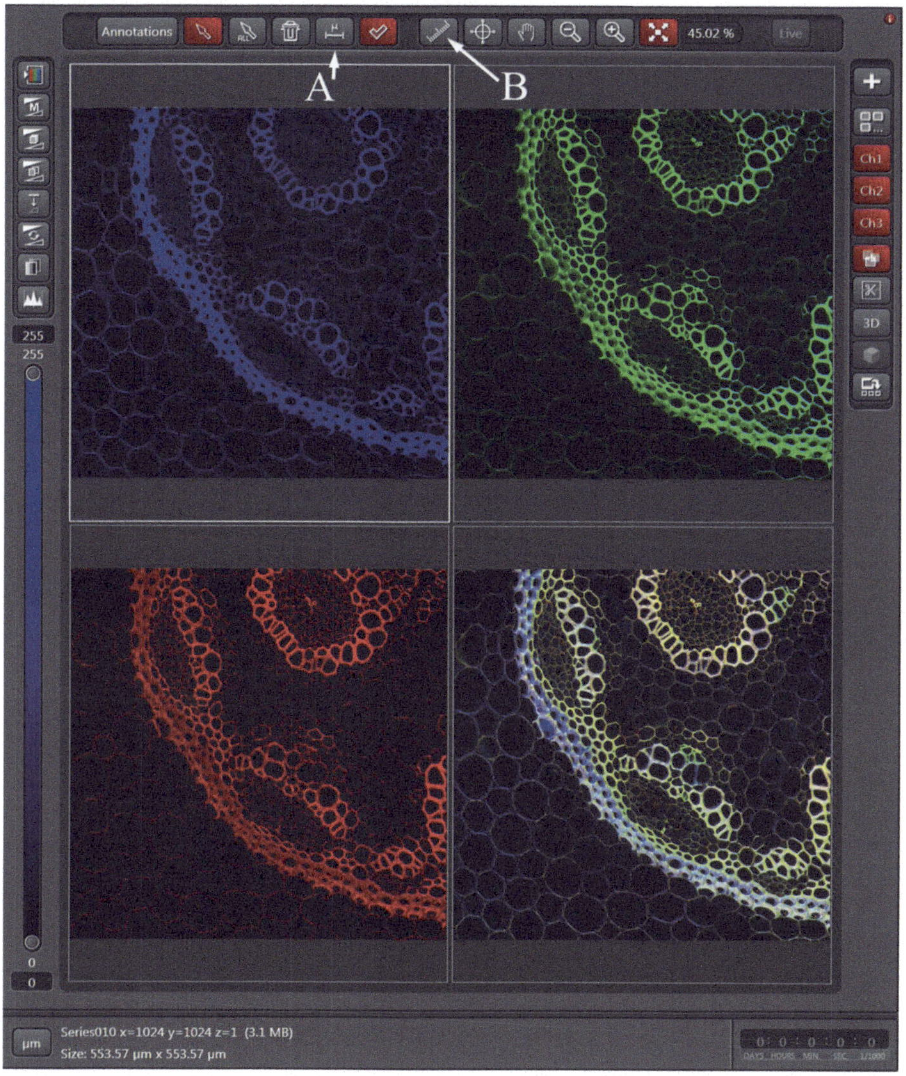

Fig. 9.47 An overview of the "Display Window" showing a three-channel image with the merged image in the bottom right. Arrow A indicates the scale bar icon and arrow B the ruler tool. Other icons along the right and left margins are discussed in the following figures

the common functions in this window used for displaying images will be addressed here. Figure 9.47 shows an overview of the window with an image of Convallaria collected with the multiphoton laser (blue) tuned to a wavelength of 800 nm, a green image collected with the 488 nm laser, and a red image collected with the 552 nm laser. The fourth image is a merge of the three channels. The top row of icons provides several tools for annotating images including the addition of scale bars (arrow A) and a ruler for measurement of image structures (arrow B). In practice, with a heavily used instrument, most annotation and measurement functions should be performed with other software (ImagePro, ImageJ, Metamorph, etc.) so instrument time is not tied to these basic image analysis and enhancement functions. However, for quick analysis these functions are useful.

The right column of icons provides access to the various channels in the image. As shown in Fig. 9.47, "CH1," "CH2," and "CH3" are highlighted red so all channels, in addition to the merged image, are shown. If additional channels were available, more icons would be present. To display a single channel, click on the channel of choice and the icon above "CH1" as shown in Fig. 9.48. In Fig. 9.48 several of the functions in the top row of icons have been activated, and a scale bar and the rulers around the periphery of the image are now present.

The image in Fig. 9.48 was selected from a "Z-Stack" so additional information is now present along the right margin where the slider (arrow A) indicates this image was the fourth of 10 images in the stack. Other images can be accessed by moving the slider with the mouse or by the "Play" buttons above the slider. Under the "CH3" button, an icon indicates "MAX" which provides information about the type of projection used in projection of the Z-Stack. At the bottom of the "Display Window," the file name (Series 13), the resolution (x=1024; y = 1024), the size of the file (35.8 MB), and the "Size" width and height of the image (553.57 μm × 553.57 μm) are provided (arrow B).

In Fig. 9.49 the bottom right column icon has been activated (now red; arrow A) which displays a number of images from the "Z-Stack" as thumbnails. In this case the second image in the stack is shown, and the second thumbnail is bordered in red to indicate this is the image shown in the "Display Window." This icon also opens additional information in the bar just above the thumbnails. The red highlight "triangle" (arrow B) indicates that image data will be displayed at the bottom of the "Display Window," and the "circled i" (arrow C) displays the information for each thumbnail as shown. "Noise-based method" is the algorithm used to project the Z-Stack data. The thumbnails shown are in two rows of four. This can be adjusted with the red slider (arrow D). Other icons in this row can be used to select a region of interest (arrow icon) and to trash an image. If the three-dimensional software package is available, the "3D" icon (arrow E) can be selected to access this optional software package. Three-dimensional reconstructions are addressed in Chap. 10.

As noted previously a LUT should be used to insure use of the full dynamic range in an image. Arrow A in Fig. 9.50 indicates the icon in the "Display Window" where LUTs can be accessed. Toggling this icon will change the images from the LUT as shown here, to a grayscale image, and back to the assigned colors. The six icons below the LUT icon provide access to a series of automated functions related to use of LUTs.

The final icon in the left column (Fig. 9.51, arrow A) is a histogram function for analysis of pixel or voxel value distribution in an image. In Fig. 9.51 the blue window is selected as shown by the white bounding box. The histogram at the bottom

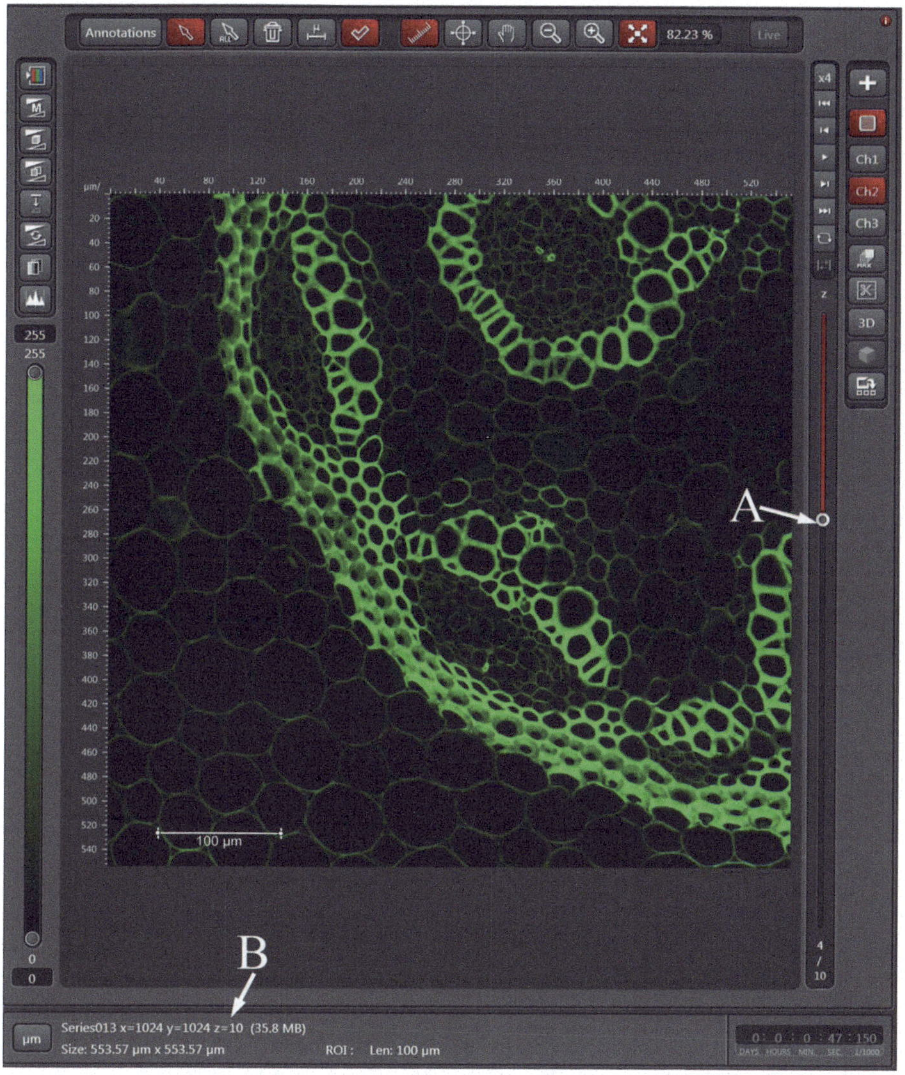

Fig. 9.48 The "Display Window" with the scale bar and ruler tool active. The single green channel was selected by choosing the "CH2" icon and the single image icon above "CH1." This image was selected from a Z-Stack as shown by the slider bar (arrow A) and information at the bottom of the window indicating "Z = 10," or 10 images in the Z-Stack

of the "Display Window" shows the distribution of voxel values (this is a Z-Stack with image 4 of 10 selected as shown by the information previously discussed). Brightness of the image can be adjusted with the slider bar on the left of the "Display Window" (arrow B) and "Gamma" with the slider bar at the bottom of the window (Arrow C).

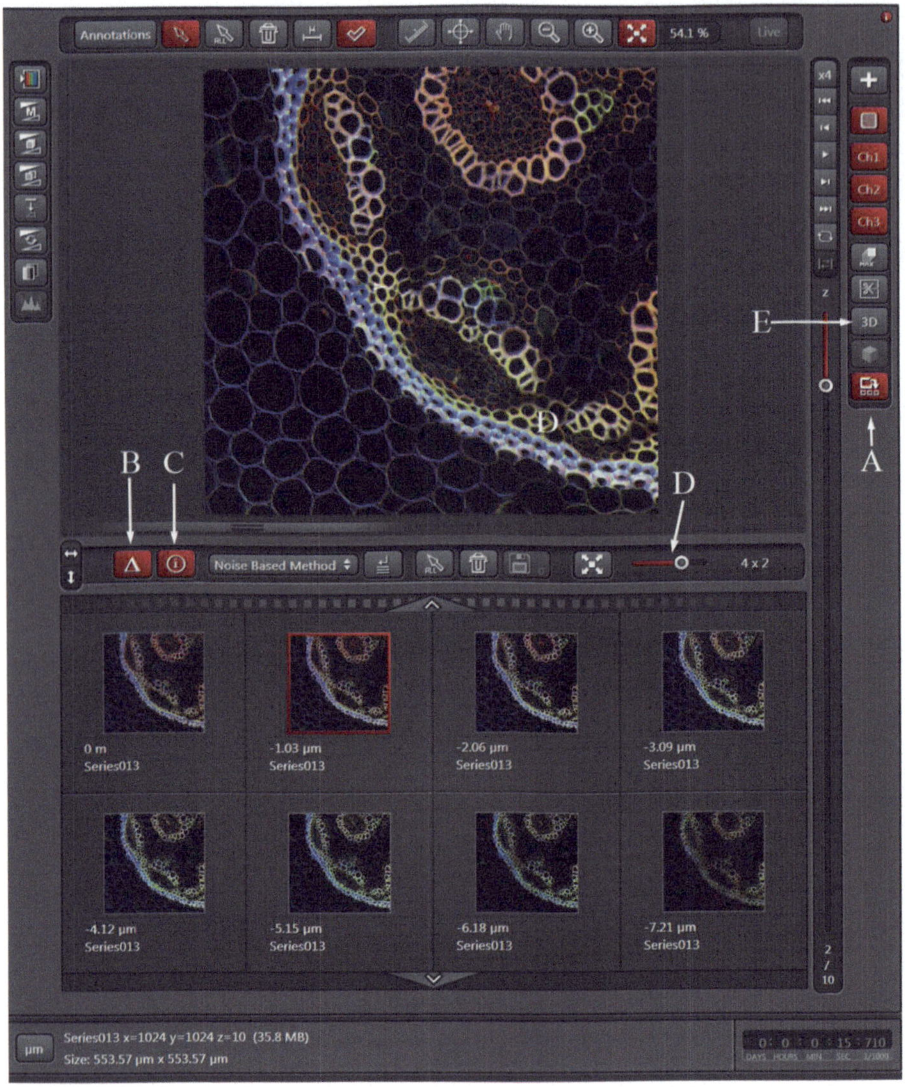

Fig. 9.49 To display additional information about the Z-Stack, select the bottom icon in the right column (arrow A). This will open the information bar on the right of the image as well as the gallery of images. Arrow B indicates information at the bottom of the "Display Window" will be shown, arrow C indicates the information for each thumbnail will be shown, and arrow D controls the number of thumbnails presented. If the optional 3D reconstruction package is available, this can be accessed by the "3D" icon (arrow E)

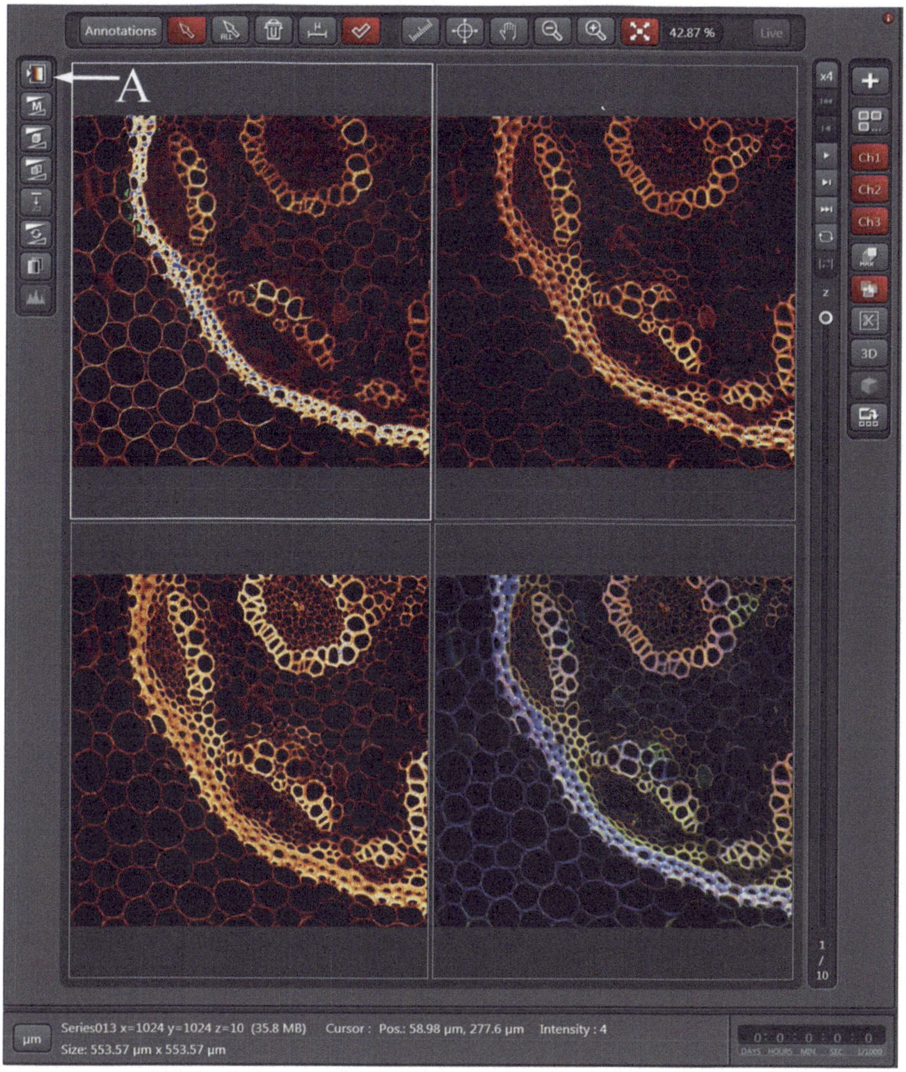

Fig. 9.50 Use of LUTs to obtain the full dynamic range in an image was discussed in Fig. 9.7. Access to the LUTs for this is with the icon indicated by Arrow A

9.11 Summary

As illustrated above the operating software for a confocal microscope is complex and includes a large number of functions. However, if the basic operating principles are learned and applied, it should be possible to collect publication quality images on any confocal microscope after a short period of time. For the most part, it is

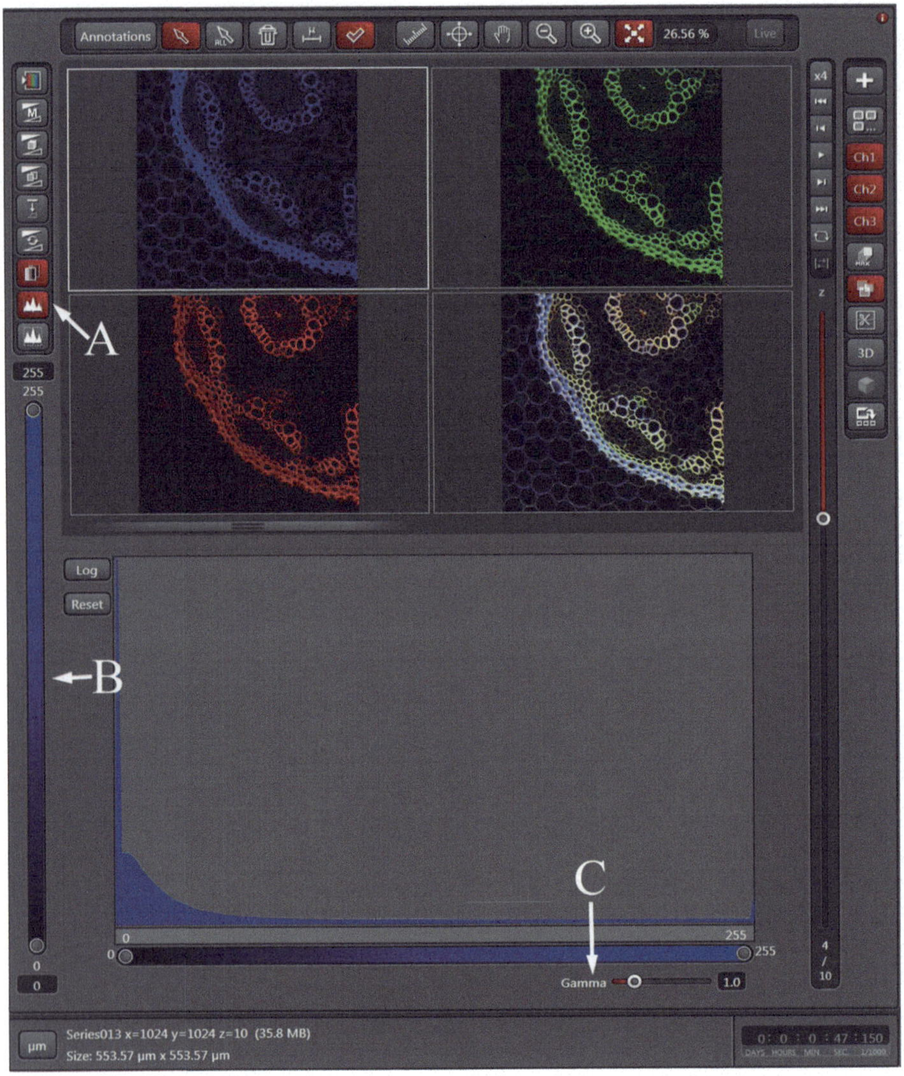

Fig. 9.51 A histogram function (arrow A) is available for analysis of pixel/voxel distribution in an image. Brightness (arrow B) and gamma (arrow C) can also be adjusted while viewing the histogram

simply a matter of learning where the programmers have placed the various functions for the proper collection and setup of images. In fact, as confocal software has evolved, operation of confocal systems has become much easier and more intuitive with the addition of functions such as the "Dye Database" and "Dye Assistant" for creating protocols, inclusion of automated algorithms for calculation of "Optical Slice Thickness" for collection of a "Z-Stack," etc. While this chapter is based on the Leica SP8 LASX version 3.1.5 software, software for many other confocal systems is now organized similarly, and moving between platforms should be relatively simple as experience is gained.

Literature Cited

Zucker RM, Price OT (1999) Practical confocal microscopy and the evaluation of system performance. Methods 18:447–458

Chapter 10
3D Reconstruction of Confocal Image Data

Thomas C. Trusk

10.1 Introduction

The main advantage of the confocal microscope is often said to be the ability to produce serial optical sections of fluorescent samples, ultimately for the purpose of reconstructing microscopic objects in three dimensions (3D) (Carlsson and Aslund 1987). There are many ways, and reasons, to reconstruct confocal image data. As an example, consider the sample of embryonic mouse heart shown in Fig. 10.1 reconstructed using a variety of 3D techniques. This chapter will introduce these methods and discuss such topics as (a) why one might want to undertake this task, (b) some definitions of the representation of 3D space using images, (c) the different types of 3D representations that can be made, and (d) the necessary steps to making useful reconstructions. Along the way, the limitations and potential pitfalls that arise will be discussed.

10.2 Why Reconstruct?

Reconstructing image data into three-dimensional representations is a fairly recent activity made necessary by the invention of devices which virtually deconstruct their subjects by scanning serial planes at different depths. Positron-emission tomography scanners, computed axial tomography scanners, magnetic resonance imagers, ultrasonic imagers, and confocal microscopes are all examples of devices which can collect these 3D image datasets. Prior to the development of those tools, building accurate 3D models of microscopic biological specimens had always been

T. C. Trusk (✉)
Department of Regenerative Medicine and Cell Biology,
Medical University of South Carolina, Charleston, SC, USA
e-mail: TruskT@musc.edu

W. G. Jerome, R. L. Price (eds.), *Basic Confocal Microscopy*,
https://doi.org/10.1007/978-3-319-97454-5_10

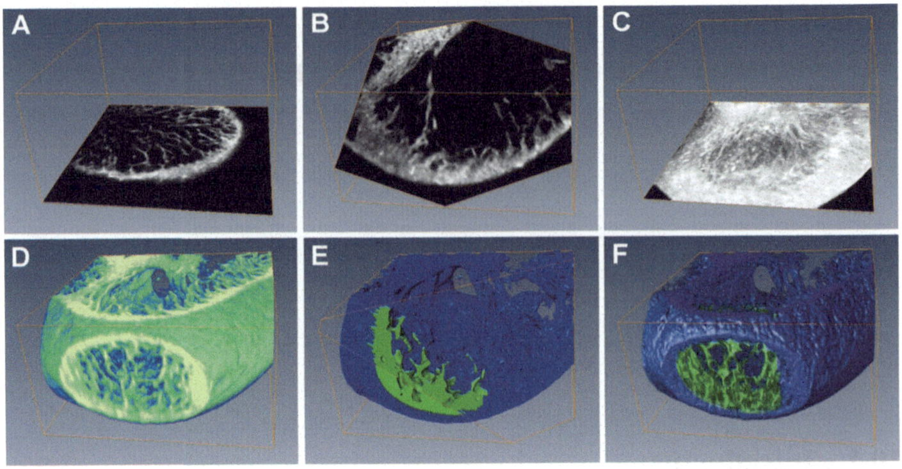

Fig. 10.1 Sample 3D reconstructions. The left ventricle of an embryonic mouse sampled in a confocal microscope. The 3D image dataset is contained inside the orange box where only the first confocal optical slice is shown (**a**). Methods which sample the 3D dataset include (**b**) an oblique slice, (**c**) a maximum projection through the z-axis, (**d**) a simple volume render, (**e**) a surface render manually segmented to highlight interior features, and (**f**) a volume render using complex lighting and segmentation information derived from (**e**)

a daunting task, that is, until powerful computers and sophisticated software were introduced. Instead of using scissors, cardboard, wax, modeling clay, or Styrofoam, it is now possible to sit in front of a computer workstation and construct, in a fraction of the time, a model that can be displayed in multiple forms and used to convey a wide variety of useful information.

Historically, the primary reason to reconstruct in 3D has been to display the morphology of the subject. In many cases, the 3D shape of the microscopic object may be completely unknown, or some experimental condition may influence its structure. It might also be helpful to reconstruct the various compartments of the subject and divide it into organs, tissues, cells, organelles, and even domains where some fluorescently labeled protein might be expressed. These arrangements often produce clues concerning function (e.g., see Savio-Galimberti et al. 2008). The value of a computerized 3D reconstruction is that the finished reconstruction can be viewed at different angles, including visualizations from perspectives impossible to achieve in the confocal microscope. Furthermore, most 3D software applications provide tools to produce digital movies, including interactive surface slicing, transparency, or subregion coloration in the models. These animations can be invaluable in conveying the structure of the subject.

3D reconstructions can be used to provide a substrate for the superimposition of other information (Hecksher-Sorensen and Sharpe 2001). For instance, the multiple channels used in confocal microscopy can be used to highlight the domain of proteins bound to fluorescent markers emitting different colors. Depending on the experimental question, these domains may be expected to be mutually exclusive or partially (or even completely) correlated within the structure of the subject. In another example, quantitative information (such as cell density or rate of cell proliferation) may be overlaid as a coded color scheme onto a reconstructed object and provide a visualization of that property throughout the sample (Soufan et al. 2007).

Finally, once a fully segmented, surface-rendered reconstruction has been made, it is a simple matter to extract quantitative information from the sample. This includes counting objects, determining volumes, measuring surface areas, or quantifying the spatial arrangements of parts (such as measuring distances or angles between features).

10.3 Defining the 3D Space Using Images

The 3D Cartesian coordinate system (Fig. 10.2) provides a convenient space within which to reconstruct image data. The Cartesian system consists of three physical dimensions – width (x), length (y), and height (z) – which are perpendicular to each other (a relationship termed orthogonal) and are typically oriented with the z dimension pointing up. Individual images are considered orthogonal slices through the z dimension, and the pixel plane of each image is defined in the x and y dimensions. Figure 10.1a is an example of an orthogonal slice. Planes cut through the Cartesian space that are not perpendicular to a major axis are said to be oblique, as in the slice shown in Fig. 10.1b.

10.3.1 Perspective

When an observer views a three-dimensional scene consisting of objects of equal size, objects that are off in the distance will appear smaller than objects closer to the viewer. An artist (and a computer) will enhance the perception of depth in rendering

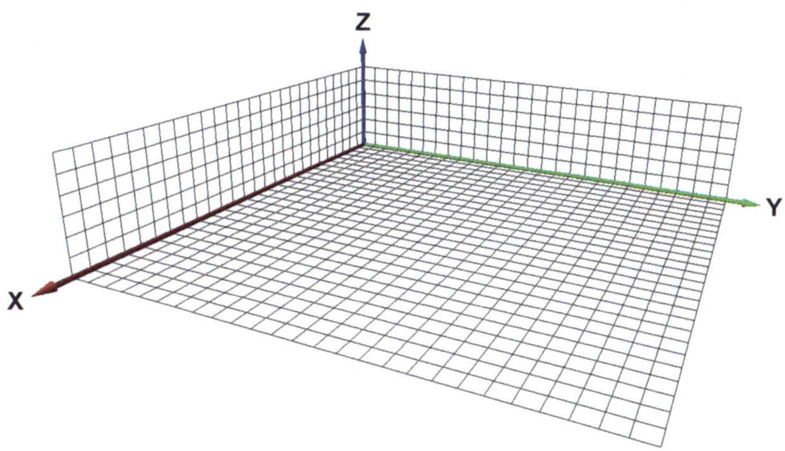

Fig. 10.2 The 3D Cartesian system. A mathematical model of three-dimensional space. Every position in the space is defined by its location along one of three axes (x, y, and z)

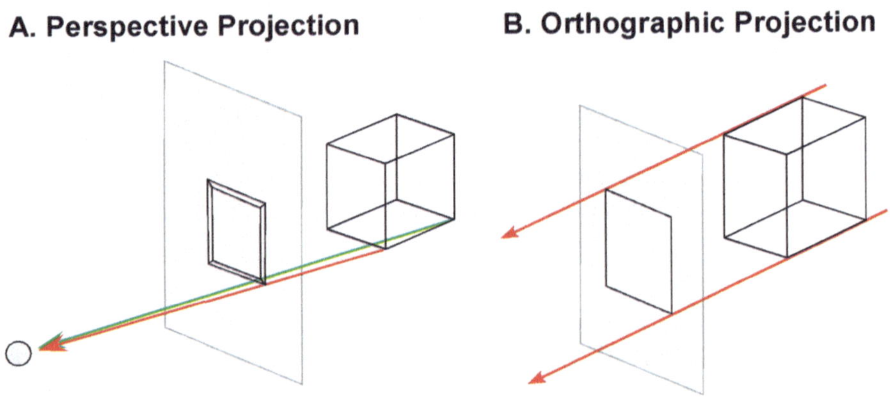

Fig. 10.3 Perspective vs orthographic projection. For objects drawn on a screen in perspective (**a**), all rays project from the object to the viewer. The more distant back face of the box appears smaller than the front face. In an orthographic projection (**b**), all rays projected from the object to the display screen remain parallel, and the perception of depth in the scene is lost

a scene by drawing objects "in perspective," as if the drawing was traced on a window through which the viewer was looking. In Fig. 10.3a, rays exit the object and project directly to the viewer. The object is drawn on the display such that the back face of the box appears smaller than the front face. In an orthographic projection (Fig. 10.3b), all rays passing through the scene toward the display are parallel, and the object on the display does not appear in perspective. These types of renderings are sometimes useful, as in architecture, where object size comparisons are more important than the perception of depth. Most 3D software suites can draw objects on the screen with or without perspective, as the viewer chooses. The perception of depth in a scene is quite helpful to the viewer, especially when objects may be moving in dynamic displays or animations. Objects drawn orthographically and rotated often appear to suddenly switch rotation direction.

10.3.2 Voxels

A single confocal "optical slice" has a "thickness" in the z dimension that is defined by the microscope optics and the confocal pinhole. Recall that increasing the diameter of the pinhole allows extra light from above and below the focal plane to contribute to the light gathered at each pixel. On the other hand, a single "image" is infinitely thin along the z dimension (see Fig. 10.4). Confocal z-series images are typically collected at evenly spaced intervals, and when the images are restacked at the correct z distance, the empty space between the images is readily apparent.

In order to fill the space between images, it is necessary to expand the concept of a two-dimensional image pixel into a three-dimensional volumetric voxel. This is demonstrated for a single voxel in Fig. 10.5. The grid of two-dimensional image

Fig. 10.4 A single image has no depth. Five evenly spaced confocal slices demonstrating the infinitely thin nature of image data in the depth (z) plane

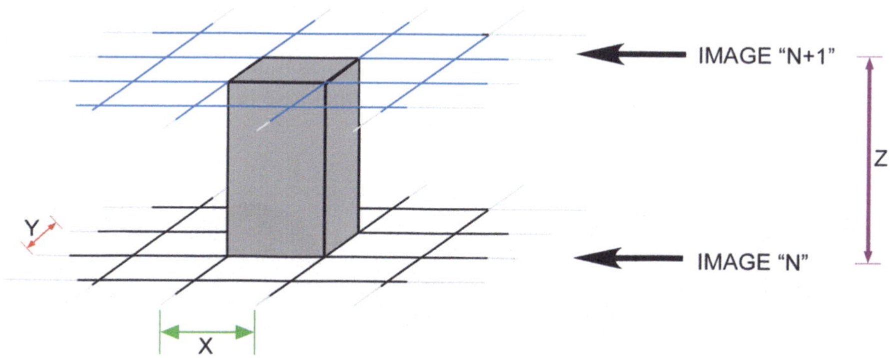

Fig. 10.5 Definition of a voxel. A portion of two consecutive images from a confocal series is shown. Image pixel dimensions were x units by y units, and images were collected every z units. A single voxel is the space defined by the pixel dimensions and the distance between consecutive images

pixels is stretched in the z dimension to produce three-dimensional voxels. Each voxel has the same dimensions of single image pixels in the x and y dimensions and the distance between consecutive images in the z dimension. Thus, the volume of each voxel is defined as the area of each pixel (x multiplied by y) multiplied by the distance between optical slices (z). This voxel volume is the smallest unit of volume that can be measured in that image dataset.

One can easily imagine that voxels are the "bricks" used to construct a 3D model. Each voxel not only has spatial dimensions, but each voxel also has a single unique brightness value inherited from its basis pixel. In an 8-bit sensitivity system, this value will range from 0 to 255 and is constant within the confines of a single voxel. Much like the bricks used to construct buildings, the voxels in a 3D image dataset are all the same shape and size, and each voxel has only one homogeneous color.

10.3.3 Resolution

The shape of individual voxels in an image dataset is determined by the resolution of the system. Some confocal microscope software controls will, by default, choose to optically slice a sample using a z distance that is as close as feasible to the pixel size. This would result in voxels that are nearly cubic, where x, y, and z dimensions are approximately equal. This is desirable for many reasons. First, cubic voxels more easily reconstruct 3D features correctly no matter what orientation they may have. For example, if you were to choose to use bricks to build a sphere, what shape brick would make the best representation? Figure 10.6 displays surface reconstructions of the same sphere rebuilt with different-shaped voxels and demonstrates the effect of increasing the voxel z dimension (or slice thickness) on the surface of reconstructed objects.

Most mathematical functions used in 3D reconstruction and visualization are much more efficient when the dataset uses isotropic or cube-shaped voxels. Using extremely anisotropic voxels can lead to inaccuracies in many measurements, especially surface area, and produce 3D visualizations that lose resolution when viewed from particular angles (see Fig. 10.7).

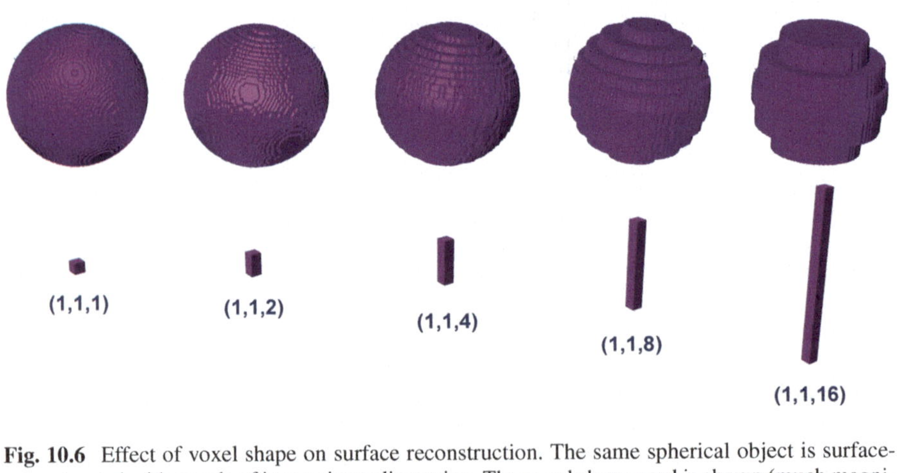

Fig. 10.6 Effect of voxel shape on surface reconstruction. The same spherical object is surface-reconstructed with voxels of increasing z dimension. The voxel shape used is shown (much magnified) below each rebuilt sphere with the voxel dimensions in x,y,z. The leftmost sphere was rebuilt using cubic voxels

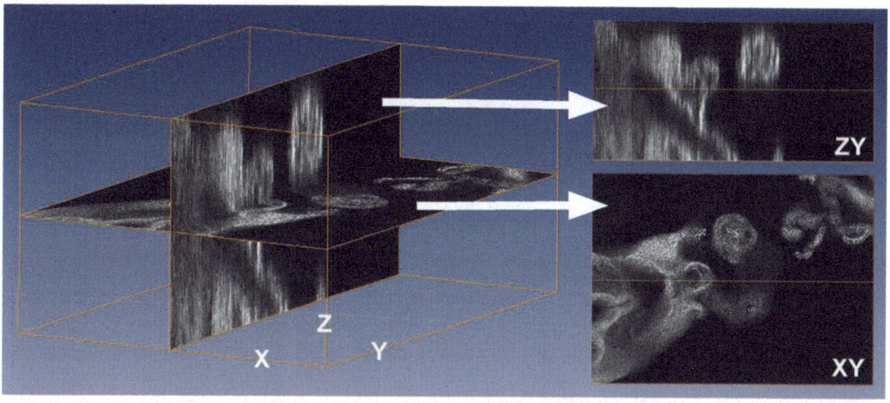

Fig. 10.7 Effect of voxel shape on 3D visualization. A confocal 3D image dataset is sliced in the XY plane (as originally collected by the confocal system) and in the ZY plane. Voxel x,y,z dimensions in micrometers are 2.4, 2.4, and 12, respectively

One solution offered in 3D reconstruction software is to resample the image data into near-cubic dimensions. That is, the original data will be interpolated to fit a cubic voxel model. However, this approach only helps in the calculation efficiency and does not improve the visualization of the sample. Resampling the data into larger voxels can be quite helpful if the computer memory is insufficient, but resampling into smaller voxel sizes will only make the volume take up more memory and not increase visible resolution at all. It is worth noting that resolution decisions are best made at the microscope when acquiring images, and this is demonstrated in Fig. 10.8. The same epithelial cell was captured using a 20× lens, a 40× lens, and a 40× lens with a digital zoom of 5 (for an effective 200× magnification). The actual collected pixel sizes were 0.73 μm, 0.37 μm, and 0.07 μm for the 20×, 40×, and 200× images, respectively. In each image, a white box identifies the same mitochondria, and the image data inside has been magnified to show individual image pixels. It should be easy to discern that the visual information obtained at 20× would be difficult to resample (or subdivide) into the higher resolution information seen at higher optical magnifications. This demonstration of resolution in the x and y dimensions also applies to the z dimension.

Resolution in the z dimension is set at the confocal microscope by choosing an appropriate distance between optical slices. For 3D reconstruction work, a good rule of thumb is to optically slice at a distance between 1 and 5 times the size of pixels in the XY images. Oversampling the object by slicing at smaller intervals is easier to correct later by resampling. If you are worried about datasets that are too large or take too long to collect, consider that the cost of data storage is essentially free. Returning the sample to the microscope later is always risky as fluorescence may have waned or the tissue may have deteriorated.

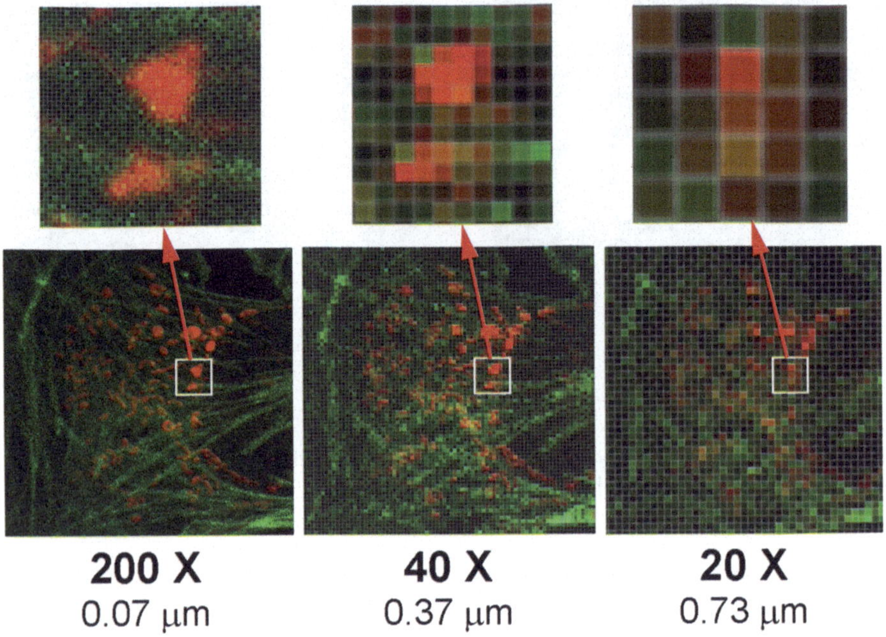

Fig. 10.8 Changing resolution. A bovine epithelial cell acquired on a confocal microscope using increasing objective power to improve resolution. The actual pixel dimension resolved is given for each image in micrometers. The white box region in each image has been magnified to reveal actual image pixels

10.4 Types of 3D Reconstruction

3D image datasets can be reconstructed and visualized in a variety of ways. These 3D models range in complexity from projections, a common function within most confocal system software suites, to surface reconstructions, which often require additional software. More recent confocal microscopes now include a sophisticated selection of reconstruction routines as part of the operating software, or possibly as optional additions. There are three basic methods used to present a 3D dataset: projections, volume renders, and surface reconstructions. The following will briefly describe how each type is calculated.

10.4.1 *Maximum Projections*

A projection is a single two-dimensional image generated using data from the entire 3D image set. In the case of fluorescent images, this consists of finding the brightest, or maximum pixel value at each pixel address from each image in the entire dataset, and projecting it onto a result image (see an example in Fig. 10.1c). While the

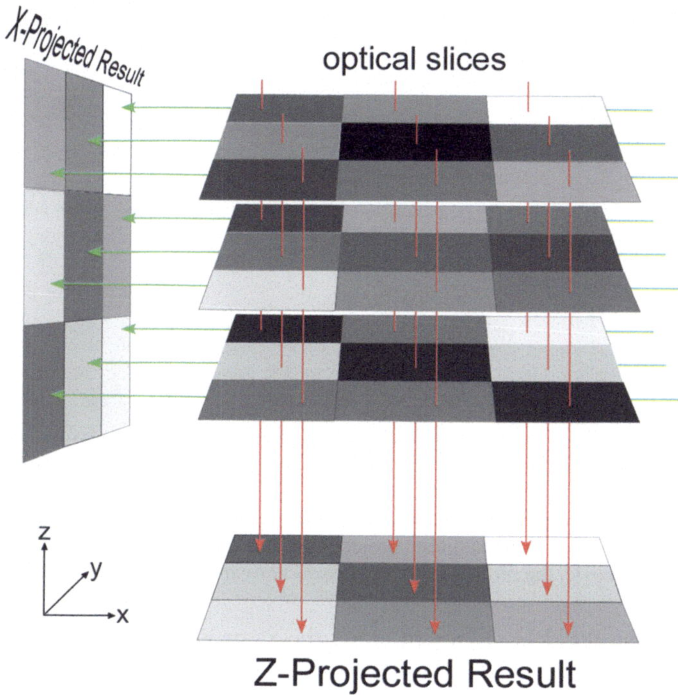

Fig. 10.9 Projected images. Three optical slices, each 3 × 3 pixels, are projected along the z-axis (red lines) and along the x-axis (green lines). The maximum (or brightest) pixel values along each projection line are placed in the resulting projected image

obtained image itself is not a 3D object, it conveys a visualization of the entire object from a particular viewpoint. Typically, the viewpoint used has been orthogonal, or along a principal axis (x, y, or z), which simplifies the mathematics of calculating a projected image.

Consider the simple 3D dataset in Fig. 10.9. Three optical slices, each a 3 by 3 pixel image, are projected along z-axis (red) and x-axis (green) projection lines. The brightest pixel encountered along each line is placed in the result image. Note that these projection lines are parallel to each other; thus they produce an orthographic rendering of the data, as previously discussed. As such, projected images tend to lack perceived depth.

An illusion of depth can be partially elicited by making a projection animation (or multi-frame movie) where the object appears to rotate in place, either in complete circles, or by alternately rocking the object clockwise and counterclockwise. These animations are generated by mathematically rotating the object a few degrees and recalculating the maximum pixels encountered along parallel rays projected toward the result image (see Fig. 10.10). When the frames are collected into a movie and played in sequence, the projected object appears to rotate. Recall that the resolution along the z-axis (as seen through the microscope) is always best and that resolution

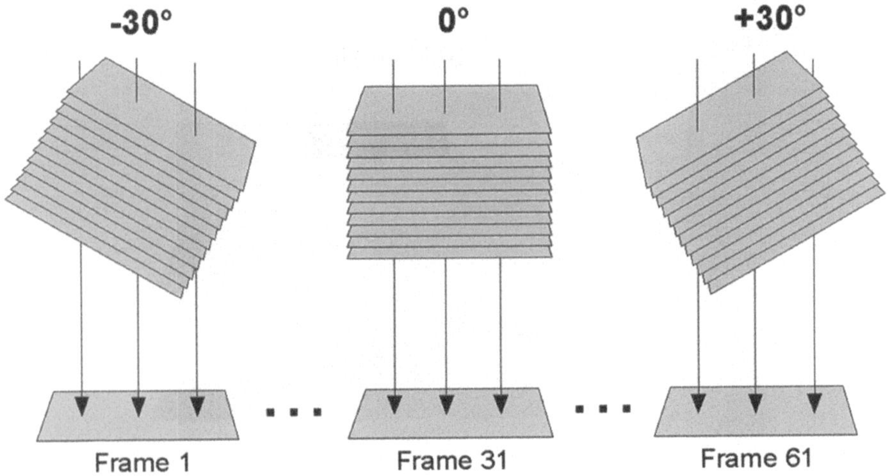

Fig. 10.10 Maximum projection animation. A 61-frame animation is made by rotating a 3D image dataset one degree before each maximum projection result frame is calculated. When the frames are collated into a movie, the projected image appears to rotate in place

is reduced when the data is viewed along the x- or y-axis. This is why projected images tend to appear to lose clarity as the object rotation exceeds 30–45° away from perpendicular to the z-axis. For this reason, most animations are made to show the object shifting rotation between ±30° of tilt.

A common mistake is to slice the confocal sample at step sizes 10 or more times the size of individual pixels. This will produce a 3D dataset that will lose resolution too rapidly as the object is rotated. The illusion of depth is also lost when the number of slices is very small when compared to the pixel dimensions of each XY image. For example, 10 z-slices taken at 1024 × 1024 resolution will make disappointing rotation projection animations. The best projection animations use 3D datasets with z-slice intervals close to xy pixel size (i.e., near-*cubic voxels*) and small rotation factors between projection animation frames (about 1° of rotation for each frame).

Maximum projections are the most commonly seen 3D reconstructions, and all confocal software systems routinely generate orthogonal views at the press of a button. Procedures for making projection animations are also widely available. Projections work best as orthogonal visualizations of 3D fields with little depth and provide the "all-in-focus" micrographs difficult to obtain on widefield fluorescent microscopes. However, be warned that maximum projections can also be misleading. When the object is hollow with internal structure, like the heart ventricle shown in Fig. 10.1c, the external signal will mask the interesting internal features. More importantly, maximum projections of multichannel image sets will mix colors by addition. Red and green channel signals that lie along similar projection lines, but are not otherwise colocalized, will produce yellow objects in the result image. This could lead to misleading conclusions.

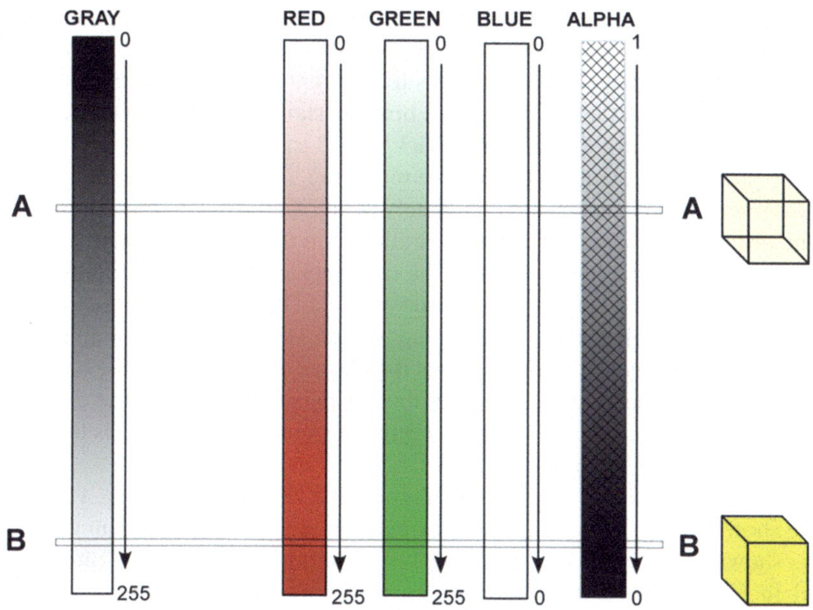

Fig. 10.11 RGBA lookup table. An 8-bit lookup table is used to convert gray values from a voxel into the corresponding red, green, blue, and alpha (transparency) values used for display. Red and green table values range from 0 to 255, blue is 0 across the range, and alpha ranges from 1 (transparent) to 0 (opaque). The gray values at A and B are converted in the mixed RGBA values and displayed in voxels at the right. The darker value at A mixes red and green to produce a transparent weak yellow voxel, while the brighter value at B produces a more saturated and opaque yellow voxel

10.4.2 Volume Rendering

As acquired in the confocal microscope, each voxel can be described with certain x,y,z spatial coordinates and a fluorescent intensity recorded in each channel. In order to visualize a 3D dataset in proper perspective, we need to render an image on the display in a manner which will allow internal structure to be correctly observed. One way to accomplish this is to arbitrarily assign each voxel an additional attribute of transparency. This attribute is termed alpha in imaging technology, and it has an inverse commonly used in 3D rendering called opacity, i.e., a voxel of 100% opacity has no alpha transparency. Alpha ranges from 0 to 1; thus opacity is 1-alpha.

By now you should be familiar with the concept of pseudo-coloring images using color lookup tables. For 3D renderings, alpha becomes a fourth attribute, and voxel pseudo-color values are often referred to as RGBA tuples, referring to the four values assigned to each voxel. A typical RGBA lookup table is shown in Fig. 10.11 for two possible voxel values. In fluorescent images, the background is typically dark (not fluorescent). Since we wish to see through nonfluorescent regions, we assign the most transparent alpha values to the darkest image pixels. In this particular example, we choose to make brighter image values to produce more saturated yellow voxels that also become less transparent (or more opaque) as intensity increases. This RGBA table will produce brightly stained fluorescent objects as yellow structures and leave the unstained background transparent.

In the process of calculating a volume rendering, imaginary projection rays proceed from the display screen and completely penetrate the 3D dataset. If the display is to be orthographic, the rays remain parallel from the display through the object. For displays remaining in perspective, the rays appear to originate from the viewer and are not parallel. The number of rays possible can be quite high, but only enough are needed to fill each pixel on the display screen. As this limited number of rays pass through the object, the software has the choice of using all of the data, where calculations are made from every voxel the ray passes through, or the data can be interpolated by averaging across a select number of neighboring voxels to speed up the process for more dynamic displays.

The first steps in the process of calculating voxel values used along a projection ray are shown in Fig. 10.12. Consider a small sample, a 3 × 3 × 3 volume. The volume is shown in the brick model in (A) and, in another conceptual format, the cell model in (B). In the cell model, voxels are viewed as the vertices of a three-dimensional grid. A single projection ray is shown in (C) as it passes through the volume and intersects the display pixel where the result will be seen. As the ray passes through the volume, it passes through several cells, each defined by the 8 voxels in the cell corners. The intensity value used for each cell is calculated using trilinear interpolation, as shown in (D). In the first approximation, four pairs of corner voxels are averaged, then those four values are averaged, and then finally the last pair is averaged. The final value is converted using the lookup table into an RGBA tuple. This process is repeated for all cells along the projection ray, and the final step is to composite all of these tuples into a single value to be displayed.

In compositing these values, calculations usually proceed from the back of the object space to the front, toward the viewer. As the ray first strikes the object, it has the color or intensity value of the background. The intensity passing through each cell in turn is modified according to the formula:

$$\text{Intensity}_{\text{out}} = \text{Intensity}_{\text{in}} \times \left(1 - \text{alpha}_{\text{cell}}\right) + \left(\text{Intensity}_{\text{cell}} \times \text{alpha}_{\text{cell}}\right),$$

and this process continues until the ray leaves the object, where the final intensity$_{\text{out}}$ is placed in the display pixel.

Most modern computer video display adapters are now capable of performing much of these 3D rendering computations very rapidly. This makes it possible for most modern desktop computers to perform volume renders as quickly as expensive graphic workstations of the recent past. The addition of hardware accelerators dedicated to 3D texture mapping, mainly used in modern gaming software, has also made it possible to generate even more complex volume texture renderings. For these methods, each cell can be given attributes beyond color intensity and transparency, such as texture mapping, specular or diffuse reflection, shadowing, and other lighting effects (as shown in Figs. 10.1f and 10.20c).

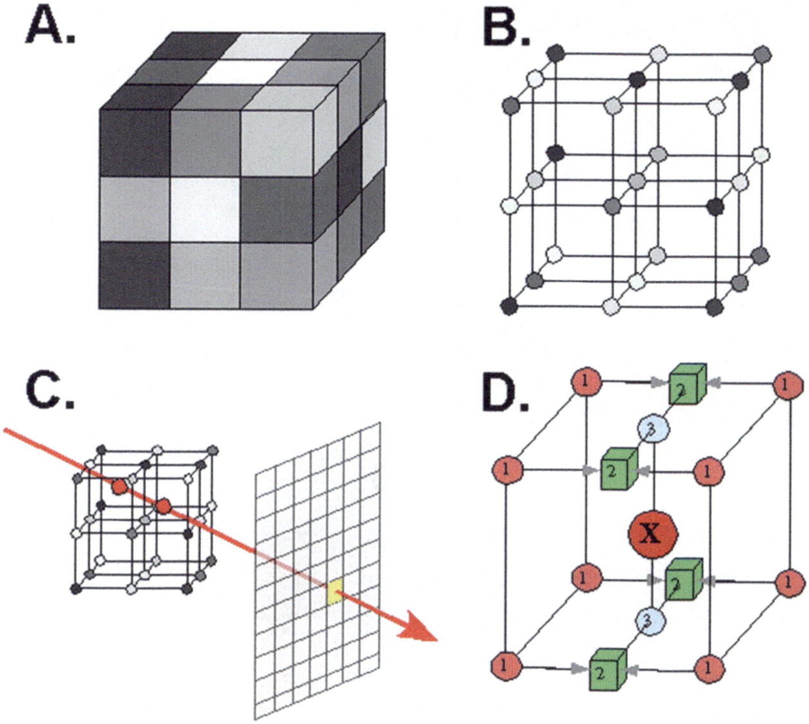

Fig. 10.12 Voxel averaging. A simple 3 × 3 × 3 volume shown using the brick model (**a**) and the cell model (**b**). In the cell model, voxels are viewed as the vertices of a three-dimensional grid. A single projection ray is shown in (**c**) as it passes through the volume and intersects the display pixel where the result will be displayed. As the ray passes through the volume, it passes through cells, each defined by the eight voxels in the corners. The intensity value used for each cell is calculated using trilinear interpolation, as shown in (**d**). In the first approximation, four pairs of corner voxels are averaged, then those four values are averaged, and then finally the last pair are averaged

Volume render animations more faithfully display internal structure as compared to maximum projection animations and are less likely to generate false positive conclusions regarding co-localization. However, a single frame volume render of a multichannel 3D dataset can be just as misleading regarding co-localization as a maximum projection of the same dataset. Depending on the software, multichannel volume renders will often mix RGBA tuples for the final display image or may choose to display one channel as masking another. It is best not to rely on one visual aspect when ascertaining co-localization.

10.4.3 Surface Reconstruction

In image-based reconstruction, a surface is a three-dimensional polygonal boundary surrounding voxels of special interest. These voxels can be defined by setting a single intensity threshold value or by including a specific range of intensity values.

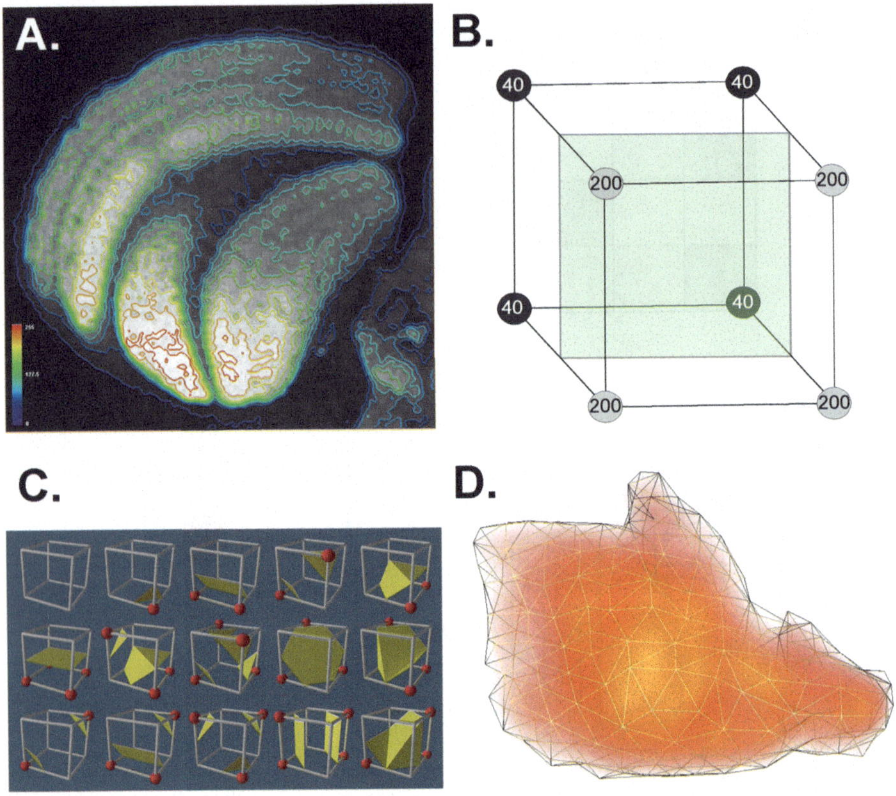

Fig. 10.13 Surface reconstruction. (**a**) Intensity isolines in a two-dimensional image. (**b**) A single 8-voxel cell is subdivided by a green surface boundary based on voxel intensities above 100. (**c**) The 15 possible solutions to the marching cube algorithm. (**d**) A 3D polygonal wireframe surface bounding volume-rendered voxels

Alternatively, individual voxels can be selected manually (in a process called segmentation) for inclusion into a surface boundary. A mathematical routine called the marching cube algorithm is typically used to define a surface in 3D space.

As an introduction to 3D surfaces, consider the simpler two-dimensional isosurfaces used to map intensity values in single images. We are used to seeing contour lines mapping altitude or isotherm lines showing temperature differences on weather maps. In Fig. 10.13a, the color-coded isolines surround pixels of a narrow range of intensity values. In 3D space, the isolines become surfaces which divide eight-voxel cells (as described above in volume rendering) to isolate voxels either by intensity or by segmentation. For example, given a threshold value of 100, the voxels of the single cell shown in Fig. 10.13b require the green surface to divide the 4 voxels of intensity 200 from the 4 voxels of intensity 40. This is how the marching cube algorithm functions, evaluating every cell in the volume so as to place boundaries sufficient to isolate voxels which either exceed a selected threshold or which have been

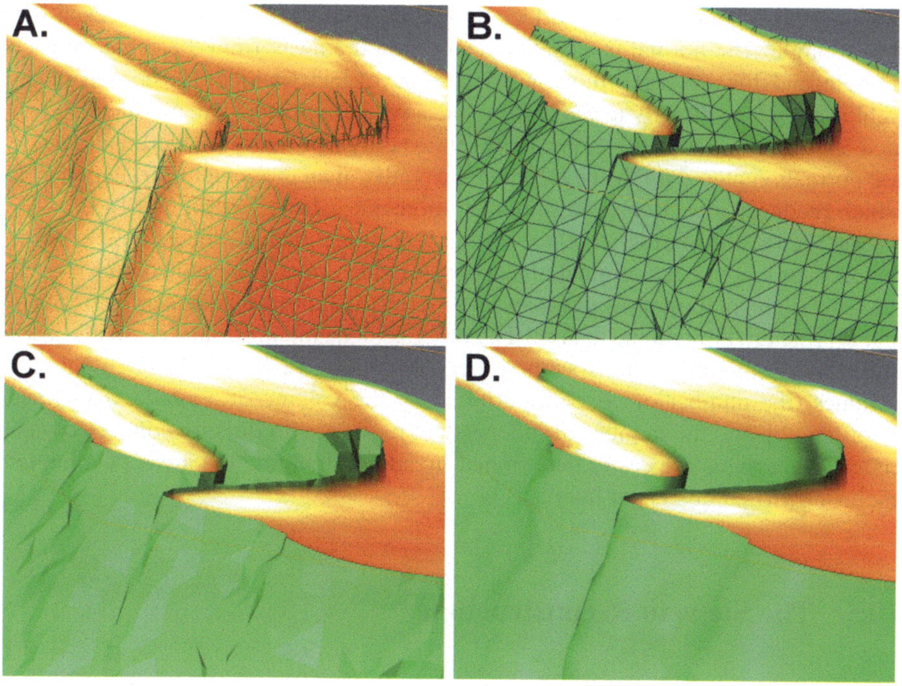

Fig. 10.14 Surface rendering. A portion of a folded surface reconstruction rendered as a wireframe (**a**), a wireframe filled with green (**b**), a green surface without the wireframe (**c**), and a green surface with reflection normals moved to the center of vertices (**d**)

manually selected. There are 256 possible solutions to an eight-voxel cell analysis, but given rotational symmetries, this number can be reduced to only 15 potential answers (Fig. 10.13c). After cleaning up some inconsistencies and imposing some rules governing maximal surface folding, the algorithm produces a set of vertices defining a mesh of triangles which enclose the selected voxels (D).

The number of triangle faces produced by the marching cube algorithm depends on the voxel resolution. An ideal solution contains just enough faces to efficiently enclose the 3D region of interest. However, the speed of rendering a surface is directly related to the number of faces to draw. Most efficient surfaces have less than 40,000 faces, yet the first approximation of the marching cube algorithm will often yield over a million faces. The face count can be reduced by either averaging across two or more voxels before executing the marching cube algorithm or by using a surface simplification routine on the final surface.

A variety of choices are available when rendering a surface to the display (Fig. 10.14). Drawing the mesh of triangles which connect the vertices defining the surface produces a wireframe (Fig. 10.14a) which is transparent. Triangles can be filled with a chosen solid color (Fig. 10.14b), and the wireframe triangles can be hidden (Fig. 10.14c). This reveals the faceted reflections of the light bouncing off the triangles. This facet effect is caused by the 3D lighting calculations of the display system, which depend upon a surface normal that has been generated for each surface triangle. The normal is a spatial vector pointing away from the center of each triangle, and this vector is orthogonal to the plane of each triangle. The normal guides light in the scene directly to the triangle face, where it is reflected, producing a faceted appearance. This reflection can be removed by forcing the rendering engine to use normals centered on the triangle vertices rather than the center of triangle faces, which produces reflections of a smoother, more natural surface (Fig. 10.14d).

Surface reconstructions produce virtual representations of objects in 3D. While they appear less like the original object compared to projections and volume renderings, the segmentation process for surface reconstructions is a necessary first step to quantifying anything in the dataset. The process of segmenting dataset voxels based on intensity (or some other specific criteria) provides a framework for calculating volumes (the number of voxels multiplied by the volume of a single voxel) and counting objects (the number of independent regions with directly connected voxels). Surface areas are easily determined by adding together the areas of triangles enclosing virtual objects of interest.

10.5 The Steps to Reconstruction

Three possible reconstruction pathways are shown in Fig. 10.15. Each path is based on the main types of reconstruction that can be produced, and the flow chart points out the main steps that will be required to achieve the desired result. Note that results obtained in more complicated reconstructions may contribute to other types of visualizations. The steps to reconstruction can be reduced to seven basic processes that must be followed in order, but it should be clear that not every step is required to produce a 3D visualization.

Most 3D reconstruction software systems perform similar functions (Clendenon et al. 2006; Rueden and Eliceiri 2007), but it is beyond the scope of this chapter to produce a guide to reconstruction for each system. As is generally true with software, the more you pay, the more power you get. Each system should have a user guide to the functions, and with luck, some tutorials designed to reduce the learning curve for novice users. This section will present a more philosophic approach to 3D reconstruction and attempt to describe the generic steps to performing this complicated task.

10.5.1 Planning

When building (or rebuilding) any structure, it is always helpful to have a plan. In 3D reconstruction, this means gathering as much information about the sample as is feasible and to have some idea of what the final product will include. For example, what is the approximate real size of the largest and smallest features to be

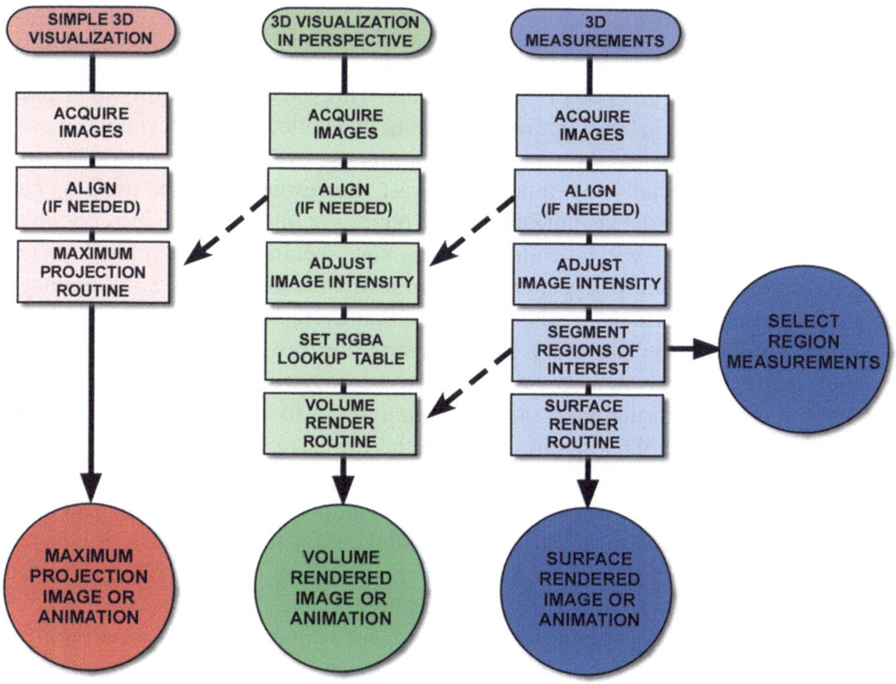

Fig. 10.15 The reconstruction process flowchart. The steps required to reconstruct will depend upon the type of result desired. (Red) Simple 3D visualizations generated using maximum projection methods only require a confocal z-series. (Green) Volumes rendered in perspective may need some intensity adjustment and the assignment of a suitable RGBA lookup table. (Blue) Surface reconstructions and any 3D measurements will require segmentation of the volume into regions of interest. The dashed arrows between pathways indicate that results from one path can contribute to other types of reconstruction. For example, the segmentation output of a surface reconstruction can be used to assign textures to multiple regions of a volume render (instead of using an RGBA lookup table)

reconstructed? That information will guide the microscopist to select an objective with enough resolving power and sufficient field of view. In some cases, there is no objective that will work well for both resolution and view field. This typically is true when the entire object is too large to fit in the view of an objective needed to resolve the smallest features of interest. In those situations, it's best to use the higher power objective for the best resolution of the smaller features. The microscopist should then plan to collect overlapping fields of view in acquisition and then realign all the parts into one complete dataset.

There should be specific goals in place. Ask yourself which products will result from the reconstruction? Are you attempting to visualize the sample from unique 3D perspectives? Will you need to subdivide the structure into components to study how they fit together? Do you wish to quantify some spatial characteristics of your sample? The answers to these questions will determine which tasks you must perform and will lead you to an efficient path to your goals.

10.5.2 Acquisition

Most of the content of this book is aimed at improving the readers' ability to acquire quality confocal images, but it is worth repeating that collecting the best dataset possible will greatly influence the reconstruction process. All of the issues previously discussed apply, especially maximizing contrast and resolution in the images. The basic rule is to enhance the visualization of the regions of interest as much as possible.

Your raw image data will be collected using your specific model confocal microscope's software system. Every manufacturer's software suite saves z-series image stacks in a different format, so it is important to understand how your system stores these images. Furthermore, every 3D reconstruction program reads these image stacks in a different manner, so you must also be prepared for the possibility that you might need to reformat your original data in order to move it into the 3D software. Hopefully your 3D software can read the raw data format of your confocal microscope, but if not, the best solution is to export z-series into TIFF format images using filenames that are in the correct order when alphabetically sorted. If you must export TIFF images, you must also keep different confocal channel images in separate, appropriately named folders. As explained elsewhere in this book, it is best to avoid using JPEG format when moving 3D datasets.

Another consequence of exporting TIFF images is that spatial information may not be read with the data into the 3D software. If you plan to produce quantitative information using physical units, be prepared to know the pixel size and z-slicing distance. Most 3D systems ask for the size of a single voxel as an x,y,z input. Be sure to report all of these values in the same units (e.g., micrometers).

10.5.2.1 Deconvolution

Deconvolution of confocal image data may seem unnecessary, but this process can actually remove a noticeably large proportion of the background noise, especially in the z-plane, and greatly assist in later reconstruction steps. Deconvolution should really be considered an adjustment method, but because its implementation has more stringent image acquisition requirements, it is best to mention it here. Any discussion of the process of deconvolution would be beyond the scope of this chapter, so the reader is greatly encouraged to consider recent papers for a general introduction (Biggs 2004, 2010; Feng et al. 2007; McNally et al. 1999).

10.5.3 Alignment

One of the more difficult and tedious tasks in reconstruction is aligning the serial slices of the 3D image set. When serial tissue slices from a microtome are mounted, stained, and photographed, the slices must be returned to their original positions by transforming the images (moving in x and y directions and rotating). Often, the process of cutting the tissue will change the shape of serial slices as the sample is dragged across a sharpened knife. For confocal data, this step is completely unnecessary. The sample remains intact, the serial slices are obtained optically, and every image is collected in perfect alignment to the entire sample.

However, circumstances do arise when it becomes necessary to align multiple confocal datasets into one large dataset. Consider the case where larger features of interest cannot fit into the field of view of the objective required to resolve the smallest regions of interest. In these situations the entire range of features can be obtained by tiling 3D image sets together. Many modern confocal microscopes come equipped with automated x-y stages and software which will stitch together neighboring z-series into large 3D image datasets.

10.5.4 Adjustment

Typically, if you were carefully setting the gain and offset during acquisition, your confocal images should need little adjustment beyond minor changes in brightness or contrast. Please keep in mind the guidelines for scientific data as applied to images, such as those defined by the Microscopy Society of America (Mackenzie et al. 2006). However, it might be necessary to adjust the intensity values in the images to enhance the differentiation of regions of interest for thresholding or manual segmentation procedures to follow. Always remember that it is important to keep a record of these procedures and to report them in any scientific reports so that your results are repeatable.

Most 3D software suites provide image adjustment functions that will alter the brightness data in the images either in individual slices, across all slices, or even using 3D kernels as image filters. These functions are strictly to be used to help differentiate regions of interest when necessary. Some "automatic" segmentation routines work best when regions of interest have sharp boundaries based on intensity, and these functions can often help the efficiency of those routines. When manual segmentation procedures will follow, these adjustment routines greatly assist the observer in finding regional boundaries.

As an example, consider the application of the histogram equalization function. Histogram equalization is often used to increase contrast in an image by "equalizing" the frequency of pixel intensities across the available range. The effect of typical unrestrained histogram equalization is shown in Fig. 10.16a and b. For confocal images, this type of equalization tends to emphasize the out-of-focus epifluorescence, which tends to defeat the purpose of optical slicing. A better solution is to apply the histogram equalization function repeatedly over smaller, more contextual regions of the image in a procedure called local adaptive histogram equalization. This greatly maximizes contrast, making hidden regions more visible, but also greatly enhances the background noise and produces shadows at image intensity edges (Fig. 10.16c).

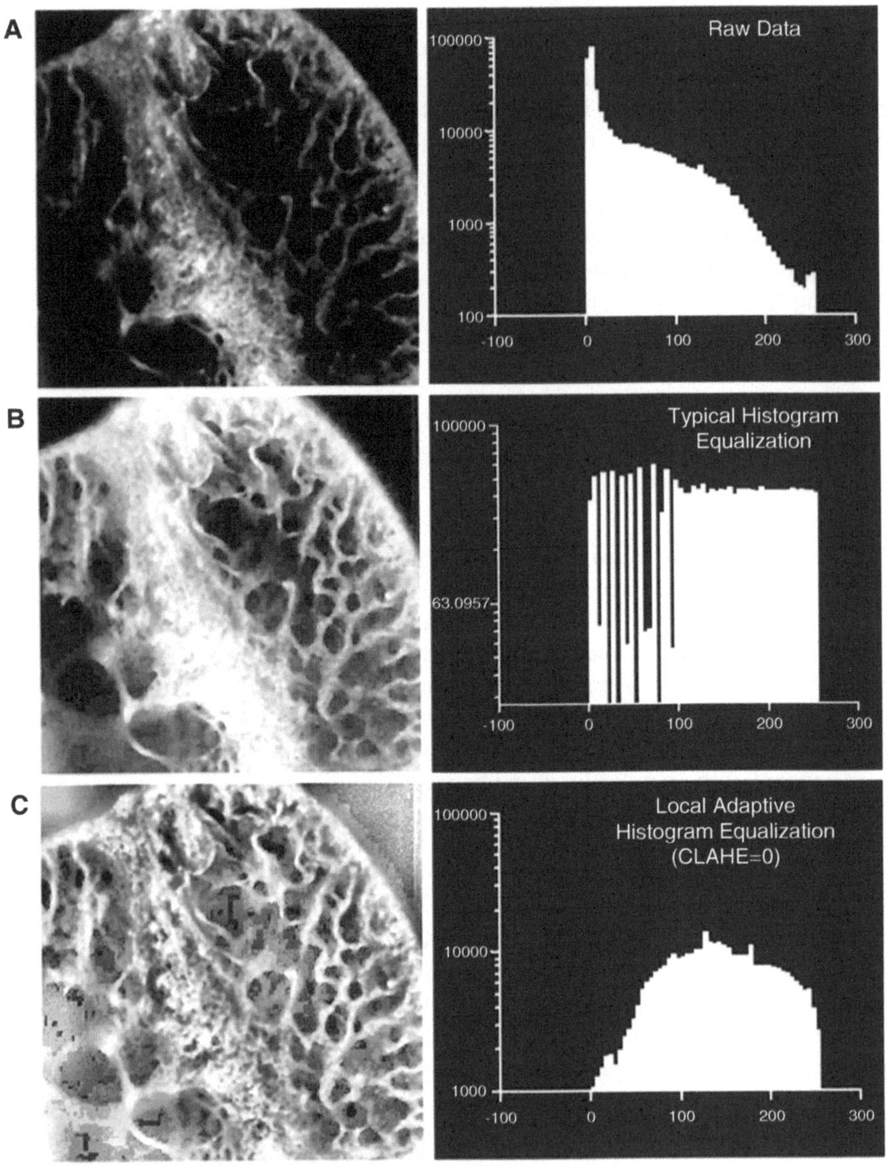

Fig. 10.16 Image adjustment. Histogram equalization. (**a**) A raw image and the corresponding histogram. (**b**) The result of a typical unrestrained histogram equalization. (**c**) Local adaptive histogram equalization without contrast limit. The effect of increasing the contrast limit in adaptive histogram equalization to 5 (**d**), 10 (**e**), and the maximal 15 (**f**)

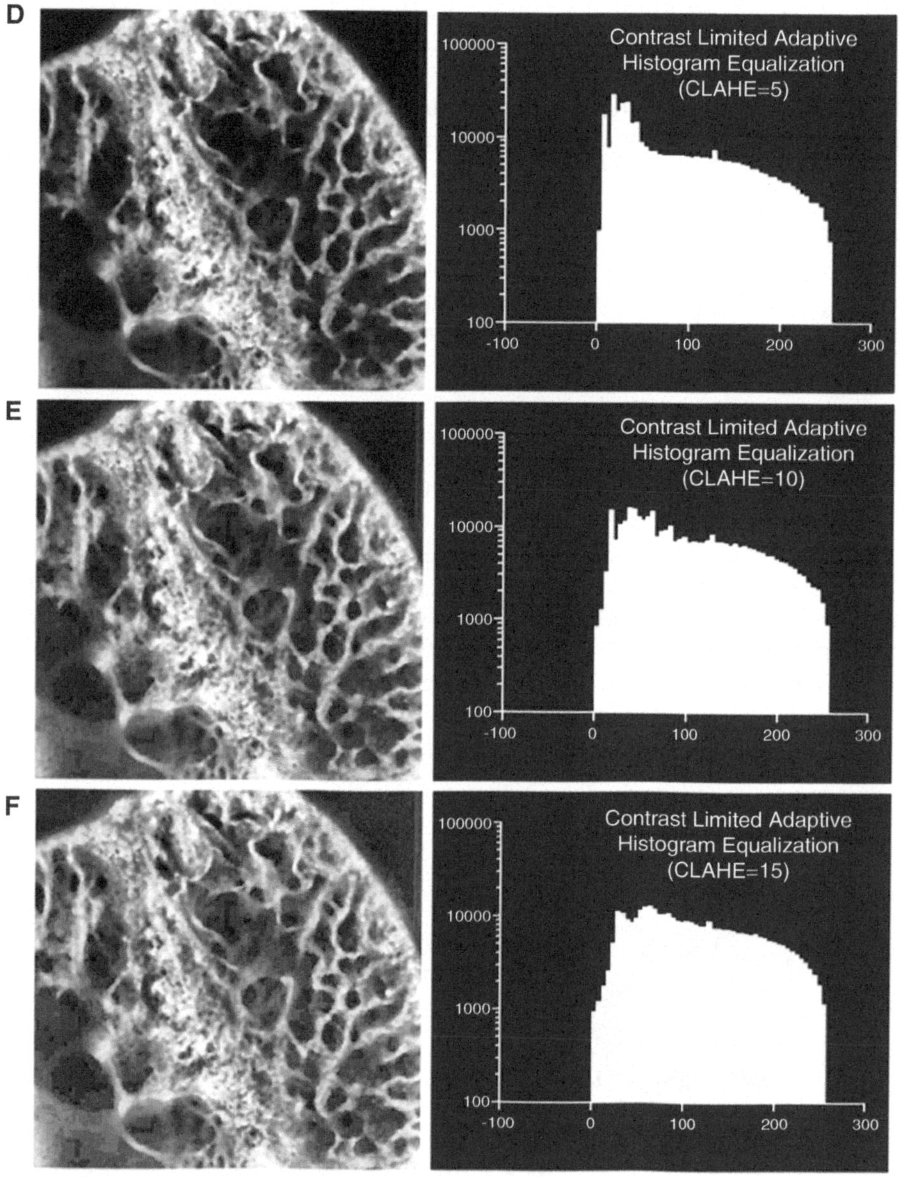

Fig. 10.16 (continued)

10.5.4.1 Contrast-Limited Adaptive Histogram Equalization (CLAHE)

This is a procedure originally developed for medical imaging and can be quite helpful in enhancing low-contrast images without amplifying noise or shadowing edges. Briefly, CLAHE works by limiting the amount of contrast stretching allowed in small local areas based on the contrast already present. So in regions where intensity is uniform, the amount of equalization is reduced, and the noise is not emphasized. The CLAHE effect is typically set by variably adjusting the degree of contrast stretching to be applied in local areas, and this is demonstrated in Fig. 10.16d–f. For the purpose of regional segmentation, the adjusted image in Fig. 10.16d would be far easier to differentiate than the original for both automatic and manual procedures.

10.5.4.2 Z-Drop

The sample itself will often interfere with both excitation light on the way in and emission light on the way out (Guan et al. 2008; Lee and Bajcsy 2006; Sun et al. 2004). The effect of this is to produce an image set where the average intensity of each optical slice appears to be reduced in deeper z-sections (a phenomenon termed z-drop, see Fig. 10.17). Most automated segmentation routines used threshold-based rules based on voxel intensity; thus it would be desirable that voxel intensity within regions of interest be uniform throughout the sample. Many 3D reconstruction systems offer routines to correct for z-drop as an adjustment procedure.

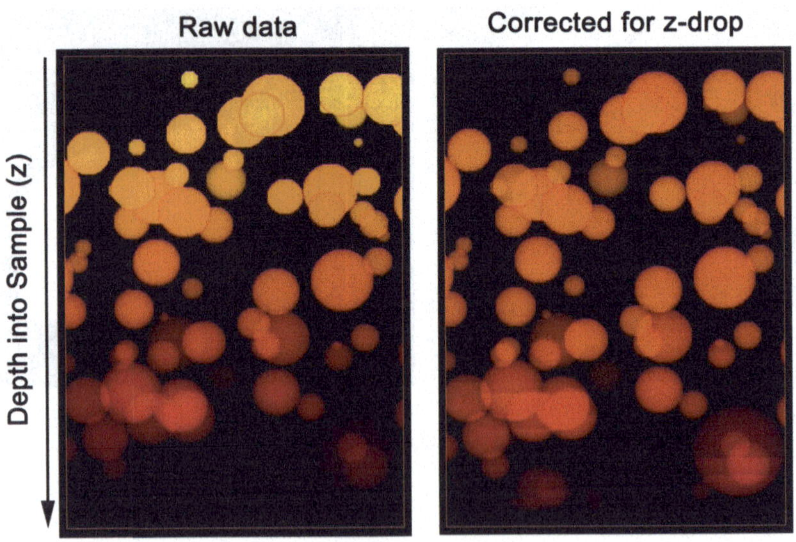

Fig. 10.17 Correcting for z-drop in confocal data. Volume rendering (XZ view) of confocal dataset showing the effect of z-drop through the depth of the sample (right). Following correction (left), voxel intensities have been adjusted so that average intensity is more nearly constant with depth

It is often possible to correct z-drop in the acquisition process by systematically adjusting laser intensity or detector gain (or both) as deeper optical slices are collected. Many confocal microscope control systems include a routine which will adjust these settings with optical depth, usually based on linear models of the z-drop effect.

10.5.5 Segmentation

Segmentation is the process of dividing the image dataset into regions of interest. This is often the most time-consuming task and typically requires a great deal of expertise. It is worth noting here that segmentation is required in order for any measurements to be collected. In the simplest case, a confocal dataset collected in a single fluorescent channel can be divided into two regions: fluorescent signal and background. Segmenting this channel using a fast, automated method is a matter of choosing a threshold intensity value that discriminates between these regions. However, the selection of a single threshold value that will apply across every optical slice is often thwarted by effects such as noise or z-drop. When such problems cannot be otherwise corrected in acquisition or by adjustment, manual segmentation methods are usually necessary. In this simple case, that may involve selecting a different threshold appropriate for each optical slice (to subvert conditions like z-drop) or manually editing the results of an automatic segmentation to remove artifacts.

An example of a segmentation editor is shown in Fig. 10.18. This routine is from the 3D software Amira (www.visageimaging.com) and offers a slice-by-slice view of the data from any perspective in addition to a selection of automated and manual selection tools. In the sample shown, a threshold has been set in the Display and Masking section to select all voxels (in the entire confocal dataset) with an intensity greater or equal to 64. Those voxels were then placed in a material named "myocardium" and outlined in green. This material data is saved in a separate label data file associated with the image data.

Once the nonfluorescent background has been isolated from the fluorescent signal, further refinements are possible. For the sample data shown in Fig. 10.18, it was decided to further segment the fluorescent signal into two histologically relevant regions: trabecular myocardium and compact myocardium. Obviously these regions cannot be isolated based on voxel intensity, so manual segmentation by an experienced observer is required. In Fig. 10.19, the "myocardium" material was renamed "compact myocardium," and a second material "trabecular myocardium" has been defined and color-coded red. The background material "exterior" was locked (to prevent any changes), and a large paintbrush was used on each optical slice to select voxels from the green regions that belong in the red material. For 170 optical slices and a trained observer, this procedure took about 2 hours.

More sophisticated approaches to automatically segmenting confocal data have been reported (Lin et al. 2003; Losavio et al. 2008; Rodriguez et al. 2003;

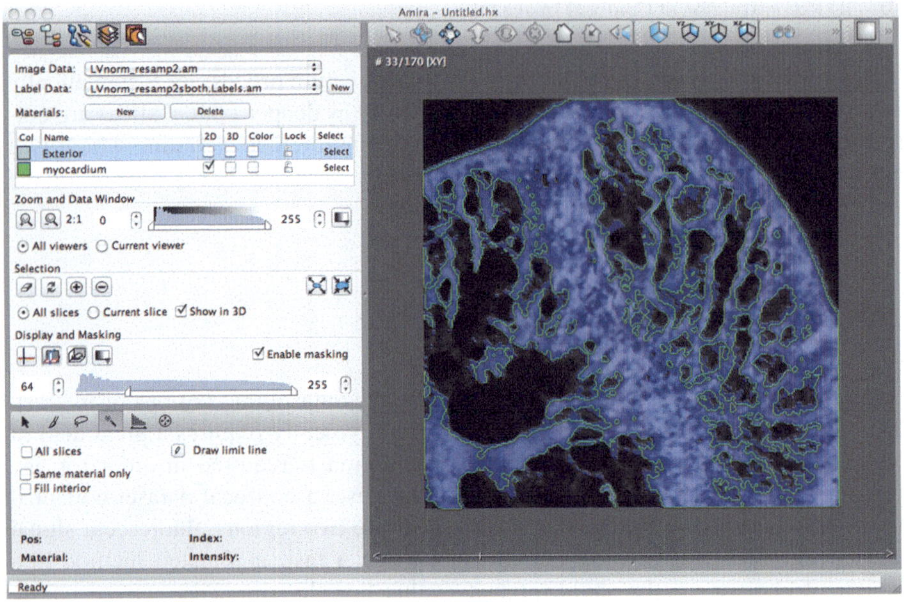

Fig. 10.18 Segmenting confocal data based on threshold. The segmentation editor in Amira (www.visageimaging.com) was used to locate voxels with intensity values between 64 and 255 (see the Display and Masking section on the right). The selected voxels are assigned to the material "myocardium" and outlined in green. Although only one slice is shown, the selection was completed on all 170 optical slices simultaneously

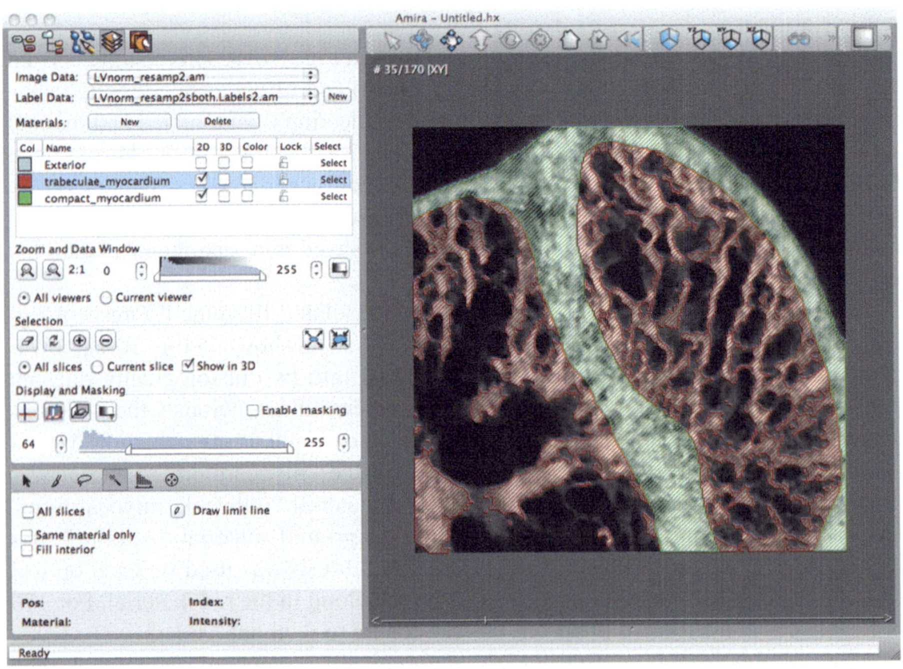

Fig. 10.19 Segmenting confocal data based on manual selection. The automatically segmented data in Fig. 10.18 was further separated into trabecular (red) and compact (green) myocardium on each optical slice by manual selection using the drawing tools

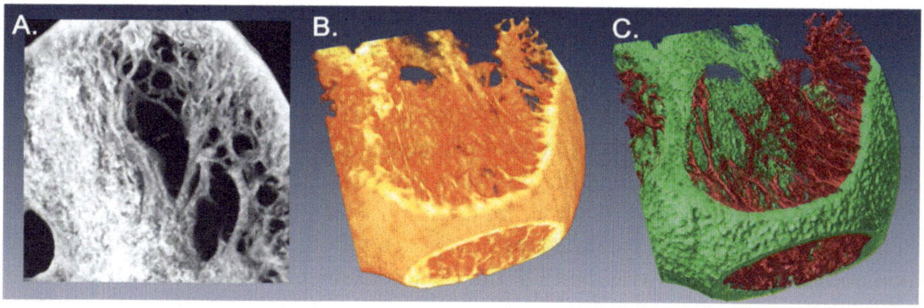

Fig. 10.20 Reconstruction models. (**a**) Projected data from the middle 50 slices (of 170). (**b**) Volume rendering of all 170 optical slices. (**c**) Volume rendering using manual segmentation to highlight trabecular (red) and compact myocardium (green)

Yi and Coppolino 2006), and some have even been made available in some specialized 3D software programs. These methods are usually based on finding the edges of regions based on voxel intensity, and some use filters based on the shape and size of the regions they expect to locate. As such, these routines require the image data to have the best contrast available (the data uses most of the available brightness spectrum), and the resolution should be optimal (and probably even deconvolved). While these routines can save time, the results will still contain errors and require some manual editing.

10.5.6 Modeling and Visualization

Most modern computers can now render 3D models practically in real time; thus visualized results are often rendered repeatedly until desired results are obtained. These computational speeds also make it easier to render animation frames, where the subject changes position or rotates in the field of view in a series of predetermined steps. Nearly all 3D software programs offer methods of generating animations in AVI or MPEG format or can generate a folder full of single frames suitable for import into digital movie applications such as Adobe Premiere.

The amount of information conveyed in a 3D visualization can be augmented with just a little extra work. Consider the data from the confocal dataset from the previous segmentation example now portrayed in Fig. 10.20. In (A), the middle 50 optical slices are shown as a maximum projection image generated immediately after acquisition. If all 170 optical slices had been included in the projection, the interior detail of the trabecular myocardium inside the ventricle would have been obscured (see Fig. 10.1c). In Fig. 10.20b, all of the data is visualized in a volume render using an RGBA lookup table which renders the background noise as transparent. The more intense voxels are rendered more opaque. The 3D nature of the sample is more apparent, and the trabecular myocardium can now almost be differentiated from the compact myocardium. In (C), the volume render has made use of the segmentation data to color code the two histological types of myocardium, and the discrimination of trabeculae inside the ventricle wall composed of compact myocardium is clear. Note that (C) is not a surface reconstruction but a special volume render utilizing the segmentation data to color-code voxels and adding diffused lighting and specular effects.

While single images of 3D models are easy to publish, often it is of value to share a model which can be viewed from multiple angles. This has typically involved supplying a supplementary animation file with a guided tour of the object as directed by the filmmaker. It is now possible to include 3D objects entirely within the Adobe Acrobat PDF file format (www.adobe.com) such that the data can be interactively manipulated by anyone with the PDF file (Ruthensteiner and Hess 2008). Note that only surface reconstruction data (not volume renders) can be imported into PDF files as of the writing of this chapter.

10.5.7 Measurement

Measurements obtained from 3D datasets are based on the spatial dimensions of single voxels in the data. For example, if the size of a single voxel is 1 mm × 1 mm × 1 mm, then the smallest volume that can be measured is a single voxel, or 1 mm^3. To determine the volume of a fluorescent object in an image dataset, it is necessary to identify (segment) and then count the voxels in that object. The total volume will then be the voxel count multiplied by the volume of a single voxel.

The calculation of surface area will require a surface reconstruction. In this case, the individual triangles within the surface will each have unique areas determined by the x,y,z coordinates of the three vertices of each triangle. The planar areas of each triangle are calculated and summed to produce a total surface area. (One can appreciate why surface calculations proceed more slowly when there are millions of triangles.)

The segmentation data of Fig. 10.19 was used to make the surface reconstructions shown in Fig. 10.21. Figure 10.21a is a surface model of the voxels contained in both segments. For segmented data and surface reconstructions, it is possible to model the segments together (Fig. 10.21b) or independently (Fig. 10.21c–d). The volume and surface area measurements for the defined segments are shown in Table 10.1.

Most 3D software systems with measurement capabilities will further subdivide each material into regions based on voxel connectivity. That is, an isolated island of contiguous voxels is considered a region separate from other voxel regions. Material statistics can be gathered for materials based on regions, which are useful for counting cells, or nuclei, or any other objects that exist as isolated fluorescent entities.

Figure 10.22a shows a field of fluorescently stained nuclei that were segmented using an automatic threshold and then surface rendered (Fig. 10.22b). Material statistics generated on this segmentation yields the data in Table 10.2, where the nuclear material has been analyzed based on region (the table is abbreviated).

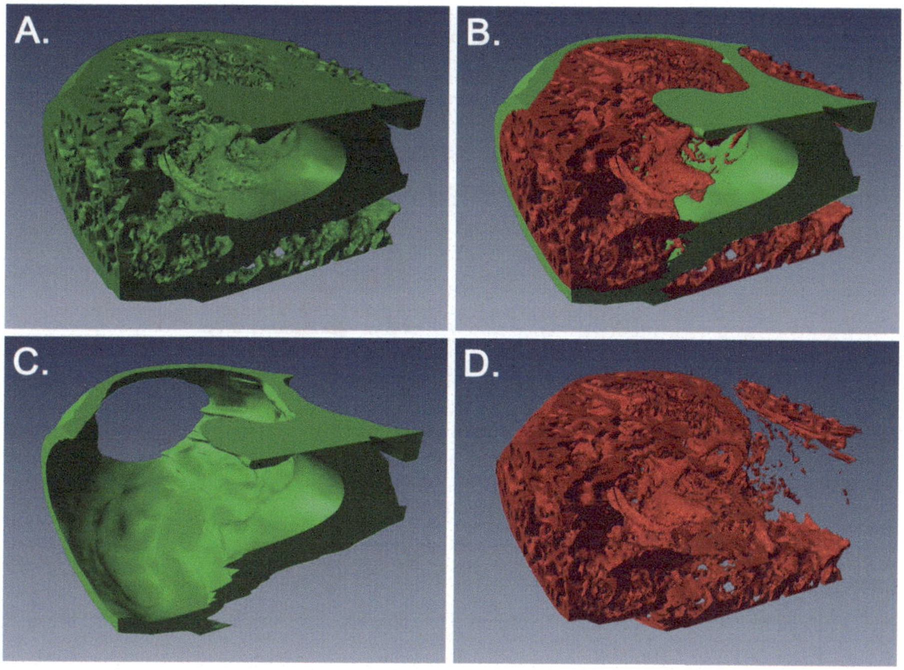

Fig. 10.21 Surface reconstructions from segmented confocal data. (**a**) Surface from threshold-based automatic segmentation. (**b**) Surface from data in (**a**) manually segmented into trabecular (red) and compact (green) myocardium. Segmented regions of interest allow separate rendering (**c**, **d**) and measurement

Table 10.1 Measurements from 3D reconstructed confocal data

Material	Voxel count	Volume (m^3)	Surface area (m^2)
Exterior	7,083,350	0.760	
Trabecular myocardium	1,838,729	0.197	21.4
Compact myocardium	2,219,041	0.238	7.3
Total	***11,141,120***	***1.196***	

Voxel size = 5.08 × 5.08 × 4.14 μm
Total dataset is 256 × 256 × 170 voxels

The analysis provides a total count of the regions revealing that 139 isolated regions were identified. The volume; voxel count; x,y,z center coordinates; and mean voxel intensity within each region are provided. The data can be exported for more complex analyses. Note that this data from Amira is sorted by volume. It is a task left to the experimenter to decide if the smallest regions (nuclei with volumes of 1 μm^3) are noise

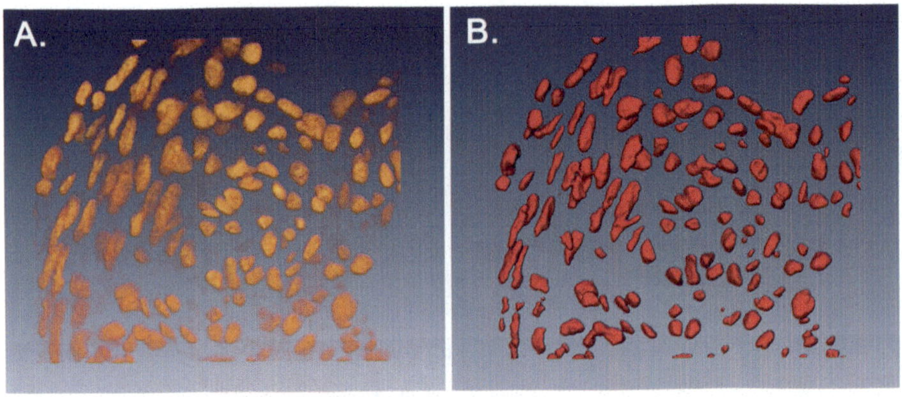

Fig. 10.22 Counting nuclei in segmented confocal data. (**a**) Voxel rendering of a field of stained nuclei. (**b**) The nuclei were segmented using an automated threshold for a surface reconstruction. Material statistics for the regions in this segmentation report volume, spatial location, and fluorescence intensity data for each nucleus (see Table 10.2)

Table 10.2 Regional measurements

Nr	Region	Voxel count	Volume	CenterX	CenterY	CenterZ	Mean
1	Exterior 1	2,196,590	116823.1	93.6	119.4	5.3	13.0
2	Nuclei 139	13,248	704.5	44.7	133.3	6.2	78.6
3	Nuclei 138	13,229	703.5	122.8	96.4	5.5	81.3
4	Nuclei 137	9868	524.8	104.0	111.3	5.6	87.6
5	Nuclei 136	9134	485.7	93.1	92.6	6.2	89.2
6	Nuclei 135	8511	452.6	62.9	114.3	5.5	71.9
7	Nuclei 134	8500	452.0	80.5	105.9	5.1	79.1
8	Nuclei 133	8015	426.2	72.0	83.7	5.7	82.0
9	Nuclei 132	7334	390.0	78.5	92.4	5.6	84.9
10	Nuclei 131	7048	374.8	77.6	123.0	5.6	73.6
…	…	…	…	…	…	…	…
138	Nuclei 3	19	1.0	41.4	149.6	6.83	59
139	Nuclei 2	19	1.0	93.7	72.6	2.9	57
140	Nuclei 1	19	1.0	82.3	91.4	8.8	62

The measurement routines in most 3D software systems will provide the basic data described here. Some provide more sophisticated tools, but these systems tend to be more expensive and will have highly specialized functions of little use to everyone. Thus, the last advice this chapter can offer is that the user should have some knowledge of the type of quantitative measures needed and to carefully select software that will fulfill the research requirements.

Literature Cited

Biggs DS (2010) 3D deconvolution microscopy. Curr Protoc Cytom. Chapter 12: p. Unit 12 19 1–20

Biggs DSC (2004) Clearing up deconvolution. Biophoton Int (February):32–37

Carlsson K, Aslund N (1987) Confocal imaging for 3-D digital microscopy. Appl Opt 26(16):3232–3238

Clendenon JL, Byars JM, Hyink DP (2006) Image processing software for 3D light microscopy. Nephron Exp Nephrol 103(2):e50–e54

Feng D, Marshburn D, Jen D, Weinberg RJ, Taylor RM II, Burette A (2007) Stepping into the third dimension. J Neurosci 27(47):12757–12760

Guan YQ, Cai YY, Zhang X, Lee YT, Opas M (2008) Adaptive correction technique for 3D reconstruction of fluorescence microscopy images. Microsc Res Tech 71(2):146–157

Hecksher-Sorensen J, Sharpe J (2001) 3D confocal reconstruction of gene expression in mouse. Mech Dev 100(1):59–63

Lee SC, Bajcsy P (2006) Intensity correction of fluorescent confocal laser scanning microscope images by mean-weight filtering. J Microsc 221(Pt 2):122–136

Lin G, Adiga U, Olson K, Guzowski JF, Barnes CA, Roysam B (2003) A hybrid 3D watershed algorithm incorporating gradient cues and object models for automatic segmentation of nuclei in confocal image stacks. Cytometry A 56(1):23–36

Losavio BE, Liang Y, Santamaría-Pang A, Kakadiaris IA, Colbert CM, Saggau P (2008) Live neuron morphology automatically reconstructed from multiphoton and confocal imaging data. J Neurophysiol 100(4):2422–2429

Mackenzie JM, Burke MG, Carvalho T, Eades A (2006) Ethics and digital imaging. Microscopy Today 14(1):40–41

McNally JG, Karpova TS, Cooper JA, Conchello J-A (1999) Three-dimensional imaging by deconvolution microscopy. Methods 19(3):373–385

Rodriguez A, Ehlenberger D, Kelliher K, Einstein M, Henderson SC, Morrison JH, Hof PR, Wearne SL (2003) Automated reconstruction of three-dimensional neuronal morphology from laser scanning microscopy images. Methods 30(1):94–105

Rueden CT, Eliceiri KW (2007) Visualization approaches for multidimensional biological image data. BioTechniques 43(1 Suppl):31 33-6

Ruthensteiner B, Hess M (2008) Embedding 3D models of biological specimens in PDF publications. Microsc Res Tech 71(11):778–786

Savio-Galimberti E, Frank J, Inoue M, Goldhaber JI, Cannell MB, Bridge JHB, Sachse FB (2008) Novel features of the rabbit transverse tubular system revealed by quantitative analysis of three-dimensional reconstructions from confocal images. Biophys J 95(4):2053–2062

Soufan AT, van den Berg G, Moerland PD, Massink MMG, van den Hoff MJB, Moorman AFM, Ruijter JM (2007) Three-dimensional measurement and visualization of morphogenesis applied to cardiac embryology. J Microsc 225(Pt 3):269–274

Sun Y, Rajwa B, Robinson JP (2004) Adaptive image-processing technique and effective visualization of confocal microscopy images. Microsc Res Tech 64(2):156–163

Yi Q, Coppolino MG (2006) Automated classification and quantification of F-actin-containing ruffles in confocal micrographs. BioTechniques 40(6):745–746, 748, 750 passim

Chapter 11
Analysis of Image Similarity and Relationship

Jesse Aaron and Teng-Leong Chew

11.1 Introduction

The ability of fluorescence microscopy to simultaneously image multiple specific molecules of interest has allowed biologists to infer macromolecular organization and, in the case of live cell imaging, even transient molecular interaction. Many consequential conclusions are drawn based on the various approaches to display or quantify these images. These include merged colorimetric display of two monochrome images and colocalization analyses. Collectively, these techniques may yield information about relative molecular abundance, spatiotemporal co-occurrence of molecules within a given cellular space, biological functions (in the case of biosensors or ionic probes), as well as other more complex examples of coupled variables. Unfortunately, when implemented without in-depth understanding, these approaches are often fraught with problems. Advances in computer technologies and software development have made the implementation of these techniques appear, at first glance, so deceivingly straightforward and intuitive that the various caveats and the underlying quantitative aspects of these methods are frequently overlooked. This chapter will discuss the underlying principles of how these techniques quantify their corresponding coefficients as well as their strengths and limitations. It will subsequently explore the practical applications of these methods.

J. Aaron · T.-L. Chew (✉)
Advanced Imaging Center, Howard Hughes Medical Institute Janelia Research Campus, Ashburn, VA, USA
e-mail: aaronj@janelia.hhmi.org; chewt@janelia.hhmi.org

W. G. Jerome, R. L. Price (eds.), *Basic Confocal Microscopy*,
https://doi.org/10.1007/978-3-319-97454-5_11

11.2 Colocalization: The Analysis of Similarity in Two Grayscale Images

One of the most common questions in life sciences is to interrogate the extent of biological association – whether a biomolecule or structure of interest is associated with a given organelle, compartment, another protein, or other structure within a cell (Dunn et al. 2011). This analysis of coupled variables forms the foundation of a "colocalization" study. Colocalization is often used by biologists as a proxy for molecular interaction. However, this analytical approach is fraught with potential problems. One rather surprising fact is that none of the so-called colocalization indices actually measures "colocalization" per se in the strictest biological sense (Ramírez et al. 2010). This is especially the case when the biological question involves the study of interaction at the molecular level, as colocalization is a technique of measuring relative proximities within the limitation of the spatial resolution of the component images. In short, "colocalization" is a misnomer, and the use of the term should indeed be discouraged and phased out. In general, colocalization analysis methods tackle the problem from one common angle – that is, to compare "image similarity" by comparing the coupled variables, which are the signals from two monochrome channels. The outcome of these analyses is affected by the resolution of the optical instrument, observer's color perception, autofluorescence and other background noise in the images, and image processing strategies. The accuracy of image similarity analysis ultimately hinges on implementation of the optimal method to tackle the biological questions in hand. We will begin by addressing these issues individually.

11.3 Resolution

One of the most important factors that would immediately affect the measurement of image similarity is the resolution of the optical instrument. In fact, the accuracy of the image similarity analysis can only be as precise as the resolution of the imaging instrument. This limitation, unfortunately, is often not well considered, resulting in biologists erroneously equating any readout from these quantitative indices as "colocalization" or even molecular interaction. There is a fundamental mismatch between normal optical resolution (on the order of 300 nm or larger) and the truly meaningful associative distance between biomolecules as defined by the Pauli exclusion principle (Pauli 1925), which is usually 10 nm or less. Yet, because of diffraction, even a single fluorescent molecule will appear as an "Airy disk" as described by the point spread function in a conventional optical image (Sheppard 2017). Figure 11.1 shows an idealized representation of this concept. While there is clearly signal overlap between green and red channels, as shown by the yellow pixels, the actual positions of the two molecules, indicated by a white + and × signs, respectively, are separated by >150 nm in both the lateral and axial planes – far

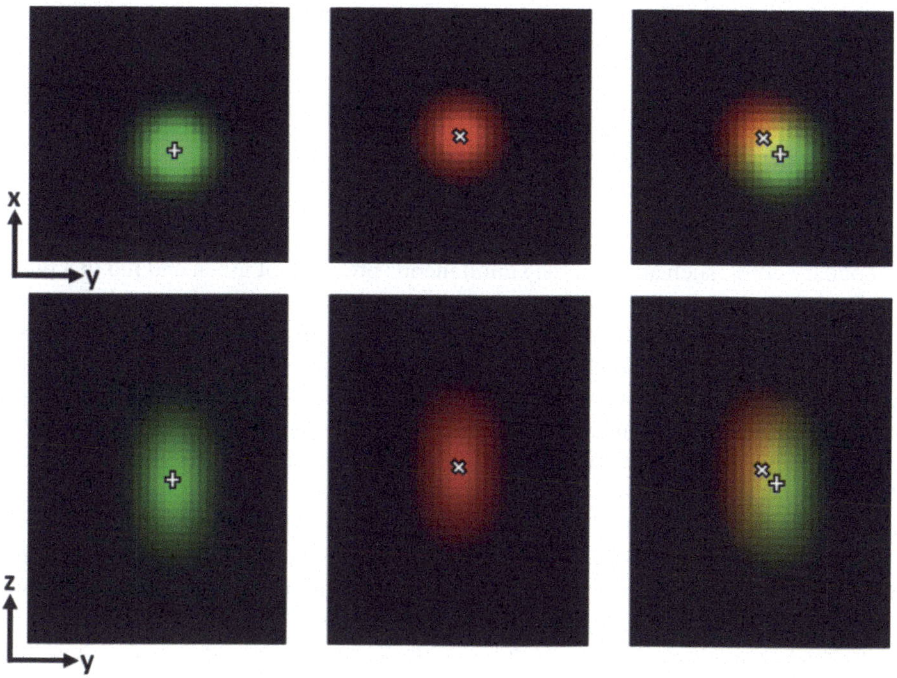

Fig. 11.1 A single molecule is imaged in the green channel, while another single molecule is imaged in the red channel. Each molecule is imaged in three dimensions, resulting in both a lateral plane image and an axial plane. Their merged display is shown in the right column

beyond meaningful molecular interaction distance. We therefore cannot claim molecular interaction based on pixel overlap, and the appearance of a yellow pixel (indicative of overlap of red and green pixels) should never be used as a quantitative measure. The only commonly used optical technique to directly measure molecular interactions makes use of Förster resonance energy transfer (FRET) (Chew and Chisholm 2005), which has an effective proximity range of <10 nm, or fluorescence cross-correlation spectroscopy. Yet, these techniques have their own limitations, and FRET calculations are themselves frequent victims of poorly performed channel bleed-through correction.

Likewise, the development of localization-based super-resolution fluorescence microscopy capable of resolving molecular separation at the range of 10–20 nm (reviewed by Schermelleh et al. 2010) has highlighted one of the most glaring limitations of image similarity studies. At low magnification, objects as large as single cells can sometimes overlap, yet in PALM/STORM images with resolution of <20 nm, even single molecules rarely show any real spatial co-occurrence (i.e., the absence of "yellow" pixels when a super-resolved red image is digitally merged with a super-resolved green image). This observation challenges the notion of using colorimetric analysis, in which yellow is often rudimentarily interpreted as overlap when the corresponding pixel pair contains signals from both green and red monochrome channels, as a reliable proxy for molecular interaction.

11.4 Color Perception and Colorimetric Display

Arguably the most common colocalization analysis is the visual perception of a secondary color, such as when the simultaneous presence of green and red in a pixel makes it appear yellow. However, color perception is nonlocal: it is also influenced by the color of surrounding regions in the image as well as the brightness and color of the lighting in the room. To illustrate these issues, Fig. 11.2a–c shows three pseudo-colored variations on the same three-channel biological image. In each variation, each channel's monochrome intensity values are identical: Only the pseudocolor assigned to each channel has been changed. If human perception of color was local and accurate, each image should look about the same, except for the differences in color. In fact, each variation *appears* to be almost an entirely different image. Objects easily visible in one variation are virtually invisible in others and vice versa, all depending on the colors and combinations of colors present. More disturbingly, objects that appear "colocalized" in one image seem totally noncorrelated in another. Thus, it is impossible to accurately judge if and how the objects in the three channels overlap. For these reasons, colorimetric methods should generally be avoided for all but the most qualitative analyses.

Among the many underlying factors that contribute to such display discrepancy in Fig. 11.2 is the fact that most of the color schemes used conventionally to display the merged images from two monochrome channels do not have uniform luminosity.

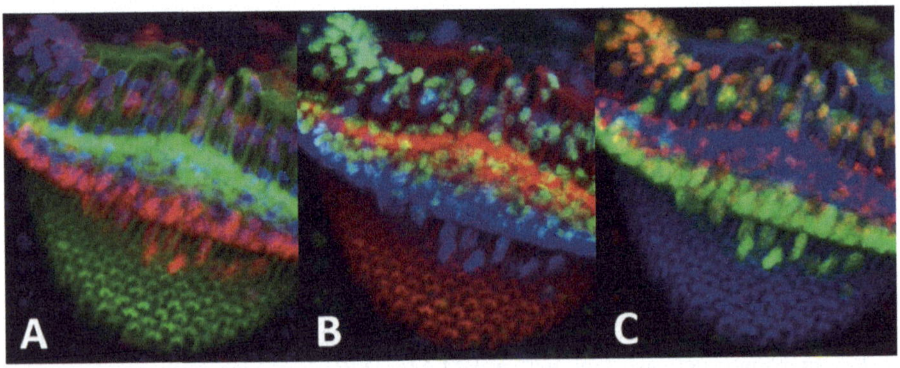

Fig. 11.2 Human visual perception can be misleading. The optic lobes from third-instar *Drosophila melanogaster* larvae were tripled-stained showing photoreceptor intricate spatial relationship of axons and glia. In panels **a**–**c**, the same combination of three monochrome images was displayed without manipulation of pixel intensities. However, the look-up table (LUT) assignments for the three channels were scrambled. Image courtesy of Dr. Vikki Weake, Purdue University

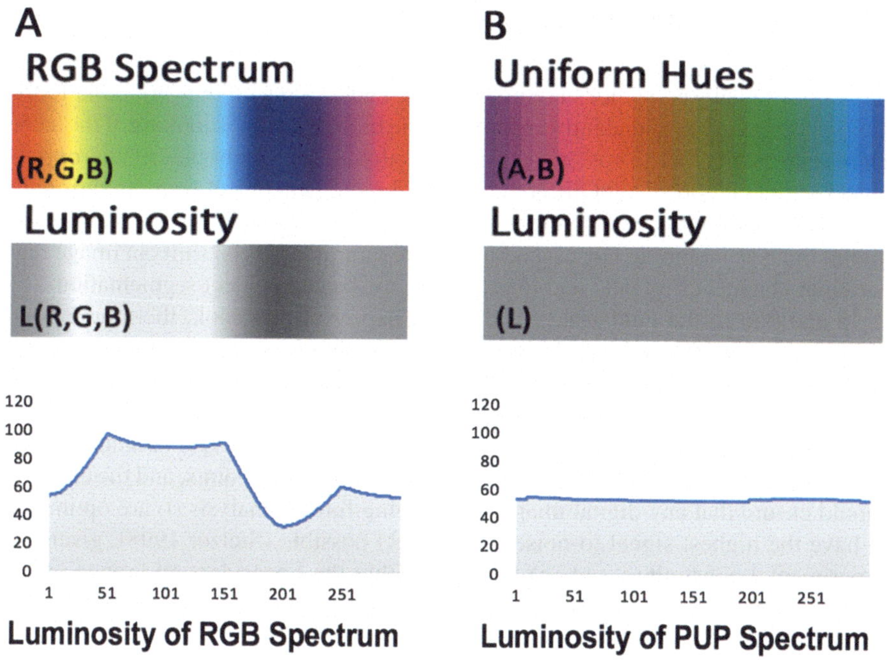

Fig. 11.3 Comparing the RGB and PUP color spaces. (**a**) RGB spectrum and the corresponding luminosity (perceived brightness) of color scheme. The profile plot highlights the irregularity in the luminosity in the RGB color space. (**b**) Perceptually uniform hues used in the PUP display. Note the uniform luminosity level throughout the color space

As a result, even though the pixel values of a merged image are faithfully maintained (whose corresponding information can be easily extracted in most image processing software simply by pointing the cursor to any pixel), ratiometrically accurate pseudo-color and grayscale intensity values are not necessarily perceptually equivalent for human vision (Taylor et al. 2017) (Fig. 11.3).

In fact, in a 24-bit standard RGB scheme (red, green, and blue) merged image, the most intense green color (intensity value equivalent to 0, 255, 0) is perceived to be approximately twice as bright as the most intense red color (intensity value equivalent to 255, 0, 0). While human visual perception should never be trusted to perform image quantification, the decision to explore certain biological features or molecular relationships is still frequently based on perceptual impressions. In fact, quite often the decision to implement further quantitative image analysis is driven by a visually perceived outcome. We therefore argue that, as underappreciated as it is, a perceptually accurate display should be treated as an important component of the investigative process in biological science. In light of that, as the first exploratory step, it is important to turn to a color scheme in which the luminosity values of the hues in the spectrum are equalized: the PUP (Perceptually Uniform Projection) display method (Taylor et al. 2017). This color scheme is available as a download from https://tinyurl.com/yc9daskb.

11.5 Optimization of Image Quality

Most image analysis methods are pixel-based, and as a result, they are blind to the biological structures abundantly apparent to the biologist who is looking at the same set of images. In other words, image analysis software, in the absence of appropriate object segmentation, cannot differentiate a real biological object from its surroundings. These programs are therefore sensitive to any contaminating signals in the digital images, including noise, shading errors, saturated pixels, shifts in image registration, channel cross talk, and improper (or the lack of) object segmentation.

In addition to the guidance provided in Chap. 9 of this book, there have been several excellent tutorials to aid readers in collecting high-quality digital fluorescence images (North 2006; Waters 2009). North describes the many optical considerations that must be optimized for high-quality microscopy images (North 2006). Waters continues by outlining important acquisition parameters for obtaining quantitative data (Waters 2009). This chapter builds off these key points, and the end user should ensure that any digital images undergoing further analysis (i) are optimized to have the highest signal-to-noise ratio (SNR) possible (Stelzer 1998), given the experimental constraints, and (ii) fall well within the linear dynamic range of the microscope detector, with no image saturation (Nakamura 2005; Stelzer 1998). (See also Chap. 9.)

There are several image corrections and manipulations that may be necessary before further image analysis can be performed. Firstly, it is important to subtract the image offset, as failing to do so will inflate the apparent signal level. Offset refers to the constant intensity value added to all pixels, regardless of the detected signal, and can generally be provided by the camera manufacturer or be measured directly. Further, many microscopes do not evenly illuminate the entire field of view. In such cases, it is imperative to obtain a correction image of a highly homogeneous sample with which to account for such imperfections. This so-called shading correction (Leong et al. 2003) is especially important when performing ratiometric imaging (discussed later). Further, it is important to assess the amount of signal from each fluorophore that is detected in the opposite color channels. The presence of such channel "bleed-through" requires subtraction of a proportion of one image from the other. The exact proportions are measured using control samples containing each fluorophore individually, under identical conditions as the experimental samples (Piston and Kremers 2007). Other image corrections are equally imperative. For instance, all images should be corrected for fluorophore photobleaching if time-course experiments are being performed (Vicente et al. 2007). Microscopes with multiple cameras, as well as the chromatic aberrations present in nearly any optical system, may require researchers to align one image with another. Multicolor sub-diffraction-sized fluorescent microspheres, such as Tetraspek® beads (Life Technologies, T-7280), can be used for this purpose in combination with affine transformation or image correlation techniques, among many other methods (Zitová and Flusser 2003).

Moreover, it is vital to subtract the unwanted background signal from the images. This signal is usually due to fluorescence from endogenous cellular components (Andersson et al. 1998), although mounting media or even the glass coverslip can contribute. One of the most commonly used techniques to remove unwanted background is the "rolling-ball" method (Dickinson et al. 2001; Sternberg 1983), which looks for minimum pixel intensity values within small neighborhoods throughout the image (Dickinson et al. 2001). More sophisticated methods based on spectral unmixing and apodization can also be employed to remove unwanted background (Haaland et al. 2009; Ojeda-Castaneda et al. 1988).

Finally, as will be discussed further, it may be necessary to determine an unbiased threshold image intensity value below which the image signal is not considered. There are numerous techniques proposed toward this end that utilize a variety of information, including image intensity distributions, image entropy, morphological features, and combinations of each (Glasbey 1993; Kapur et al. 1985; Otsu 1975; Peters 1995). Readers should explore numerous methods to determine which algorithm produces the desired results given the samples and structures being imaged. To aid in performing the preprocessing steps outlined above, we have collected a list of software plug-ins that are available for the open-source ImageJ/FIJI image processing package. These are summarized in Table 11.1. Similar functionalities are also provided in many commercially available software packages. We encourage the readers to consult with the manufacturer for more information should you wish to use them. In any case, the readers should always explain in detail any manipulations that are performed on image data featured in publications, and these manipulations should always fall well within the guidelines set forth by any publisher or funding agency such as those given by NIH in Chap. 12.

11.6 Object-Based Overlap Analysis

When the objects of interest are significantly larger than the diffraction-limited spot size, object-based overlap can be useful and is far more reliable than visual perception of secondary colors. The process of defining objects within an image is termed segmentation (Solomon and Breckon 2011) whereby a threshold is applied to create a binary image that distinguishes structures of interest from background signal. The binary images are often further morphologically filtered (selected based on size or shape), until only the objects of interest remain. As a quality control check, we strongly recommend overlaying final binary images onto the original images to verify the accuracy of the segmentation procedure. Once the binary images from each channel are acceptable, a Boolean AND operation is used to create an image of objects that represent the amount of overlap between the two channels. Figure 11.4 gives a schematic representation of this general algorithm.

The number and/or size (or shape) of these overlapped objects can be measured automatically. If the objects in the input channels are expected to be entirely coincident, the number of overlapped objects will be the most useful measurement. Other

Table 11.1 List of freely available FIJI/ImageJ plug-ins for image preprocessing, colocalization analysis, and intensity-modulated display of ratiometric images

Processing step	ImageJ/FIJI plug-in
Shade correction	http://rsb.info.nih.gov/ij/plugins/shading-corrector.html
Bleed-through/cross talk[a]	http://rsbweb.nih.gov/ij/docs/guide/146-29.html#toc-Subsection-29.9
Photobleach correction[a]	http://cmci.embl.de/downloads/bleach_corrector#bleach_corrector
Image registration	http://bigwww.epfl.ch/thevenaz/turboreg/
Rolling-ball background removal[a]	http://rsbweb.nih.gov/ij/docs/guide/146-29.html#toc-Subsection-29.14
Image threshold[a]	http://fiji.sc/wiki/index.php/Auto_Threshold
Colocalization with randomization	http:// imagejdocu. tudor. lu/doku.php?id=plugin:analysis:confined_displacement_algorithm_determines_true_and_random_colocalization_:start

[a]Pre-installed in FIJI

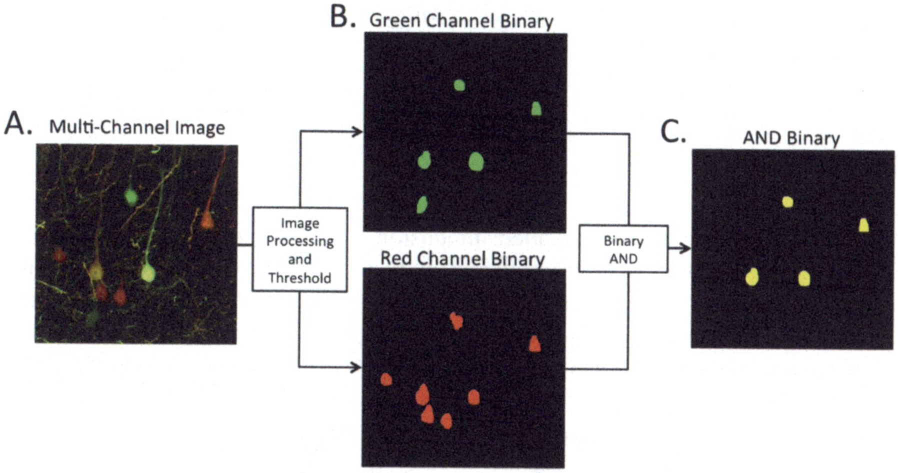

Fig. 11.4 Object-based overlap study. (**a**) Green- and/or red-labeled neurons in situ. Note that only one neuron is visually yellow. Image processing and segmentation operations are used to define the objects of interest on each channel (here, cell bodies). (**b**) The resulting binary images can be automatically counted to determine the number of green objects and the number of red objects. (**c**) Combining these binaries with an AND operation creates a new image containing objects that are both green and red, which again can be automatically counted

definitions of object overlap, such as measuring the distance between the input object's centroids, have also been devised (Lachmanovich et al. 2003).

Additional steps should be taken, however, to determine that the amount of overlap observed is greater than that expected by chance alone. One way to tackle this problem is to leverage the power of repeated random sampling in Monte Carlo-based simulations to show all the possible outcomes. Costes et al. have devised a method which is essentially a block-scrambling technique (Costes et al. 2004) – by randomly shifting blocks with the dimension of the full width at half maximum (FWHM) of the optical point spread function and then recalculating their overlaps many times until the process produces a distribution of chance outcomes. The Costes method will be discussed in greater detail in Sect. 11.8.3 when we delve further into image similarity analysis. However, one important shortcoming of the Costes method in dealing with object-based overlap analysis is that it assumes that the pixels composing the objects in each channel are independently distributed in space, when in fact their positions are often highly dependent, because they are grouped together to form a small number of larger objects. Thus the Costes method tends to inflate randomization of the fluorescent pattern. Rather than block-scrambling, a better strategy for object-based overlap analysis would be to randomize the locations of the input object through a technique called confined displacement algorithm (Ramírez et al. 2010). To implement this method properly, the random locations should be restricted to image areas that are physically plausible, e.g., the objects should always fall within cell boundaries if subcellular objects are being measured. This method can be complex to implement in practice and requires programming knowledge. However, a FIJI/ImageJ plug-in is available that can perform this analysis, as summarized in Table 11.1.

11.7 Scatterplot Analysis

Intensity correlation offers an added dimension over color-based or object-based colocalization analysis. By plotting the intensity distribution of corresponding color 1 and color 2 pixels in a scatterplot, the degree of colocalization can be qualitatively assessed (Fig. 11.5a–d). In the case with a high degree of colocalization (Fig. 11.5a), an increase in the color 1 channel pixel intensity is accompanied by a proportional

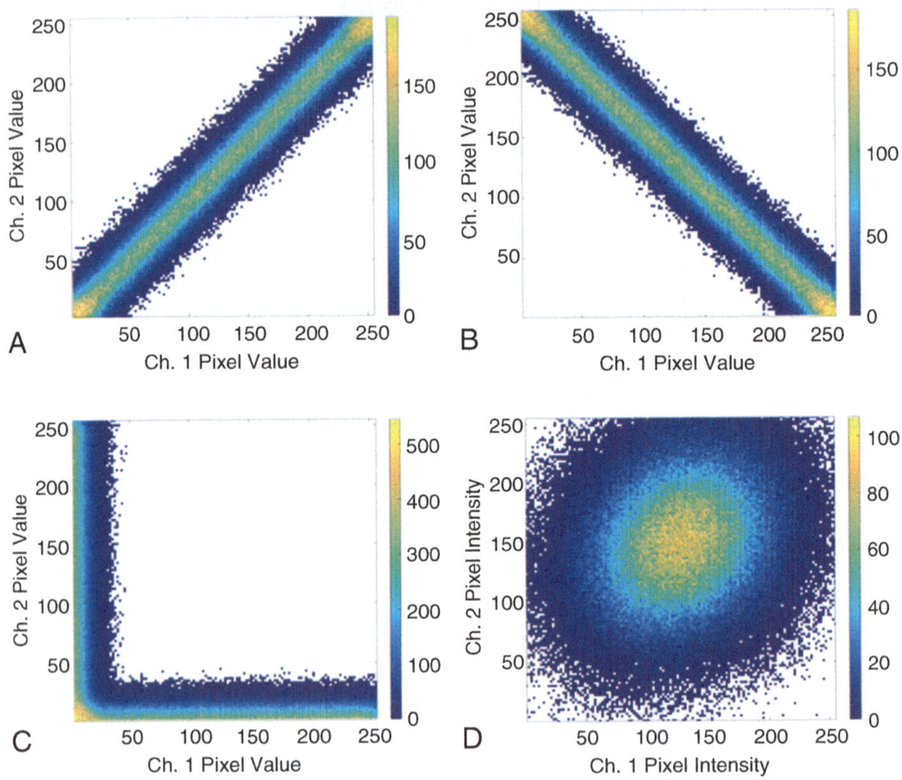

Fig. 11.5 Scatterplots for assessing theoretical pixel correlation. These four scatterplots represent four *theoretical* correlative relationships between signals from two channels. (**a**) Linear correlation. An increase in the image 1 signal intensity is accompanied by a proportional increase in the image 2 signal intensity at each pixel. (**b**) An opposite situation is illustrated. In this case, high image 1 intensity is accompanied by low image 2 intensity and vice versa. This indicates that signals in each image tend toward mutual exclusion, often termed molecular repulsion. (**c**) Zero intersection. The two signals do not interact. (**d**) No correlation between pixels in image 1 and image 2. In this case, no clear relationship between the molecules of interest can be surmised

increase in the color 2 channel pixel intensity. An opposite situation is shown in Fig. 11.5b, whereby regions of high color 1 signal are accompanied by little or no color 2 signal and vice versa. A third case is illustrated in Fig. 11.5c, displaying no intersection between color 1 and color 2. A practical example of such behavior can be seen when labeling two molecules that occupy separate cellular compartments. Lastly, Fig. 11.5d shows two signals with zero correlation, indicating that the interaction of the two signals is random and displays no discernible relationship.

Practically speaking, a scatterplot will nearly always display at least two and sometimes even all three of these cases due to biological factors such as non-specific labeling and image noise. Thus, subtle changes can be hard to observe qualitatively. Therefore, numerical coefficients have been proposed to better quantify changes in colocalization.

11.8 Image Similarity Coefficients

An image similarity coefficient describes, in numerical terms, the degree of overlap or correlation between two image channels. Two indices are commonly used for this purpose. The first measures the degree of synchrony (or correlation); the second quantifies the extent of contribution (which measures co-occurrence). These phenomena should not be confused with each other. Indeed, each can occur in the absence of the other. Before we proceed, our discussion of image similarity analysis hereafter assumes that the two images to be analyzed have been properly background-corrected and that appropriate intensity threshold has been applied.

11.8.1 Pearson's Correlation Coefficient

Pearson's correlation coefficient (PCC) evaluates image similarity by measuring *intensity correlation* between two channels (Pearson 1896). It asks: when a pixel in channel 1 deviates from the mean intensity value, how likely will the corresponding pixel intensity in channel 2 deviate in the same manner? PCC can be expressed as:

$$\mathrm{PCC} = \frac{\sum_i \left(C2_i - \overline{C2}\right)\cdot\left(C1_i - \overline{C1}\right)}{\sqrt{\sum_i \left(C2_i - \overline{C2}\right)^2 \cdot \sum_i \left(C1_i - \overline{C1}\right)^2}}$$

where $C2_i$ and $C1_i$ refer to each pixel in the "color 2" and "color 1" image, respectively, while $\overline{C2}$ and $\overline{C1}$ denote the mean pixel intensity of the entire image in each channel. Note that intensities are expressed with respect to their deviation from the mean values and have a range of −1 to 1. A coefficient of 1 is a complete synchrony, while a value of −1 is 100% anticorrelation.

It is important to note that Pearson's calculation only applies to the groups of pixels in which the two channels intersect, as shown in Fig. 11.6. It does not consider pixels that appear only in one of the channels. This has significant consequence. As can be seen in Fig. 11.6, Pearson's coefficient is insensitive to the percentage area of intersection; it merely concerns itself with how well the pixels between the two channels in the area of intersection "correlate" with one another in intensity fluctuations. While it is immensely powerful to determine how well the intensity signals are correlated between the two channels (an important proxy for "interaction" or "association"), it does not calculate the area of overlap. It is therefore important to use PCC carefully and appropriately.

The power of Pearson's coefficient thus lies in the pixel-by-pixel covariance between the two channels (Adler and Parmryd 2010). Practically speaking, Pearson's coefficient can be particularly sensitive to changes in colocalization patterns when one or both images contain relatively sparse signals across the field of view. But this sensitivity can create alarming situations for unsuspecting biologists. Put simply, if the intensity in either image channel does not vary greatly (such as when labeling a large and homogenous biological structure), the PCC will likely return a nonintuitive result. Let's consider the situation presented in Fig. 11.7. This illustration provides a good (while extreme) example of how the PCC can return an unexpected result. The diagram shows a red object overlapping with a green object. The area of intersection is 50% for each object. The most important feature in this hypothetical situation is that the two objects are saturated in their intensity; thus both objects show no intensity variations. Even if by all biological definitions these two objects are "colocalized," PCC will fail mathematically to return a coefficient. Since there is no variation in intensity, PCC simply cannot compute due to division by zero. Not only does this extreme and hypothetical situation serve as a cautionary example of why image similarity analysis should not be performed on images with saturated pixels, but it also serves as a good example of strong co-occurrence in the absence

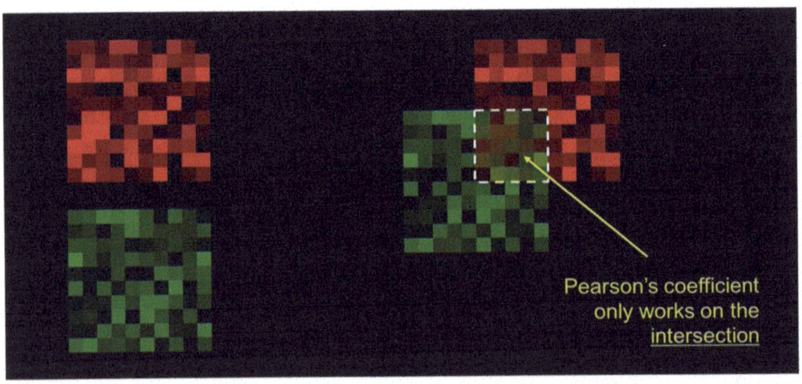

Fig. 11.6 Pearson's coefficient and object intersection. Consider two objects (red and green), each with an area of 100 square pixels, and a quarter of each of the objects intersects one another. Pearson's coefficient would only be applicable in the area in which the two segmented objects intersect

Fig. 11.7 Effects of intensity variability on Pearson's correlation coefficient. Two identical-sized objects (red and green) are shown here with 50% area overlap. These two objects have homogeneous pixel intensity with no variability, thus contributing to a zero value in the denominator for Pearson's correlation coefficient, rendering the calculation impossible. This is an example where co-occurrence of signals does not translate into correlation

of correlation. It is important to remember that PCC relies on each image containing a wide range of pixel values over which to correlate each image. Small (or the lack of) variances in signal can produce problematic PCC values, even if there is a strong signal overlap. Likewise, the inclusion of background pixels will artificially inflate PCC. This is because the background of two images can be highly correlated with each other, not to mention that they can significantly deviate from the mean intensity values. Excluding the image background via the application of threshold will generally return a more intuitive PCC value.

These situations, while extreme, show how a powerful analytical tool can be wrongly interpreted. In short, the PCC is most valuable to biologists when considering images that vary widely in pixel intensity and where the background can be ignored by applying threshold. If the main aim of the analysis is to quantify overlapping area, then PCC is not the right algorithm. However, if the goal of the experiment is to quantify how the two signals correlate with one another in the area in which the two signals intersect, then PCC is the right tool. Another drawback is that the PCC provides no channel-specific information. To help answer this question, we turn to Manders' coefficients.

11.8.2 Manders' Overlap Coefficients (MOCs)

It is common to encounter situations wherein most of the pixels from one channel contribute to colocalization, while the other channel does not. For example, almost all the signal from labeled transcription factor molecules will colocalize with a DAPI-stained nucleus, but not vice versa. A more pertinent measurement in this case may be to quantify the contribution of both fluorescent intensity and area from each channel toward the overlapping region. In such situations, one may prefer turning to Manders' overlap coefficients (MOCs) (Manders et al. 1993). The MOC examines the ratio of intensity-weighted intersecting volume to total object volume. In other words, what percentage of color 1 (and color 2) pixels and cumulative intensity contribute to the color overlap? The MOC is defined as follows:

$$\mathrm{MOC} = \frac{\sum_i C2_i \cdot C1_i}{\sqrt{\sum_i C2_i^2 \cdot \sum_i C1_i^2}}$$

where $C2_i$ and $C1_i$ are defined as previously described. Notice that pixel intensity values are now expressed in absolute terms, not as deviations from the mean as in the PCC. Thus, pixels with zero intensity are intrinsically omitted, eliminating the possibility of negative values. Manders also proposed individual channel coefficients to determine the overlap of each channel on the other:

$$M_1 = \frac{\sum_i C1_{i,\mathrm{coloc}}}{\sum_i C1_i}$$

$$M_2 = \frac{\sum_i C2_{i,\mathrm{coloc}}}{\sum_i C2_i}$$

where

$$C1_{i,coloc} = \{ \begin{matrix} C1_i \; if \; C2_i > 0 \\ 0 \; if \; C2_i = 0 \end{matrix}$$

and

$$C2_{i,\mathrm{coloc}} = \{ \begin{matrix} C2_i \; if \; C1_i > 0 \\ 0 \; if \; C1_i = 0 \end{matrix}$$

Unlike PCC which focuses on the signal correlation within the image *intersection* (Fig. 11.6), MOC is implemented by calculating the coefficients based on the *union* of the two channels, as shown in Fig. 11.8.

Consequently, this creates significant mathematical difference between MOC and PCC. First, MOC is sensitive to the number of pixels that are above threshold. In other words, the *area* covered by above-threshold pixel will affect MOC, but not PCC. MOC can therefore report the percentage of total intensity contributed by each channel to the overlap. This is something PCC cannot deliver. Consider the situation in Fig. 11.9. Figure 11.9b and c has extra green objects in the merged

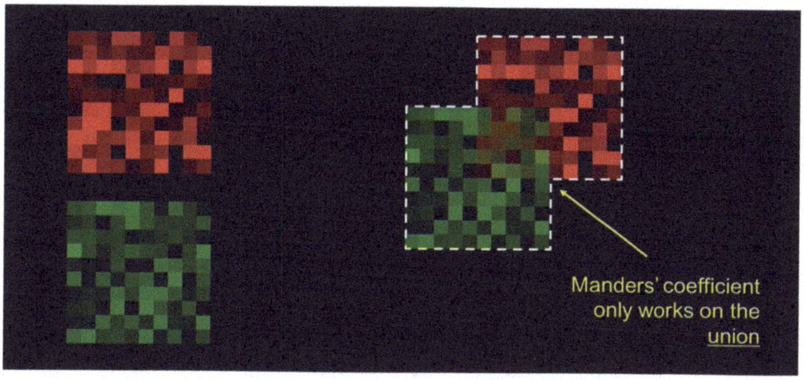

Fig. 11.8 Manders' coefficient and object union. Consider two objects (red and green), each with an area of 100 square pixels, and a quarter of each of the objects intersects one another. Manders' coefficient would be implemented in the total area covered by both signals, which is referred to as the "union" of the two signals

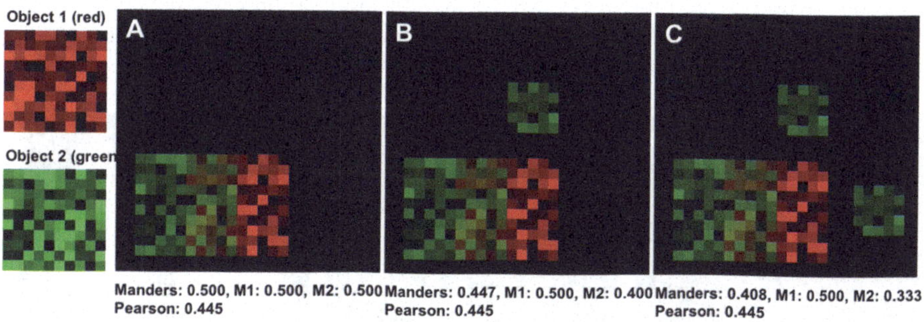

Fig. 11.9 Effects of adding "non-colocalizing" signals on PCC and MOC. Consider two objects (channel 1, red, and channel 2, green), each with 10 × 10 pixels. In panel **a**, 50% of these pixels overlap. This gives rise to an overall and channel-specific MOC of 0.500 and PCC of 0.445. In panels **b** and **c**, extra nonoverlapping green objects are added. While the degree of "colocalization" should decrease in these latter cases, only MOC reports the decrease in signal overlap accordingly. PCC, which only reports signal correlation within the area of intersection, is not affected by these extraneous objects. This is because the area of intersection has not changed in these scenarios

image that are not found in Fig. 11.9a. The extent of colocalization has decreased with the addition of extraneous, nonoverlapping objects. Yet, since the intersection has not changed, PCC remains identical throughout Fig. 11.9a–c. However, since the additional objects affect the "union" of the two signals, the MOC value drops accordingly.

In addition, since MOC is weighted for intensity value, it is therefore more than merely a calculation of area overlap. The brighter the pixels in the overlap, the higher the MOC score. Likewise, it has the intrinsic propensity to diminish the contribution of dim background noise. So MOC is relatively refractory to SNR fluctuation to a certain extent. Yet, this feature is a double-edged sword. Pixels that contain very high intensity values (such as non-specific antibody binding, large shading, or out-of-focus light) will inflate the readout of MOC. A word of caution is that none of these features of MOC preclude the need for background signal subtraction. It is important to remember that pixel intensity is still affected by unwanted signals.

Therefore, the fundamental difference between the MOC and the PCC is how each image pixel contributes to the overall coefficient value. The MOC is based on the absolute magnitude of fluorescence intensity, while the PCC is based on deviation from the mean intensity. Thus, as the intensity of a given pixel decreases, its overall contribution to the total Manders' coefficient is likewise reduced. In the same way, if the background/offset in either image is significant, it will severely skew the resulting MOC to a higher value. In addition, an abundance of high-intensity co-occurring pairs can produce Manders' coefficient that is refractory to other low-intensity pairs, whether the latter are colocalized or not.

To further understand the relationship between the PCC and MOC, we will examine how they assign weight to each pixel intensity pair. Figure 11.10a–b shows images of α-actinin and actin, respectively, in a cultured murine embryonic fibroblast. There are areas where the two proteins show strong colocalization and other locations where α-actinin decorates focal adhesion complexes devoid of actin (Fig. 11.10c). An example of low colocalization is highlighted by yellow arrows, in

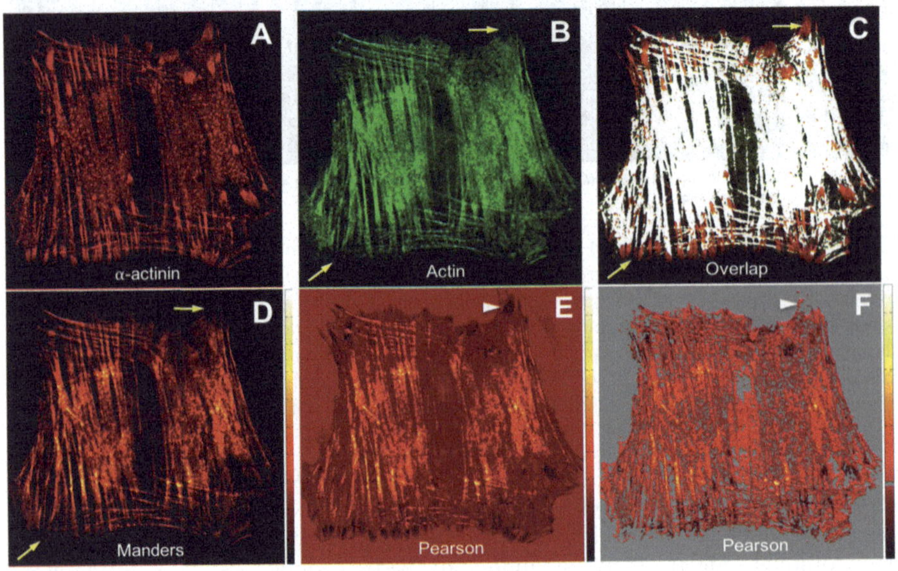

Fig. 11.10 How PCC and MOC weigh biological image. Panels **a** and **b** show α-actinin and actin immunostain in a murine embryonic fibroblast, respectively. (**c**) Overlapped pixels are highlighted in white. Yellow arrows indicate areas where the two proteins show no overlap. (**d**) Relative contribution of each pixel pair to the total MOC. (**e**) Relative contribution of each pixel pair to PCC prior to threshold application. (**f**) The same as **e** but with threshold applied. White arrows show how pixels with negative correlation respond to intensity threshold. Intensity scale bars on the right of **d**–**f** show relative weights assigned to the pixels in each calculated scenario

B and C. Figure 11.10d shows the relative contribution of each pixel pair to the total MOC. As such, areas with poor overlap (yellow arrows) are assigned minimal weight. Similarly, the coverslip area makes no contribution to Manders' algorithm due to low intensity. Most visible cytoskeletal structures, on the other hand, receive equal weight except for a few high-intensity spots where color 1 and color 2 pixels co-occur.

Figure 11.10e shows areas of both positive (red) and negative (black) correlation as calculated by PCC – offering information that is unavailable using the MOC. The white arrowhead points to an example area with negative correlation. Interestingly, however, the PCC assigns a non-negligible weight to the coverslip area. This is due to the fact that, while dim, these pixel intensities deviate from the mean image intensity value and thus increase the PCC due to their correlation. To avoid this, a threshold should be applied to the images (Fig. 11.10f). Refer to the "Image Acquisition and Preprocessing" section above for guidelines on selecting an appropriate threshold value. Note that, regardless of the method being employed, it is important to apply the same methodology to all images that are being compared in a set of experiments.

A question arises, however, as to what pixel intensity pairs contribute most to either the PCC or MOC. To elaborate this important distinction, scoring matrices for each coefficient are plotted in Fig. 11.11. This illustration is adapted from Adler and Parmryd (2010) and indicates the relative "importance" of any given pixel intensity combination, within an 8-bit range. These plots show that pixel pairs with near identical intensities receive the highest relative weighting from both algorithms. However, as the color 2 and color 1 pixel intensities diverge from each other, so do the behaviors of each coefficient. Figure 11.11a shows that the importance given to the MOC

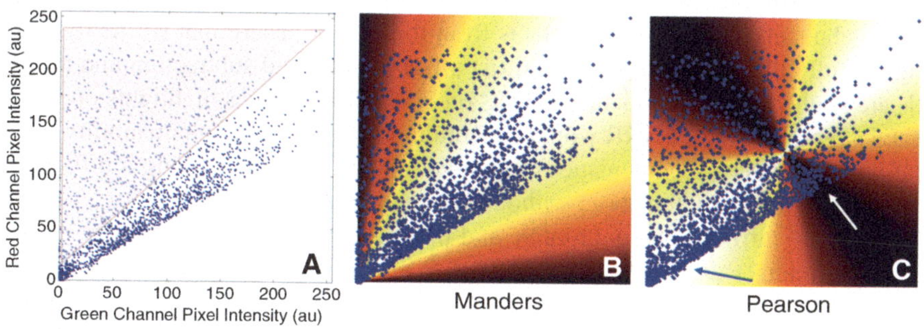

Fig. 11.11 Correlation of individual pixel pairs with the scoring scheme of MOC and PCC. (**a**) Scatterplot derived from the two channels used in Fig. 11.10. The pink area shows pixel pairs that do not fall into the linear correlation. (**b**) The underlying "heat map" indicates how MOC assigns its scores. High-intensity areas (white and yellow) receive high MOC scores, while the darker areas where two channels show decreasing overlap receive low MOC scores. The scatterplot in panel **a** is then transposed onto panel **b**, showing that the pixel pairs with linear correlation receive comparable MOC scores. (**c**) The underlying heat map shows how PCC assigns its scores. The scatterplot from panel **a** is likewise transposed onto panel **c**. Note that in this example, pixels with linear correlation receive varying scores from PCC; the blue arrow shows pixels receiving high PCC scores, while the white arrow shows pixels within the same linear correlation receiving low PPC

decreases proportionately as one channel's intensity changes relative to the other. On the other hand, the weightings of the Pearson's coefficient (Fig. 11.8b) follow a more complex pattern that is determined by two factors: (i) the intensity differences between pixel pairs and (ii) the deviation from each channel's mean intensity.

Figure 11.11a shows a scatterplot derived from the images in Fig. 11.10a and b. It indicates that a portion of the α-actinin signal (color 2 channel) is correlated well with actin (color 1 channel). However, a portion of the α-actinin does not correlate with actin, as highlighted by the pink bounding box. Now let's superimpose this colocalization scatterplot onto the scoring matrix of PCC and MOC (Fig. 11.11b–c). It is clear that the pixels outside of the bounding box receive similar importance from the Manders algorithm (Fig. 11.11b), regardless of their absolute intensities. This linear relationship is intuitive; as long as the molar ratio of actin and α-actinin remains consistent, an equal importance is assigned toward the final MOC.

However, if one traces the linearly correlated pixels within the scatterplot from low-intensity pairs to high-intensity pairs, one would notice that PCC assigns strong significance to the very dim and the very bright pixel pairs (blue arrow) but low significance to the pixels near the mean intensity value (white arrow), as shown in Fig. 11.9c.

11.8.3 Setting Appropriate and Unbiased Intensity Threshold Level

Pearson's coefficient is therefore highly sensitive to a pixel pair's deviation from their respective mean intensity values and also to the difference in pixel intensity between the two channels. To minimize artifacts from this effect, ensure that (i) the dynamic range is fully filled in each image and (ii) offset/background subtraction and thresholds are applied. The dependence of PCC on SNR of the images highlights one of its weaknesses. A decreased SNR would concomitantly decrease the predictability of the relationship between the intensities of the two images, likewise making it difficult to set the proper intensity threshold without introducing observer bias.

In order to introduce a quantitative and unbiased method to set intensity threshold for correlative analysis, Costes et al. (2004) devised a progressive method by calculating the PCC scores across a range of threshold values. In this approach, thresholds for the two images are first projected at near the maximum pixel values for each. The PCC is computed for pixels both above and below the threshold values, and the process is reiterated with incrementally lower threshold values that fall along a linear regression of the scatterplot. This process is repeated until the PCC for the subthreshold pixels approaches zero. This is considered the Costes threshold value for segmentation from the background. It is important to note that this method may not work optimally when the signal of interest is not well correlated in comparison to the background. The Costes thresholding method is not able to identify the clear distinction in the PCC values between the two.

11.8.4 Expanding Correlation Analysis with Spearman's Rank Correlation Coefficient

PCC makes a frequently underappreciated assumption, in that it expects the signals in the intersection to exhibit a linear correlation. PCC assigns its highest magnitude score (+1 or − 1) only when the pixel-intensity relationship is linear. As a result, in a situation wherein two signals are clearly correlated, but with a varying degree of proportions, PCC tends to underestimate the degree of correlation. This shortcoming in PCC is addressed by Spearman's rank correlation coefficient (SRCC) (Adler et al. 2008). In essence, the SRCC computation is equivalent to PCC, except that it is applied to pixel intensity *ranks*, whereby PCC is applied to the intensities themselves (Spearman 1904). SRCC converts pixel value to pixel rank by giving the lowest above-threshold pixel intensity in the image a rank of 1, the next lowest intensity value would then receive a rank of 2, and so on until every intensity value in the image has iteratively received a rank. In cases where multiple pixels have the same intensity, that particular intensity value would be assigned an average score. For instance, if two pixels are tied for the fifth and the sixth lowest value, they would both be ranked with the average value of 5.5. This ranking approach thus linearizes a scatterplot, making the PCC applicable to nonlinear correlation.

Examples of linearly and nonlinearly correlated image pairs can be found in Fig. 11.12. In Fig. 11.12a, two nearly identical images are displayed in columns 1 and 2. The linear correlation is likewise reflected in the intensity scatterplot presented in column 3. The *ranked* scatterplot as implemented by SRCC is shown in column 4. In this example, the SRCC and PCC show an almost identical result as the original intensity values are linearly correlated, so the SRCC ranking makes negligible impact. However, in a situation where the two signals are well correlated but in a nonlinear fashion, such as that presented in Fig. 11.12b, the benefit of SRCC becomes apparent. In Fig. 11.12b, the channel 2 image (red) has been altered from panel 11.12a. This modification produces a very well-correlated, albeit nonlinear relationship between the two images, as indicated by the scatterplot in column 3 of panel 11.12b. As a result of the nonlinearity, the PCC value is lower than expected, even though the two signals show near-perfect correlation. On the other hand, by linearizing the relationship through ranking the pixel intensity values in the two images, the SRCC restores the near-perfect correlation. It is therefore important to note that by eliminating its reliance on a linear relationship, the fidelity of SRCC is not dependent on the assumption that the signals must obey a linear relationship and therefore is more practically versatile in assessing biological signals, whose association rarely exhibits clean and straightforward linearity. SRCC should therefore be preferred over PCC for all practical purposes.

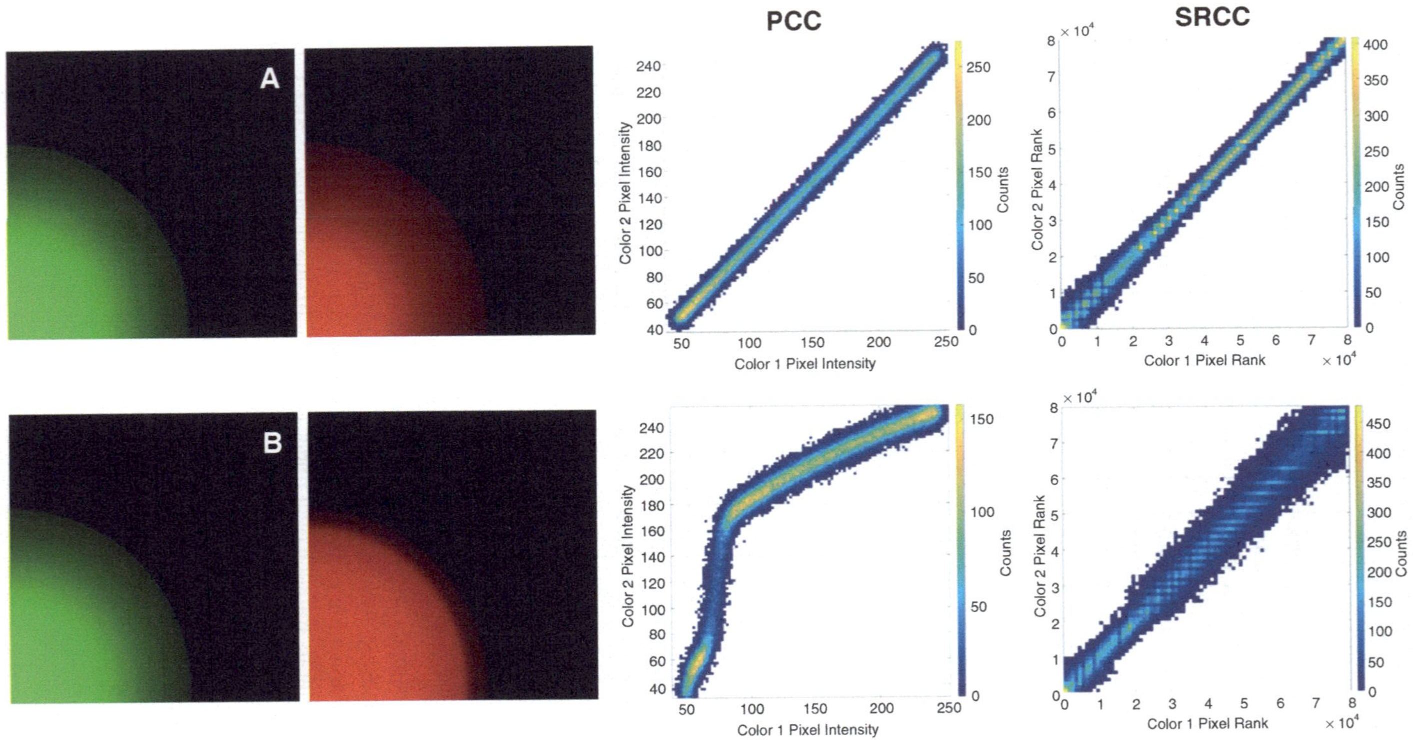

Fig. 11.12 Pearson's and Spearman's rank correlation coefficients. (**a**) Two nearly identical images (except for random noise introduced into the images) are shown in columns 1 and 2. The images in panel **a** exhibit a linear correlation that produces comparable results for PCC and SRCC. (**b**) Another pair of nearly identical images is shown in columns 1 and 2. The green image is identical to the one in panel **a**, while the red image has been slightly altered compared to its counterpart in panel **a** to generate a highly correlated but nonlinear relationship with the green image. This relationship is displayed in column 3 of panel **b**. Due to the nonlinear correlation, PPC score decreases to 0.875, demonstrating the tendency of PCC to underestimate perfectly correlated but nonlinear signals. However, SRCC (column 4) converts the pixel intensity values of the two images into ranks, essentially "linearizing" the correlation. In doing so, this reflects the near-perfect correlation more accurately

11.9 Global Factors Affecting Molecular Clustering

The study of image similarity, however, is more complex than obtaining readout from colocalization coefficients, especially when these indices are often considered a representation of spatial relationship of biomolecules. Even with the appropriate implementation of the optimal quantitative index, these measurements, which are statistical and probability analyses, do not consider other biological variables that will impact the outcomes. It is not uncommon that a global factor can vastly affect the pattern of molecular clustering. These global factors can result from (but are not limited to) drastic cell shape changes, molecules being confined to structural constraints during transportation, cellular polarization, macromolecular realignment, rapid variations in molecular intensity due to expression levels, etc. These factors, while biologically significant, may skew the *apparent* image similarity measurements such as colocalization and ratiometry, leading to misinterpretation of the *intrinsic*, local molecular interaction.

Unfortunately, these "global biases" are rarely decoupled from the local molecular relationship prior to making a data interpretation (Adler and Parmryd 2010; Bolte and Cordelieres 2006; Costes et al. 2004; Dunn et al. 2011; Tambe et al. 2011; Yannis et al. 2015). These confounding global effects often skew, if not inflate (or conceal), the underlying molecular interaction at the regional level. For example, a metastatic cancer cell squeezing itself through tight spaces in between the extracellular matrix and the endothelium during invasion tends to "indiscriminately" squeeze a lot of biomolecules within the narrowest part of itself, as it actively changes its shape to overcome the size constraint of the obstacle. This process will inevitably increase the "colocalization" readout due to molecular crowding, regardless of what image similarity coefficient (be it PCC, MOC, SRCC, etc.) is used. This increase may not be a result of underlying local molecular forces but merely due to the drastic change of cell shape. So some mathematical, heuristic approach must be devised to uncouple that global bias. There are several approaches to this problem (Helmuth et al. 2010; Lagache et al. 2015). One way is to normalize the readout to simulated data and identify the global bias as a confounding factor for subsequent elimination from the calculation (Vander Weele and Shpitser 2013).

Yet, eliminating the global effect outright would also mean throwing away equally essential biological data. These global biases frequently are the real biological effects (e.g., cell shape changes) that define the molecular events being interrogated. A more desirable method would be one that could simultaneously decouple the global bias from the local interaction yet is capable of scoring both factors. In light of that, Zaritsky et al. (2017) have proposed an elegant algorithm, *DeBias*. The underlying assumption of DeBias is that the apparent spatial relationship between two variables is the sum effects of a global bias and a local interaction component. Briefly, to decouple the two factors, the algorithm randomizes the two variables that carry orientation information (Drew et al. 2015; Nieuwenhuizen et al. 2015) and then resamples the distribution of interactions ("resampled"). In this case, in a scenario with neither global bias nor local interaction contributions, the randomized alignment would be uniformly distributed ("uniform"). The power of the randomization step is that it allows the global bias factor to be easily extracted, as it decouples the effect of any local interactions from the global bias. In this case, the global bias is defined as the difference between the "uniform" and the randomized (and "resampled") distributions. On the other hand, the effects of local interaction are represented by the difference of dissimilarity between the observed and "uniform" distributions and dissimilarity between the "resampled" and "uniform" distributions. This freely available (https://debias.biohpc.swmed.edu) mathematical tool should therefore be in the repertoire of anyone interested in analyzing image similarity. It is also important to note that in addition to biological factors, global bias may also be introduced by non-biological factors, including spatially correlated noise and/or detector offset.

11.10 Take-Home Message

The seemingly simple concept of measuring image similarity is complex. As discussed, no colocalization coefficient is perfect, as none of them really measure the misnomer term "colocalization" per se. While various arguments exist in favor of one coefficient over another, such discussions may fail to consider practical experimental issues – unequal antibody affinities, naturally imbalanced protein stoichiometry, and the association of low-abundance proteins with a large, bright biological structure. Overall, it is important to remember that image similarity analysis never directly measures molecular interaction. It gives a scoring system to evaluate the relationship between different molecular images of the same sample. Image similarity studies are only meaningful if these coefficients reproducibly show change that can be related to experimental intervention or compared to good controls.

Another important point discussed in this chapter is how the resolution of the microscope impacts the analysis of image similarity. While recent advances may offer biological details unattainable prior to the advent of super-resolution microscopy (discussed in Chap. 8), they may also create confusion for end users. However, there are commercial instruments that merely enhance the resolution approximately 1.5–2 fold. These categories of instruments or techniques, which include structured illumination microscopy (SIM), image scanning microscopy, and the closely related pixel reassignment techniques (Sheppard et al. 2013; York et al. 2013), would indeed improve the accuracy of image similarity analyses such as SRCC and MOC. But beyond that, localization-based super-resolution techniques (Betzig et al. 2006; Rust et al. 2006) and, also to a lesser degree, stimulated emission depletion (STED) microscopy (Hell and Wichmann 1994; Klar et al. 2000) offer resolution near the molecular level and call into question the entire exercise of analyzing colocalization at the realm described by the Pauli exclusion principle. In fact, super-resolution microscopy raises questions about whether conventional analyses on image similarity can be implemented in images with sub-diffraction resolution. Unfortunately, an in-depth discussion of various new techniques to characterize and quantify signal relationships is beyond the scope of this chapter. We point readers to several original papers and reviews that deal specifically with leveraging spatial statistics-based techniques to quantitatively assess molecular interactions (Coltharp et al. 2014; Lagache et al. 2015; Nicovich et al. 2017).

In summary, the analysis of image similarity hinges on five important decisions on the part of the user: (i) appropriate image acquisition and processing parameters, (ii) correct object segmentation, (iii) informed use of the optimal coefficient for image similarity analyses, (iv) confidence in the statistical significance in the read-out of the coefficient, and (v) decoupling of global bias from local interaction. Only when all five decisions are made wisely is one able to deduce meaningful biological information from the analysis of image similarity.

References

Adler J, Parmryd I (2010) Quantifying colocalization by correlation: the Pearson correlation coefficient is superior to the Mander's overlap coefficient. Cytometry A 77(8):733–742. Retrieved from http://doi.wiley.com/10.1002/cyto.a.20896

Adler J, Pagakis SN, Parmryd I (2008) Replicate-based noise corrected correlation for accurate measurements of colocalization. J Microsc 230(1):121–133. https://doi.org/10.1111/j.1365-2818.2008.01967.x

Andersson H, Baechi T, Hoechl M, Richter C (1998) Autofluorescence of living cells. J Microsc 191(1):1–7. https://doi.org/10.1046/j.1365-2818.1998.00347.x

Betzig E, Patterson GH, Sougrat R, Lindwasser OW, Olenych S, Bonifacino JS et al (2006) Imaging intracellular fluorescent proteins at nanometer resolution. Science (New York, NY) 313(5793):1642–1645. Retrieved from http://www.sciencemag.org/cgi/doi/10.1126/science.1127344

Bolte S, Cordelieres FP (2006) A guided tour into subcellular colocalisation analysis in light microscopy. J Microsc 224(3):13–232. https://doi.org/10.1111/j.1365-2818.2006.01706.x

Chew T-L, Chisholm R (2005) Monitoring protein dynamics using FRET-based biosensors BT - live cell imaging: a laboratory manual. In: Goldman RD, Specter DL (eds) *Live cell imaging: A laboratory manual*. CSHL Press, pp 145–157. Retrieved from http://www.worldcat.org/title/live-cell-imaging-a-laboratory-manual/oclc/803364869

Coltharp C, Yang X, Xiao J (2014) Quantitative analysis of single-molecule superresolution images. Curr Opin Struct Biol. https://doi.org/10.1016/j.sbi.2014.08.008

Costes SV, Daelemans D, Cho EH, Dobbin Z, Pavlakis G, Lockett S (2004) Automatic and quantitative measurement of protein-protein colocalization in live cells. Biophys J 86(6):3993–4003. https://doi.org/10.1529/biophysj.103.038422

Dickinson ME, Bearman G, Tille S, Lansford R, Fraser SE (2001) Multi-spectral imaging and linear unmixing add a whole new dimension to laser scanning fluorescence microscopy. BioTechniques 31(6):1272–1278. https://doi.org/citeulike-article-id:2534922

Drew NK, Eagleson MA, Baldo DB, Parker KK, Grosberg A (2015) Metrics for assessing cytoskeletal orientational correlations and consistency. PLoS Comput Biol 11(4). https://doi.org/10.1371/journal.pcbi.1004190

Dunn KW, Kamocka MM, McDonald JH (2011) A practical guide to evaluating colocalization in biological microscopy. Am J Physiol Cell Physiol 300(4):C723–C742. Retrieved from http://ajpcell.physiology.org/cgi/doi/10.1152/ajpcell.00462.2010

Glasbey CA (1993) An analysis of histogram-based thresholding algorithms. Graphical Models Image Process 55(6):532–537. https://doi.org/10.1006/gmip.1993.1040

Haaland DM, Jones HDT, Van Benthem MH, Sinclair MB, Melgaard DK, Stork CL et al (2009) Hyperspectral confocal fluorescence imaging: exploring alternative multivariate curve resolution approaches. Appl Spectrosc 63(3):271–279. https://doi.org/10.1366/000370209787598843
Hell SW, Wichmann J (1994) Breaking the diffraction resolution limit by stimulated emission: stimulated-emission-depletion fluorescence microscopy. Opt Lett 19(11):780. https://doi.org/10.1364/OL.19.000780
Helmuth JA, Paul G, Sbalzarini IF (2010) Beyond co-localization: inferring spatial interactions between sub-cellular structures from microscopy images. BMC Bioinformatics 11. https://doi.org/10.1186/1471-2105-11-372
Kapur JN, Sahoo PK, Wong AKC (1985) A new method for gray-level picture thresholding using the entropy of the histogram. Comput Vis Graph Image Process 29(3):273–285. https://doi.org/10.1016/0734-189X(85)90125-2
Klar TA, Jakobs S, Dyba M, Egner A, Hell SW (2000) Fluorescence microscopy with diffraction resolution barrier broken by stimulated emission. Proc Natl Acad Sci 97(15):8206–8210. https://doi.org/10.1073/pnas.97.15.8206
Lachmanovich E, Shvartsman DE, Malka Y, Botvin C, Henis YI, Weiss AM (2003) Co-localization analysis of complex formation among membrane proteins by computerized fluorescence microscopy: application to immunofluorescence co-patching studies. J Microsc 212(2):122–131. https://doi.org/10.1046/j.1365-2818.2003.01239.x
Lagache T, Sauvonnet N, Danglot L, Olivo-Marin JC (2015) Statistical analysis of molecule colocalization in bioimaging. Cytometry A 87(6):568–579 https://doi.org/10.1002/cyto.a.22629
Leong FJWM, Brady M, O'D McGee J (2003) Correction of uneven illumination (vignetting) in digital microscopy images. J Clin Pathol 56(8):619–621. https://doi.org/10.1136/jcp.56.8.619
Manders EMM, Verbeek FJ, Aten JA (1993) Measurement of co-localization of objects in dual-colour confocal images. J Microsc 169(3):375–382. https://doi.org/10.1111/j.1365-2818.1993.tb03313.x
Nakamura J (2005) Image sensors and signal processing for digital still cameras. CRC Press. https://doi.org/10.1201/9781420026856
Nicovich PR, Owen DM, Gaus K (2017) Turning single-molecule localization microscopy into a quantitative bioanalytical tool. Nat Protoc. https://doi.org/10.1038/nprot.2016.166
Nieuwenhuizen RPJ, Nahidiazar L, Manders EMM, Jalink K, Stallinga S, Rieger B (2015) Co-orientation: quantifying simultaneous co-localization and orientational alignment of filaments in light microscopy. PLoS One 10(7). https://doi.org/10.1371/journal.pone.0131756
North AJ (2006) Seeing is believing? A beginners' guide to practical pitfalls in image acquisition. 172(1):9–18. Retrieved from http://eutils.ncbi.nlm.nih.gov/entrez/eutils/elink.fcgi?dbfrom=pubmed&id=16390995&retmode=ref&cmd=prlinks
Ojeda-Castaneda J, Ramos R, Noyola-Isgleas A (1988) High focal depth by apodization and digital restoration. Appl Opt 27(12):2583–2586. https://doi.org/10.1364/AO.27.002583
Otsu N (1975) A threshold selection method from Gray-level histograms. Automatica, 11, 23-27. - references - scientific research publish. Automatica 11:23–27. Retrieved from http://www.scirp.org/(S(lz5mqp453edsnp55rrgjct55))/reference/ReferencesPapers.aspx?ReferenceID=1244209
Pauli W (1925) Über den Zusammenhang des Abschlusses der Elektronengruppen im Atom mit der Komplexstruktur der Spektren. Z Phys 31:765–783. Retrieved from http://adsabs.harvard.edu/cgi-bin/nph-data_query?bibcode=1925ZPhy...31..765P&link_type=EJOURNAL
Pearson, K. (1896). Mathematical contributions to the theory of evolution. III. Regression, heredity, and Panmixia. Philos Trans R Soc Lond A, Containing Papers of a Mathematical or Physical Character, 187:253–318. Retrieved from http://www.jstor.org/stable/90707
Peters RI (1995) A new algorithm for image noise reduction using mathematical morphology. IEEE Trans Image Process 4(5):554–568. https://doi.org/10.1109/83.382491
Piston DW, Kremers G-J (2007) Fluorescent protein FRET: the good, the bad and the ugly. Trends Biochem Sci 32(9):407–414. https://doi.org/10.1016/j.tibs.2007.08.003

Ramírez O, García A, Rojas R, Couve A, Härtel S (2010) Confined displacement algorithm determines true and random colocalization in fluorescence microscopy. J Microsc 239(3):173–183. https://doi.org/10.1111/j.1365-2818.2010.03369.x
Rust MJ, Bates M, Zhuang X (2006) Sub-diffraction-limit imaging by stochastic optical reconstruction microscopy (STORM). Nat Methods 3(10):793–795. Retrieved from http://www.nature.com/nmeth/journal/v3/n10/abs/nmeth929.html
Schermelleh L, Heintzmann R, Leonhardt H (2010) A guide to super-resolution fluorescence microscopy. 190(2):165–175. Retrieved from http://www.jcb.org/cgi/doi/10.1083/jcb.201002018
Sheppard CJR (2017) Resolution and super-resolution. Microsc Res Tech. Retrieved from http://doi.wiley.com/10.1002/jemt.22834
Sheppard CJR, Mehta SB, Heintzmann R (2013) Superresolution by image scanning microscopy using pixel reassignment. Opt Lett 38(15):2889–2892. Retrieved from http://eutils.ncbi.nlm.nih.gov/entrez/eutils/elink.fcgi?dbfrom=pubmed&id=23903171&retmode=ref&cmd=prlinks
Solomon C, Breckon T (2011) Fundamentals of digital image processing: a practical approach with examples in matlab. https://doi.org/10.1002/9780470689776
Spearman C (1904) The proof and measurement of association between two things. Am J Psychol 15(1):72. https://doi.org/10.2307/1412159
Stelzer EHK (1998) Contrast, resolution, pixelation, dynamic range and signal-to-noise ratio: fundamental limits to resolution in fluorescence light microscopy. J Microsc 189(1):15–24 https://doi.org/10.1046/j.1365-2818.1998.00290.x
Sternberg (1983) Biomedical image processing. Computer 16(1):22–34. https://doi.org/10.1109/MC.1983.1654163
Tambe DT, Corey Hardin C, Angelini TE, Rajendran K, Park CY, Serra-Picamal X, et al. (2011) Collective cell guidance by cooperative intercellular forces. Nat Mater 10(6):469–475. https://doi.org/10.1038/nmat3025
Taylor AB, Ioannou MS, Watanabe T, Hahn K, Chew T-L (2017) Perceptually accurate display of two greyscale images as a single colour image. J Microsc 268(1):73–83. Retrieved from http://eutils.ncbi.nlm.nih.gov/entrez/eutils/elink.fcgi?dbfrom=pubmed&id=28556922&retmode=ref&cmd=prlinks
Vander Weele TJ, Shpitser I (2013) On the definition of a confounder. Ann Stat 41(1):196–220. https://doi.org/10.1214/12-AOS1058
Vicente NB, Diaz Zamboni JE, Adur JF, Paravani EV, Casco VH (2007) Photobleaching correction in fluorescence microscopy images. J Phys Conf Ser 90(1):0–8. https://doi.org/10.1088/1742-6596/90/1/012068
Waters JC (2009) Accuracy and precision in quantitative fluorescence microscopy. J Cell Biol 185(7):1135–1148. Retrieved from http://eutils.ncbi.nlm.nih.gov/entrez/eutils/elink.fcgi?dbfrom=pubmed&id=19564400&retmode=ref&cmd=prlinks
Yannis K, Inna K, Marino Z (2015) A probabilistic method to quantify the colocalization of markers on intracellular vesicular structures visualized by light microscopy. *AIP Conference Proceedings* 1641:580–587. Retrieved from http://aip.scitation.org/doi/pdf/10.1063/1.4906025
York AG, Chandris P, Nogare DD, Head J, Wawrzusin P, Fischer RS et al (2013) Instant super-resolution imaging in live cells and embryos via analog image processing. Nat Methods 10(11):1122–1126. Retrieved from http://www.nature.com/doifinder/10.1038/nmeth.2687
Zaritsky A, Obolski U, Gan Z, Reis CR, Kadlecova Z, Du Y et al (2017) Decoupling global biases and local interactions between cell biological variables. eLife 6 Retrieved from http://eutils.ncbi.nlm.nih.gov/entrez/eutils/elink.fcgi?dbfrom=pubmed&id=28287393&retmode=ref&cmd=prlinks
Zitová B, Flusser J (2003) Image registration methods: a survey. Image Vis Comput. https://doi.org/10.1016/S0262-8856(03)00137-9

Chapter 12
Ethics and Resources

W. Gray (Jay) Jerome and Robert L. Price

12.1 Introduction

In the opening sections of this text, we noted that often students and technologists are sent to a confocal microscope to collect images without a full understanding of how the system works. Unfortunately, an additional factor affecting image collection is that often the students and technologists are given a preconceived idea of what the final image or data should look like. This may put pressure on a confocal microscope user to produce data to match expectations rather than that which is seen through the microscope. It is fairly easy with a confocal system to increase the apparent labeling by changing the sensitivity of a detector or increasing the laser intensity so more photons are generated. These factors can make an image appear brighter than that seen through the microscope and result in the wrong interpretation of the data. Several other factors such as summing images and slowing the scan speed have similar effects.

It is also possible to mishandle digital images resulting in the inadvertent creation of artifacts. It is essential that accepted practices and care be taken when processing confocal images and that guidelines presented below are followed.

W. G. Jerome (✉)
Department of Pathology, Microbiology and Immunology,
Vanderbilt University School of Medicine, Nashville, TN, USA
e-mail: Jay.Jerome@Vanderbilt.edu

R. L. Price
Department of Cell Biology and Anatomy,
University of South Carolina School of Medicine, Columbia, SC, USA
e-mail: Bob.Price@uscmed.sc.edu

W. G. Jerome, R. L. Price (eds.), *Basic Confocal Microscopy*,
https://doi.org/10.1007/978-3-319-97454-5_12

12.2 Imaging Ethics

The problem of biased collection of images due to a preconceived idea of how data should appear is not a new problem as illustrated by a quote from the 1742 book *The Microscope Made Easy* (shortened title) by Henry Baker: "When you employ the microscope, shake off all prejudice, nor harbour any favorite opinions; for, if you do, 'tis not unlikely fancy will betray you into error, and make you see what you wish to see." With the sophistication of today's microscopes and the power of digital imaging techniques, accurate presentation of data and avoiding the temptations of seeing what we wish to see and of "cleaning up" images have never been more important. As noted by North (2006), "All data are subject to interpretation," and many errors are introduced in complete innocence. A perfect example, as discussed throughout this text and by North, is whether the presence of yellow in a red/green merged image represents true colocalization. Hopefully at this point, it is recognized that many factors affect and alter the colors (signal) in a fluorescence image.

Unfortunately, it is not always in complete innocence that images are used inappropriately. The number of cases involving questionable images has been on a consistent upward trend since the first Department of Health and Human Resources Office of Research Integrity (ORI) Annual Report reporting period of 1989–1990 when only 2.5% of the cases involved questionable images (Krueger 2005). Perhaps the most famous case of unethical use of digital images is that of Woo Suk Hwang's manipulation of digital images of stem cells in his 2005 science paper (Hwang et al. 2005). As noted in the first edition of this book, other examples of questionable image manipulation, including confocal images, have been reported as detailed in the 2008 ORI Annual Report (Federal Register Volume 73 Number 196 page 58968). The 2009 ORI Annual Report further indicates that 68% of the cases opened by the ORI involve cases of questioned images (Krueger 2009). Unfortunately the trend for research misconduct resulting from image manipulation continues. From 2011 to 2015, there were 45 cases investigated by the ORI where findings of research misconduct associated with image manipulation were confirmed. While these were not categorized by type of image manipulation, they did involve cases of deleting or inserting parts of micrographs or reusing and relabeling unrelated images (http://ori.hhs.gov March 2017 newsletter 24:1).

With the ease of image collection in confocal microscopy, continual diligence in handling confocal digital images through the various steps of collection and processing in Photoshop, AMIRA, and other programs is essential. Our images are our data, and just as it would not be acceptable to vary or alter pipetting volumes when loading a Western blot, it is not acceptable to vary or alter pixel data. Equally important is that early in training all are taught the importance of maintaining detailed laboratory notebooks for experiments, but few are taught the importance of the proper handling and archiving of digital images. An article by Goldenring (2010) emphasizes the importance of keeping original image files collected from the microscope and for maintaining non-flattened archival Photoshop files of all image manipulations. It was only through proper archiving of all images and image processing that Dr. Goldenring and members of his laboratory were able to disprove reviewer and editorial charges of misconduct regarding confocal images.

Maintaining records of all original files and instrument parameters emphasizes a very important point concerning archiving of confocal data. While the editor-in-chief of *Microscopy and Microanalysis*, the journal of the Microscopy Society of America, Dr. Price had a case where a very qualified reviewer refused to review a manuscript until all metadata collected with the images in the manuscript were provided. Once the metadata were provided, the paper was accepted for publication emphasizing the importance of having this information available. Although often cumbersome, it is essential when collecting a confocal image that an original copy of the data in the proprietary format of the manufacturer that includes all collection parameters such as laser intensity, detector settings, scan parameters, etc. is stored and available for review. Should questions concerning the data integrity arise, this is the only mechanism to demonstrate all data has been processed properly. Many institutions are now requiring the archiving of all data in centralized data banks, and this has become an important component of image publication (Price 2014).

12.3 Journal and Office of Research Integrity Guidelines

Most journals have published guidelines for acceptable processing of digital images. The ORI has also published a series of guidelines for handling digital images that are essential for acquisition and publication of confocal images (http://ori.dhhs.gov/products/RIandImages/guidelines). While many of these topics have already been discussed, because of the importance of the topic, some redundancy is justified, and the full list of 12 guidelines is given below. Expanded discussions of each topic are available on the ORI website as well as in a paper by Cromer (2010) entitled Avoiding Twisted Pixels: Ethical Guidelines for the Appropriate Use and Manipulation of Scientific Digital Images. Cromey's paper gives many specific examples of how digital images should be processed and manipulations reported in a manuscript.

1. Treat images as data: Scientific digital images are data that can be compromised by inappropriate manipulations.
2. Save the original: Manipulations of digital images should always be done on a copy of the raw image. The original must be maintained.
3. Make simple adjustments: Simple adjustments to the entire image are usually acceptable. Reasonable adjustments using software tools like brightness and contrast, levels, and gamma are usually appropriate.
4. Cropping is usually acceptable. Legitimate reasons for cropping include centering an area of interest, trimming empty space around the edges of an image, and removing debris from the edge of an image. Questionable forms of cropping include editing which can create bias such as removal of dead or dying cells leaving only healthy cells. Motivation for cropping should always be a primary consideration. Is the image being cropped to improve its composition or to hide something?

5. Comparison of images should only involve images that have been collected under identical conditions of preparation and acquisition, and any post imaging processing should be identical for all images involved.
6. Manipulation should be done on the entire image. It is not acceptable to alter one area of an image to enhance a specific feature.
7. Filters such as smoothing and sharpening functions degrade data and are not recommended. If filters are used, this needs to be reported in the Methods and Materials for the paper.
8. Cloning or copying pixels from other images or a different area of the same image should not be done. Copying pixels to create structures in an image which did not exist is research misconduct.
9. Intensity measurements are difficult to perform and must be done on image pixels collected and processed in an identical manner. Measurements should always be performed on the raw data.
10. Avoid the use of lossy image compression formats. TIFF (Tif) is the most widely accepted format for images, but always check the journal format prior to submitting images. In general, the JPEG format should never be used for collection of scientific images.
11. Confocal images include X, Y, and Z dimensions, and digitally altering the size (magnification) in any of these directions will alter the data. Care must be used to sample or collect images according to the Nyquist Theorem. If doubt exists concerning Nyquist sampling, then oversampling should be performed.
12. Altering the number of pixels in an image to make images fit a page can result in software interpolation of data which will create a new resolution and possibly intensity value for pixels. This can result in aliasing artifacts.

As noted by Rossner and Yamada (2004), each image should be an accurate representation of what was observed through the microscope. Manipulating images to make them more convincing can change data that others might be interested in or interpret differently.

12.4 Microscopy Society of America Statement on Ethical Digital Imaging

All of the above considerations have led the Microscopy Society of America to issue a Statement on Ethics in Digital Imaging. Although some of the terminology concerning storage media is outdated due to the rapid development of technology and very large "Big Data" files, the basic premise of the statement provides guidance on how to properly handle confocal images:

> Ethical digital imaging requires that the original uncompressed image file be stored on archival media (e.g. CD-R) without any image manipulation or processing operations. All parameters of the production and acquisition of the files, as well as any subsequent processing steps, must be reported to ensure reproducibility.

> Generally acceptable (non-reportable) imaging operations include gamma correction, histogram stretching, and brightness and contrast adjustments. All other operations (such as unsharp masking, Gaussian blur, etc) must be directly identified by the author as part of the experimental methodology. However, for any image data that is used for subsequent quantification, all imaging operations must be reported.

Even when using the simplest and generally acceptable imaging operations, one should always be aware of the changes in pixel and voxel values. Figure 12.1 illustrates the changes in pixel value when performing a simple adjustment of contrast and brightness, an image manipulation function most, if not all of us, routinely use. These functions involve grouping a range of values at the low or high end of the histogram and reassigning all of the values within the range a "0" or "255," respectively. This creates more black or white in the image. Since values are changed, it is essential that any quantitative analysis is completed prior to performing image enhancement functions. Cromer (2010) provides a number of other examples illustrating the effects of acceptable imaging operations on pixel and voxel values and how these may change the data.

If care is taken in the preparation and collection of confocal images and all of the above guidelines are followed for processing the images, few problems concerning the ethics of how your data was collected and processed should arise. Always remember the Confocal Commandments and keep an original unaltered archived file of your data and perform all image manipulations using a copy of the original file.

12.5 Available Resources

There are a number of websites maintained by manufacturers that are excellent resources for fluorescence and confocal microscopy. Among these are:

Nikon Microscopy U: http://www.microscopyu.com/articles/confocal/
Olympus Microscopy Resource Center:
http://www.olympusmicro.com/primer/techniques/fluorescence/fluorhome.html
Zeiss Online Campus: http://zeiss-campus.magnet.fsu.edu/tutorials/index.html

The Confocal Microscopy List Serve (confocalmicroscopy@lists.umn.edu) is dedicated to topics in confocal imaging, while the Microscopy Society of America also maintains a List Serve (http://www.microscopy.com/) addresses topics in confocal as well as all other forms of microscopy.

There are also a number of books available which cover many aspects of confocal imaging. However, many of these are relatively old and/or cover various techniques and applications rather than basic information on how to operate a confocal system. The rate at which confocal technology is being developed also makes it difficult to keep up with advances in confocal microscope hardware. A chronological listing of some of the books we have found very useful is as follows:

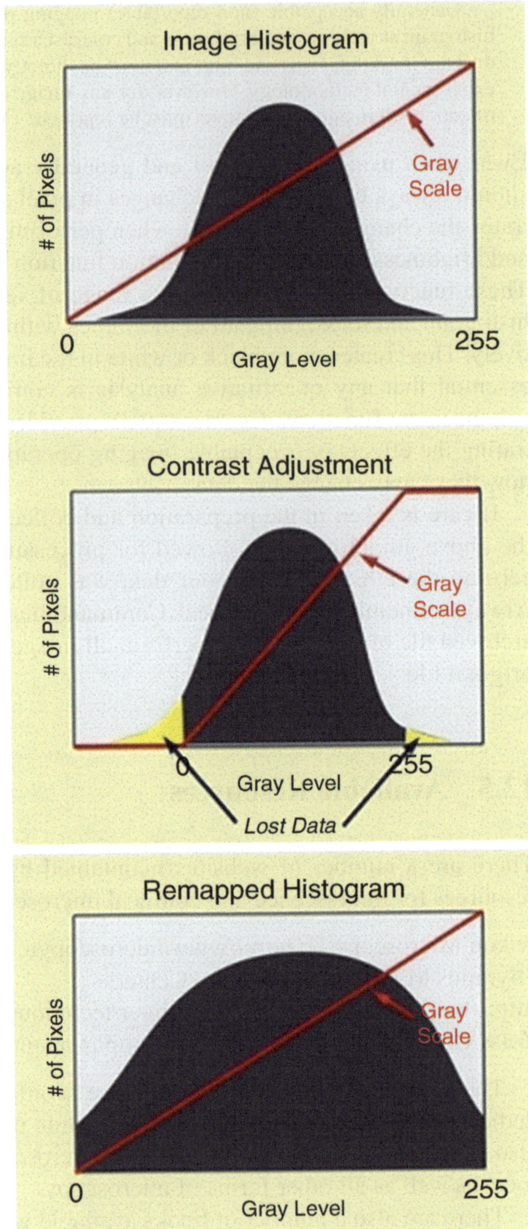

Fig. 12.1 (Top) Histogram representing an image that does not use the full dynamic range with few pixel values near "0" or "255" so few black or white values would be present. (Middle) By using the contrast and brightness controls the gray scale can be changed to group a large number of values near the low and high ends of the image. (Bottom) When these grouped values are remapped to "0" or "255" values, data is lost on both ends of the histogram

Stevens JK, Mills LR, Trogadis JE (eds) (1994) Three dimensional confocal microscopy: volume investigation of biological systems. Academic Press, 507 pp. (Good practical information; nice section on spinning disk microscopes)

Gu M (1996) Principles of three-dimensional imaging in confocal microscopes. World Scientific Publishing Company, Inc., 352 pp. (Very technical but good for advanced students)

Paddock SW (ed) (1999) Confocal microscopy (Methods in molecular biology volume 1212). Humana Press, 464 pp. (Very good practical protocols as well as basics)

Alberto D (ed) (2001) Confocal and two-poton microscopy: foundations, applications and advances. John Wiley and Sons, Inc., 576 pp. (Excellent treatise of some advanced confocal imaging techniques)

Matsumoto B (ed) (2002) Methods in cell biology volume 70: cell biological applications of confocal microscopy, 2nd edn. Academic Press, 507 pp. (Information on system hardware and applications to some specific biological organisms and systems.)

Hibbs A (2004) Confocal microscopy for biologists. Springer, 474 pp. (Good for beginners and advanced; great appendix of information; live cell imaging)

Pawley JB (ed) (2006) Handbook of biological confocal microscopy, 3rd edn. Springer, 988 pp. (A very good comprehensive review of advanced confocal microscopy).

Michael Conn P (2010) Techniques in confocal microscopy, 1st edn. Academic Press, 544 pp. (Presents of range of uses for confocal imaging).

Price RL, Gray Jerome W (2011) Basic confocal microscopy. Springer, 302 pp. (The first edition of the current book)

Paddock S (ed) (2014) Confocal microscopy. Methods and protocols. Humana Press, 375 pp. (Presents a range of techniques using various biological samples)

Liu J, Tan J (2016) Confocal microscopy. Morgan and Claypool, 90 pp; eBookISBN 9781681743387 (Primarily covers techniques used in industrial metrology and scale resolution in bio-imaging)

Gonzalez S (2017) Reflectance confocal microscopy of cutaneous tumors, 2nd edn. CRC Press, 535 pp. (Describes the use of reflectance confocal imaging in the examination of skin tumors)

Hopefully the book you are currently reading will be added to your list of useful resources for confocal imaging.

Additional Literature Cited

Baker H (1742) The Microscope Made Easy; or I. The nature, uses, and magnifying powers of the best kinds of microscopes described… for the instruction of such, particularly, as desire to search into the wonders of the minute creation… II. An account of what surprising discoveries have been already made by the microscope… And also a great variety of new experiments and observations… London, R. Dodsley. 311pp

Cromer DW (2010) Avoiding twisted pixels: ethical guidelines for the appropriate use and manipulation of scientific digital images. Sci Eng Ethics 16:639–667

Goldenring JR (2010) Innocence and due diligence: managing unfounded allegations of scientific misconduct. Acad Med 85(3):527–530

Hwang et al. (2005) Patient specific embryonic stem cells derived from human SCNT blastocysts. Science 308:1777–1783

Krueger J (2005) Confronting image manipulation of digital images in science. ORI Newsletter 13(3):8–9

Krueger J (2009) Incidences of ORI cases involving falsified images. ORI Newsletter 17(4):2–3

North AJ (2006) Seeing is believing? A beginners' guide to practical pitfalls in image acquisition. J Cell Biol 172(1):9–18

Price RL (2014 Aug) 2014. Editorial: data repositories, meeting proceedings, open access/online publications, and first reporting issues. Microsc Microanal 20(4):996–998. https://doi.org/10.1017/S1431927614012926

Rossner M, Yamada KM (2004) What's in a picture? The temptation of image manipulation. J Cell Biol 166(1):11–15

Glossary (Terms Are Defined with Respect to Confocal Imaging)

Acousto-optical Beam Splitter (AOBS) A form of AOTF which replaces the dichroic filters in Leica confocal systems making it possible to work with a larger variety of excitation and emission wavelengths without the changing of multiple filters.

Acousto-optical Tunable Filters (AOTF) Typically a tellurium dioxide or quartz crystal whose optical properties (refractive index) can be controlled by varying the acoustic waves it is exposed to. As the refractive index is altered, the wavelengths of light present are split or combined into specific bandwidths which can be selected for use in the confocal microscope.

Airy Disk Light passing through an aperture is diffracted. When this diffracted light is focused, it produces an Airy pattern of a central bright spot surrounded by sequentially concentric rings of decreasingly bright light interspersed between dark rings. The Airy disk is the central bright circle. The diameter of the spot is related to the wavelength of the light forming the pattern.

Alexa Fluor Dyes A family of dyes produced by molecular probes that has excellent photostability for use in confocal imaging. Dyes are available for a wide range of excitation and emission ranges that span the visible spectrum.

Alpha Value Measure of voxel opacity or transparency when creating and analyzing 3D reconstructions. An alpha value of 0 creates opaque voxels, while an alpha value of 1 creates transparent voxels.

Analog Image An image in which gray or color tones are continuous in infinitely small increments.

Anti-fade Reagent Various reagents such as DABCO and other mounting media that inhibits or stops the photobleaching of fluorochromes when they are excited by high-intensity lasers.

Antibody A protein (immunoglobulin) produced as an immune response by a host which recognizes and can attach to the antigen which elicited the immune response. By attaching fluorescent markers to the antibody, specific labeling of the antigen can be achieved.

W. G. Jerome, R. L. Price (eds.), *Basic Confocal Microscopy*,
https://doi.org/10.1007/978-3-319-97454-5

Antigen A specific molecule or group of molecules which is capable of eliciting an immune response from a host resulting in the production of antibodies. Attachment of fluorochromes to the antigen, or to antibodies which recognize the antigen, allows the precise localization of the antigen in cells and tissues.

Antigen Retrieval Physical and/or chemical means of unmasking hidden antigen epitopes to allow or enhance antibody labeling.

Autofluorescence Fluorescence in a sample that may arise from endogenous substances (NADH) or frequently from fixation with aldehydes. In fluorescence microscopy of biological samples, autofluorescence is most often a problem when using an excitation wavelength around 488 nm.

Avalanche Photodiode Detector (APD) High-sensitivity photon detection device exploiting the avalanche effect (impact ionization) to increase the electrical signal generated by a photon interaction with the detector. They are useful in some specialized confocal and other fluorescent imaging situations.

Avidity A measure of the number of binding sites for the target molecule a label possesses.

BABB (Benzyl Alcohol/Benzyl Benzoate) Solvent-based immersion tissue-clearing technique that used fairly harsh chemistry.

Background Spurious information in an image that may result from autofluorescence or from noise introduced by the detector and electronics of the system. Autofluorescence typically shows as non-specific fluorescence across the range of the specimen, while electronic noise typically shows as bright pixels in areas where no labeling or sample fluorescence is present.

Band-Pass or Emission Filter See Emission Filter

Binning Grouping together of adjacent imaging elements (photodiodes) in a CCD camera to increase speed of imaging or intensity of the signal. Binning results in the loss of resolution which creates a compromise situation between performing high resolution and fast imaging.

Bit Depth An expression of the number of gray tones or colors available in an image. An 8 bit image (2^8) has 256 available values, and a 12 bit image (2^{12}) has 4096 available values.

Bleed-Through Term used to describe spectral overlap of photons emitted from two or more fluorochromes labeling the same sample. This may make it difficult to determine which fluorochrome contributed the signal affecting localization and quantification of sample fluorescence.

Brightness A measure of how light or dark a pixel appears. A brightness value of 0 is black, and a brightness of 255 in an 8 bit image is white.

Cartesian Coordinate System System used for reconstruction of data sets in which the X, Y, and Z axes are all perpendicular to each other.

Charge-Coupled Device (CCD) A photodiode-based photon detection system frequently used for wide-field fluorescence and spinning disk confocal microscopes. CCD cameras are slower than CMOS cameras because processing of data from each photodiode is done sequentially.

Chromatic Aberration The result of shorter wavelengths of light (blue) being focused closer to the surface of a lens than longer (red) wavelengths of light. If objective lenses are not corrected for chromatic aberration, significant artifacts in interpretation of confocal co-localization data may result.

Clarity (X-CLARITY™) Hydrogel-based tissue-clearing method often enhanced by the use of electrophoresis to decrease the amount of time required to clear tissues.

Clearing Agents Reagents that help equalize the refractive index throughout a cell to minimize light scattering within a sample. A clearing agent allows light to penetrate into and escape from deeper depths within a cell or tissue.

CMYK Color Space A subtractive color scheme comprised of cyan, magenta, yellow, and key (black). When no colors are present, a pixel is white. When all colors are present, the pixel is represented as black.

Co-localization The presence of signal from two or more fluorochromes in the same pixel or voxel of information.

Complementary Metal-Oxide Semiconductor (CMOS) A photodiode-based photon detection system used for wide-field fluorescence and spinning disk confocal systems. Like charge-coupled devices (CCDs), CMOS cameras use an array of photodiodes to collect photons from a sample to produce an image of the sample. CMOS cameras are generally faster than CCD cameras because post-collection processing is done in parallel at each diode.

Confocal Microscope A microscope in which an aperture is placed in front of the detection device to block out-of-focus light from contributing to the final image. When used properly, this results in all planes of the final image being in focus.

Contrast Transfer Function (CTF) Describes how contrast in the original object is transferred to an image as a function of resolution of the image.

Correction Collar A collar on a microscope lens which allows the optical elements to be slightly rotated to correct for spherical aberration resulting from refractive-index mismatch due to incorrect coverslip thickness or differences between sample and mounting media refractive indices.

Cyanine Dyes A family of fluorescent synthetic dyes named based on the number of methine of trivalent functional CH groups present in the structure of the molecule. Thus, Cy3 has three methine groups and Cy5 has five methine groups. The photostability and specific excitation and emission values associated with each member of the Cy family make the dyes very useful in multi-labeling confocal microscopy experiments.

DABCO (1,4-diazabicyclo[2,2,2]octane) A common anti-fade reagent which can be mixed into mounting media such as PBS glycerine to protect the sample from photobleaching.

Deconvolution A mathematical method to recover lost signal and resolution in an image based on the pattern of light spread from a point source of emission.

Depth of Field The range the specimen can be moved in the Z-direction while remaining in acceptable focus.

Depth of Focus The distance (depth) in the Z-direction that information around a point in the sample is still in acceptable focus. Depth of focus is dependent on the wavelength of the imaging source, numerical aperture, and refractive index.

Dichroic (Dichromatic) Filter A filter which is capable of reflecting light of one range of wavelengths while allowing other wavelengths of light to be transmitted through. In confocal imaging, dichroics are typically used to reflect excitation photons toward the sample while allowing emission photons to pass through to the detector.

Differential Interference Contrast (DIC) A high-contrast, high-resolution imaging mode based on the use of polarized light for working with live cells and unstained tissues.

Diffraction-Limited Resolution The limits imposed on resolution by the diffraction of light. Diffraction-limited resolution is a function of the wavelength of the electromagnetic waves being used.

Digital Image An image that consists of discrete elements (pixels or voxels) that have values of one unit or a multiple of that unit (1, 2, 3, etc.). For example, it is not possible to have an element with a value of 1.1 because this value is not a whole multiple.

Digitization The process of converting an analog image which is comprised of a continuous series of numerical values into a digital image that has discrete whole-number values.

Diode Laser A laser in which the lasing medium is a semiconductor material. Like gas lasers, monochromatic light of different colors can be created depending upon the lasing medium. Diode lasers are more compact and less expensive than comparable gas lasers. However, their beams are more divergent than those from gas lasers.

Dipping Lens A microscope objective designed to be used without the presence of a coverslip. This allows the lens to be placed directly into the cell culture dish. Although dipping lenses typically have a low numerical aperture, they have a long working distance making them ideal for direct imaging of cells in a dish.

DISCO (Three-Dimensional Imaging of Solvent-Cleared Organ) Solvent-based immersion clearing technique often favored because of its relatively low toxicity compared to BABB techniques.

Dry Lens A microscope objective designed to be used without an immersion fluid such as water or oil.

Emission Filter A filter which determines the wavelengths of photons that will be allowed to pass through to the detector. Short pass – short-pass emission filters allow all photons shorter than a specific wavelength to pass through to the detector. Band pass – band-pass emission filters allow photons between two specific wavelengths to pass through to the detector. Long pass – long-pass emission filters allow photons longer than a specific wavelength to pass through to the detector.

Epifluorescence Microscope Fluorescence imaging system in which both the excitation and emission light pass through the objective lens. The various wavelengths of light are separated by excitation and emission filters.

Epi-illumination A system in which excitation photons are focused onto the specimen with an objective lens rather than a condenser lens and emitted light travels in the opposite direction than the excitation light.

Epitope The region of the antigen which conveys specificity to the antibody. Epitopes are often sensitive to fixation and processing, making identification of the antigen by the antibody difficult resulting in reduced intensity of labeling.

Excitation Filter A filter used to control the wavelength and intensity of the excitation photons.

Fixative (Cross-Linking) Most commonly members of the aldehyde group of chemicals which preserve tissue structure by forming molecular bridges between amino groups.

Fixative (Precipitating) A chemical which preserves cell structure by dehydration and precipitation of cell components, typically proteins. Most are organic solvents such as alcohols and acetone which remove lipids but significantly affect 3D architecture due to removal of water resulting in shrinkage of the sample.

FLIM (Fluorescent Lifetime Imaging Microscopy) Imaging method based on the length of time a fluorescent molecule remains in the excited state making it possible to distinguish dyes in the same wavelength range and to separate emission from autofluorescence. Lifetimes may vary and are affected by a variety of environmental factors making it possible to study many factors that affect living cells.

Fluorescence Emission of photons following absorption of energy of a shorter wavelength.

Fluorescence Lifetime The delay in emission of a photon from the time a photon is absorbed by a fluorochrome.

Fluorescence Recovery After Photobleaching (FRAP) A technique in which a specific region of a cell is exposed to a high-intensity laser resulting in photobleaching. The region is then monitored to determine the time required for fluorescence to return. This provides a measure of dynamic events that may occur within the cell.

Förster Resonance Energy Transfer (FRET) Also known as fluorescence resonance energy transfer, it is a mechanism by which energy from one excited fluorochrome (donor) is transferred by non-radiative coupling to and excites a second (acceptor) fluorochrome. FRET is confirmed by detecting fluorescence from the acceptor fluorochrome when the sample is irradiated by a wavelength that should only excite the donor fluorochrome. Since FRET is exquisitely sensitive to the space between two fluorochromes and only occurs if the donor and acceptor are less than 10 nm apart, FRET is often used to confirm that two molecules are adjacent to each other.

Fluorochrome Molecule which absorbs a photon of light of a specific wavelength and, after a brief interval, emits some of that energy in the form of a photon of a different wavelength.

Fluorophore Fluorochrome linked to a biologically active substance such as protein, DNA, or RNA.

Focal Length The distance from the surface of the lens to the point or plane in the specimen where the light is focused. Focal length is dependent on several factors including refractive index and chromatic and spherical aberrations.

Focal Plane The plane of the specimen which is in focus.

Formaldehyde A common fixative in which the aldehyde groups bind to amine groups in proteins. Paraformaldehyde is the powder form, while formaldehyde is a gaseous form.

Formalin A 40% solution of formaldehyde.

Full Width Half Maximum (FWHM) The width of an emission peak at half of its height. Often used for describing characteristics of fluorophore emission and emission filter specifications.

Gallium Arsenide Phosphide (GaAsP) detector A type of PMT in which the photon-detecting cathode is made of gallium arsenide phosphide. GaAsP have increased quantum efficiency compared to standard multi-alkali PMTs. However, they are more susceptible to damage, and their high quantum efficiency is limited to wavelengths between 400 nm and 650 nm.

Gamma The relationship between the gray level which is detected and that which is rendered in the final image. A gamma of 1 indicates that the detected and rendered gray levels are identical. Gamma relationships are not linear. Pixel values of 0 and 255 remain as 0 and 255, but an input pixel value of 50 and a gamma value of 0.5 may result in an output pixel value of 113.

Gamma Correction The inverse of gamma. Gamma correction can be used to restore the original image pixel values.

Gas Laser A laser in which a laser beam is produced by injecting an electrical charge into a gas to produce a monochromatic, highly coherent beam of photons.

Green Fluorescent Protein (GFP) One of a family of naturally occurring fluorescent proteins which can be incorporated into the genomic structure of cells and used as a locator molecule in fluorescence microscopy. Excitation of GFP is in the blue range and emission in the green range. Other fluorescent proteins such as red and yellow variants also exist with specific excitation and emission ranges.

Histogram Equalization An image processing procedure in which pixel values in a digital image are remapped so that all intensity values are equally represented in the image. The effect is to maximize contrast (and noise) in the image.

Histogram Stretch In an 8 bit image, the darkest pixel value will be assigned to 0 and the highest pixel value assigned to 255 so that all 256 available values are utilized to improve the contrast of the image. The histogram is stretched to use the full dynamic range.

Hue Defines the color of a pixel or voxel. All blue pixels have the same hue.

Hybrid Detectors A PMT that uses a GaAsP cathode for initial photon detection but then uses an avalanche photodiode to amplify the initial signal.

Immersion Lens Lens designed to be immersed in a fluid such as water, glycerin, or oil to minimize the refractive-index mismatch of components in the optical path. Immersion lenses differ from dipping lenses in that they are designed to be used with a coverslip in place.

Interpolation A method of resampling data sets by the computer. Often used to create near-cubic voxels to facilitate 3D reconstructions.

Jablonski Diagram A diagram of energy absorption and emission from a fluorochrome.

Joint Photographic Experts Group (JPEG) A compressed generic image file format useful for digital transfer of images and incorporation of images into programs such as PowerPoint. JPEG files are a lossy format and should not be used for data analysis or for final storage of scientific images.

Kernel A grouping of pixels or voxels used for image enhancement, analysis, and reconstruction. For example, a group of 3 pixels in the X and 3 pixels in the Y direction would represent a 3X3 kernel.

LASER Acronym for light amplification by stimulated emission of radiation. Lasers produce and intense monochromatic beam of coherent collimated light useful in confocal imaging.

LED (Light-Emitting Diodes (LEDs)) High-performance stable light sources with relatively narrow emission spectra which can be used to excite fluorochromes in epi-fluorescent and spinning disk confocal systems.

Ligand A substance which is capable of binding to and activating a cell receptor. With respect to confocal and fluorescence imaging, ligands may be conjugated to fluorescent molecules and attachment to receptors detected.

Light-Sheet Microscopy A fluorescent microscopy technique in which the specimen is illuminated by a thin plane of light to improve the axial (Z-axis) resolution of the image.

Look-Up Table (LUT) A table of alternate values (typically colors) that will be used to display pixels of certain values to be reassigned a specific color. Often used as an aid when setting optimal ranges of pixel values for display of a confocal image.

Luminance A measure of the perceived brightness. Luminance considers the color sensitivity of the human eye to certain colors such as yellow-green, which the eye is sensitive to as compared to blue which the eye is not very sensitive to.

Magnification The extent to which a sample is enlarged when comparing the actual size and the size in the final image.

Marching Cubes Algorithm A computer graphic algorithm used to define a surface in three-dimensional space by the use of voxels. Based on assigned criteria such as threshold values, the algorithm can be used to define specific structures in a region of interest for reconstruction of a 3D data set.

Maximum Projection A method of projecting multiple confocal slices and flattening into a single image. Pixel values at each location in all images of the Z-series are compared and the maximum value kept and projected into the final 2D image.

Molar Extinction Coefficient A measure of the quantity of light absorbed as a function of the path length of the excitation light and the concentration of the fluorochrome. A molecule with a high molar extinction coefficient is more likely to exhibit fluorescence.

Multiphoton (Two-Photon) Confocal Microscopes Systems which use the additive energy of long-wavelength low-energy photons to excite fluorochromes rather than the short-wavelength high-energy photons used in single-photon confocal systems. Multiphoton systems provide the distinct advantages of reduced specimen damage and deeper specimen imaging.

Multi-track (Sequential) Imaging See sequential imaging.

Neutral-Density (ND) Filter A filter which is used to reduce the intensity of light (number of photons) from the laser prior to interacting with the specimen. In most modern high-end systems, the ND filter is replaced with an AOTF which provides additional control over the intensity and wavelengths of light used to excite a specimen.

Nonlinear Optics Nonlinear optical techniques take advantage of the nonlinear response of light under certain specific conditions. Second-harmonic imaging and multiphoton excitation are examples of nonlinear optical techniques.

Numerical Aperture A measure of the resolving power and light gathering capacity of lens given by the equation $NA = n\sin\theta$ where n represents the refractive index of the medium and θ represents the half angle of the light entering the lens. As NA increases, the light-gathering capacity and resolution of the lens improve.

Nyquist Theorem A simplified explanation of the Nyquist theorem useful for confocal imaging is that to resolve a structure, it must be sampled at least twice in the data set. This is important in both 2D and 3D imaging spaces but becomes very important in 3D reconstructions and sampling when collecting Z-series data sets where Z-resolution is easily adjusted with the pinhole settings. Oversampling may result in very large data sets and extensive collection times, while undersampling will require computer interpolation when generating 3D reconstructions.

Objective Lens The primary image-forming lens and a key component of the optical path in a confocal microscope. High-quality objective lenses corrected for flat-field imaging and chromatic and spherical aberrations are essential for confocal imaging.

Oblique Planes Data presentation in which image planes are not perpendicular to the major X, Y, and Z axes used when the original image data was collected.

Orthographic Projection 3D data representation in which all projection rays passing through the data remain parallel and without reference to the viewer, so objects do not appear in perspective. This results in same-sized objects appearing identical regardless of the distance they may be from the viewer.

Orthogonal Planes A data presentation in which planes are perpendicular to each other. In imaging, this refers to image planes perpendicular to any of the X, Y, and Z axes used when the original image data was collected.

Pauli Exclusion Principle The quantum mechanical principle that two or more identical particles with half-integer spin cannot occupy the same quantum state. In the case of electrons, two electrons in an atom must have a different value for at least one of their four quantum values (quantum number, angular momentum quantum number, magnetic quantum number, spin quantum number).

Perceptually Uniform Projection (PUP) Display A method for displaying the fluorescent signal from two different molecular probes as a single-color image so that information about the probes' relative concentrations as well as their spatial co-relationship can be visualized. Unlike other methods of comparison, PUP encodes the information as two independent and uniform quantities – hue and luminosity – in the color image for a more accurate comparison.

Perspective Projection A 3D data representation in which rays from objects in a scene project directly toward and converge at the viewer. This results in objects appearing to decrease in size as they increase in distance from the viewer.

Phase-Contrast Microscopy An imaging mode useful for unstained cells in which small phase shifts of the light occur as it passes through the specimen. Based on optical components added to the light path, the phase shifts can be used to introduce contrast to transparent specimens.

Photo-Activated Localization Microscopy (PALM) A fluorescence microscopy technique allowing resolution below the diffraction limit by using photoswitchable probes. In a group of photoswitchable fluorophores labeling an object, a low-power activation beam will stochastically convert a few photoswitchable fluorophores to the active state. These are then imaged by a high-power excitation beam that immediately converts them back to an inactive state by photobleaching. Since in most instances single fluorophores are imaged while their neighbors remain dark, one can compute the center of the Airy disk to localize the fluorescing molecule. Multiple repetitions of this on-off cycle (usually thousands of cycles) ensuring that all fluorophores in the group have been imaged one at a time and then a computed model of the labeling can be constructed from the individual images. (See also stochastic optical reconstruction microscopy.)

Photobleaching Chemical or mechanical destruction of a fluorophore due to high-energy photons from the excitation source. Photobleaching results in an irreversible reduction in the number of photons emitted and can be a major concern for several fluorophores used in single-photon confocal imaging.

Photomultiplier Tube (PMT) The primary type of detector used in single- and multiphoton-based confocal scanning laser microscopes. PMTs are highly sensitive detectors capable of detecting and proportionally amplifying relatively few photons emitted from a sample.

Photon The elementary particle of light or quantum of electromagnetic radiation. Photons exhibit specific wavelength and particle properties that allow separation into specific bands of light useful for excitation and emission of fluorophores used in confocal imaging.

Pinhole (also referred to as an aperture or iris in some systems) A small opening that restricts the passage of light waves to only those that are in focus at the opening.

Pixel Discrete two-dimensional picture elements which comprise a digital image. Pixels have a two-dimensional address within an image (X,Y location) and an intensity value (i.e., 0–255).

Plan Objective An objective lens corrected for flat-field imaging, so no curvature of the lens is present. Plan objectives are important in confocal imaging since curvature may result in multiple focal planes and loss of signal due to the confocal effect of the pinhole.

Planck's Law The radiation energy of a photon is inversely proportional to its wavelength.

Pseudo-color Is a technique for enhancing visualization by assigning each value of a gray-scale image a specific unique color assigned by a look-up table (LUT).

Quantum Dots Photostable fluorescent semiconductor nanocrystals with high quantum yield that absorb light and emit at a longer wavelength similar to standard organic fluorochromes.

Quantum Efficiency See quantum yield.

Quantum Yield A measure of the efficiency of a fluorochrome presented as a ratio of the number of photons emitted relative to the number of photons absorbed. Thus it is a measure of the efficiency of fluorescence, and a quantum yield of 1.0 would be a perfect emitter with one photon emitted for each photon absorbed.

Quenching The reversible reduction in number of photons emitted from a fluorophore. A number of factors including excitation, changes in solution chemistry, and interaction with surrounding molecules may result in quenching.

Refractive Index A measure of the speed of light as it passes through various media. As light passes through a vacuum, the refractive index equals 1.0. As light passes through various other media, the speed, and thus the refractive index, varies and is presented as a ratio of the speed through the vacuum compared to the speed to the specific medium. For example, the refractive index through water is 1.33. With respect to confocal imaging, this term refers to a measurement of how the wavelength of photons is affected as they pass through components of the optical path.

Refractive-Index Mismatch As light passes through the various components in the optical path of a confocal microscopy, the speed, and thus the refractive index, varies. Refractive-index mismatch refers to this variation in the speed of light as it passes along the optical path and can result in significant deterioration of the image due to chromatic and spherical aberrations.

Resolution The ability to clearly distinguish two points or objects within a sample. Based on the Nyquist theorem, the two points must be sampled twice to be able to resolve them.

RGB color space An additive color mechanism in which images are comprised of red, green, and blue. When all three colors are present, a white pixel results. When none are present, a black pixel results.

RGBA Tuples (Red, Green, Blue, Alpha) Coding of voxel values in each of the four channels for display of 3D data sets and reconstructions.

Saturated Pixel A saturated pixel is a photodiode sensor that has accumulated its maximum amount of charge and thus is no longer sensitive to incoming photons. Two adjacent saturated pixels will read out the same maximum value (and so will have no contrast difference) even though the number of photons that collided with the senor may be different. Obviously, for maximum resolution, saturation of pixels should be avoided.

Saturation (Color) Determines how intense the color appears. A fully saturated color is brilliant, while a less saturated color appears faded.

Saturation (Pixel/Voxel) A pixel or voxel which has the highest value available. For an 8 bit image, this would be 255, and for a 12 bit image, this would be 4095.

Second-Harmonic Microscopy A microscopy technique that uses the phenomenon of second-harmonic generation to generate fluorescence from specific structures within a sample.

Segmentation A process of dividing an image data set into regions of interest based on selection of pixels or voxels which meet specific criteria.

Sequential Imaging A method of collecting multichannel images in which each fluorescent channel is collected separately. Hardware and software settings are changed between collection of each channel to avoid overlap or bleed-through of signal into multiple channels.

Shading Correction Methods to correct uneven background shading of a microscope image produced by uneven illumination of the image area.

Signal-to-Noise (S/N) Ratio Also abbreviated as SNR, it is a measure comparing the amount of signal (photons) detected to the level of noise (non-specific events) that is recorded by the detector. There are several sources of noise, but all obscure the ability to detect the true signal. Thus, a high level of signal above the background noise is desirable.

Simultaneous Imaging A setup of the operating hardware and software so that more than one channel of fluorescence is detected in a single pass of the laser(s). This results in a multichannel image, but care must be taken to avoid bleed-through of one channel into neighboring channels.

Single-Photon Confocal Microscopes Confocal systems which use a high-intensity laser of short wavelengths as the excitation source resulting in a single excitation photon producing an emitted photon of a longer wavelength.

Single-Track Imaging See simultaneous imaging.

Spectral Imaging An imaging mode where the available wavelengths of light are divided into several channels so that closely related fluorophores such as green and yellow fluorescent proteins can be separated. Spectral imaging is also often used to separate autofluorescence from true signal produced by a fluorophore.

Spherical Aberration Image defects introduced when light rays from the peripheral and central regions of the objective lens are focused in different planes. In confocal imaging, spherical aberrations can result in significant loss of signal from regions of the image since out-of-focus light is eliminated by the pinhole.

Spinning Disk Confocal Microscopes A group of instruments that use a rapidly spinning disk with several microlenses rather than a laser scanned point by point to form the confocal image. Spinning disk systems are usually preferred for rapid or live cell imaging protocols.

Stimulated Emission Depletion (STED) Microscopy A super-resolution fluorescent microscopy technique that depletes fluorescence in a specific area but leaves the central-most region active to fluorescence excitation. In this way, the origin of the fluorescence can be isolated to an area smaller than the diffraction-limited area of a focused beam of light.

Stochastic Optical Reconstruction Microscopy (STORM) A fluorescence microscopy technique allowing resolution below the diffraction limit by using photoswitchable probes. STORM is based on the ability of closely spaced cyano dyes (<2 nm) to photoswitch. In this way, for any single image, only a few dyes are excited and imaged separate from the surrounding, and the center of the Airy disk can be calculated. Repeated imaging (>60 cycles) allows all of the individual fluorophores to eventually be localized (see also the similar technique, photo-activated localization microscopy).

Stoke's Shift The difference in wavelength from the light absorbed by fluorochrome and the light which is emitted following excitation.

Super-resolution Microscopy Any of a number of schemes for collecting fluorescence information from a sample and using this information to construct an image that has better resolution than can be attained by traditional microscopy. Both physical and mathematical methods have been developed to obtain resolution beyond the diffraction limit. Some super-resolution techniques have been able to laterally resolve structures in the <30 nm size range.

Surface Render A technique used for visualizing the surface of a 3D data set. The Marching cubes algorithm is used to segment voxels based on a threshold intensity or selected value.

Tagged Image File Format (TIF) A very good uncompressed generic image format readable by most computers. The TIF format is probably the most accepted format for scientific images.

Threshold An assignment of a particular value to a group of pixels or voxels for the purpose of image enhancement and analysis. For example, a group of pixels that fall between the values of 0 and 50 may all be assigned a value or thresholded at a value of 50.

VaLaP A 1:1:1 mixture of petroleum jelly (Vaseline), lanolin, and histology paraffin that can be used to support and seal cover glasses for imaging.

Vital Dyes Stains which are minimally or nontoxic to living cells when used in low concentrations. Many are useful as they change emission spectra or emission intensity when the intracellular environment is altered.

Volume Render A technique used to display a 3D data set in a manner which will allow the internal structure present in the volume to be observed. Voxels are displayed after assigning transparency (alpha) values to certain voxel intensities so that less intense voxels do not hide brighter voxels.

Voxel A discrete volume element which consists of X, Y, and Z dimensions. Isotropic voxels have equal dimensions in all three planes, while anisotropic voxels have different dimensions in at least one of the planes. Voxels have a three-dimensional address within a 3D space (X, Y, and Z location) and an intensity value.

Wide-Field Fluorescence Microscope A microscope system in which fluorophores in the entire volume of the specimen are excited at the same time, and the resulting image included signal from both the focal plane of interest and the out-of-focus planes above and below the focal plane.

Working Distance The distance from the surface of the objective lens and the coverslip (or specimen if using dipping lens) when the specimen is in focus.

Z-Drop A phenomenon resulting in loss of intensity of pixels in deeper sections of a sample due to poor laser photon penetration or ability of photons emitted from a fluorochrome to pass through the sample and be collected.

Z-Series A series of images collected through the depth of the sample. Sample depth is typically along the Z-axis.

Index

W. G. Jerome, R. L. Price (eds.), *Basic Confocal Microscopy*,
https://doi.org/10.1007/978-3-319-97454-5

N

O

P

Q

R

S

The manufacturer's authorised representative in the EU is Springer Nature Customer Service Centre GmbH, Europaplatz 3, 69115 Heidelberg, Germany. If you have any concerns regarding our products, please contact ProductSafety@springernature.com

Printed and bound by CPI Group (UK) Ltd, Croydon, CR0 4YY
25/09/2026
02222853-0001